Materials Science and Engineering

Materials Science and Engineering

Edited by
Emilio McMahon

Larsen & Keller
www.larsen-keller.com

Materials Science and Engineering
Edited by Emilio McMahon
ISBN: 978-1-63549-177-7 (Hardback)

☐ Larsen & Keller

Published by Larsen and Keller Education,
5 Penn Plaza,
19th Floor,
New York, NY 10001, USA

Cataloging-in-Publication Data

Materials science and engineering / edited by Emilio McMahon.
    p. cm.
Includes bibliographical references and index.
ISBN 978-1-63549-177-7
1. Materials science. 2. Materials. 3. Engineering.
4. Technology. I. McMahon, Emilio.
TA403 .M38 2017
620.11--dc23

The publisher's policy is to use permanent paper from mills that operate a sustainable forestry policy. Furthermore, the publisher ensures that the text paper and cover boards used have met acceptable environmental accreditation standards.

Printed and bound in the United States of America.

For more information regarding Larsen and Keller Education and its products, please visit the publisher's website www.larsen-keller.com

Table of Contents

Permissions

Index

Preface

The book aims to shed light on some of the unexplored aspects of materials science and engineering. It describes in detail the various concepts and theories of this field. Materials science refers to the study, design and discovery of new materials. This subject is used in different areas like nanotechnology, metallurgy, forensic engineering, biomaterials, and failure analysis, etc. This book is compiled in such a manner, that it will provide in-depth knowledge about the theory and practice of material science. The topics included in it are of utmost significance and bound to provide incredible insights to students. Some of the diverse topics covered in this book address the varied branches that fall under this category. This textbook is an essential guide for both graduates and post-graduates in this discipline.

To facilitate a deeper understanding of the contents of this book a short introduction of every chapter is written below:

Chapter 1- Materials science studies different designs and materials. It includes elements of physics, chemistry and engineering. The chapter on materials science offers an insightful focus, keeping in mind the complex subject matter.

Chapter 2- The essential concepts of materials science are biomaterials, composite materials, acoustic emission, crystal growth, Hume-Rothery rules and hysteresis. Biomaterial is the substance that is made to interact with biological systems whereas a composite material is a material made from constituents that have different physical or chemical properties. The major components of materials science are discussed in this section.

Chapter 3- Aerogel is a material that is originated from a gel; the liquid constituent in the gel is substituted with a gas. Alternatively, the other emerging technologies in materials science are metamaterial, nanomaterial, programmable matter, quantum dot etc. The aspects elucidated in this text are of vital importance, and provide a better understanding of materials science.

Chapter 4- A crystal is a material that is in a proper structure. They can be recognized by their geometrical shape. The other atomic arrangements discussed in this chapter are crystal structure, Miller index, crystal structure prediction, crystallographic defect etc. This section serves as a source to understand the basic aspects of atomic arrangements present in solids.

Chapter 5- The bond between atoms and chemical compounds is known as chemical bonding. Ionic bonding, covalent bond, metallic bonding and electronegativity are other aspects of chemical bonding. This section is an overview of the subject matter incorporating all the major aspects of chemical bonding.

Chapter 6- Carbon nanotubes are allotropes of carbon. They have unusual properties and are very valuable for electronics, optics and nanotechnology. Topics such as optical properties of carbon nanotubes, synthesis of carbon nanotubes, carbon nanotube chemistry and selective chemistry of single-walled nanotubes have been explicated in the section.

Chapter 7- Metallurgy studies the physical and chemical behavior of metallic elements. It is the technology of metals. The process of working on these metals to create structures and assemblies is known as metalworking. The processes of metalworking that have been explained in this chapter are casting, forging, rolling, cladding, metal fabrication etc. This section elucidates the basic concepts of metallurgy.

Chapter 8- The allied fields of materials science are nanotechnology, polymer chemistry and radiation material science. Nanotechnology is the study of matter on atomic, molecular and supramolecular scale. This chapter helps the reader in developing an in depth understanding of all the fields that are allied to materials science.

I owe the completion of this book to the never-ending support of my family, who supported me throughout the project.

Editor

Introduction to Materials Science

Materials science studies different designs and materials. It includes elements of physics, chemistry and engineering. The chapter on materials science offers an insightful focus, keeping in mind the complex subject matter.

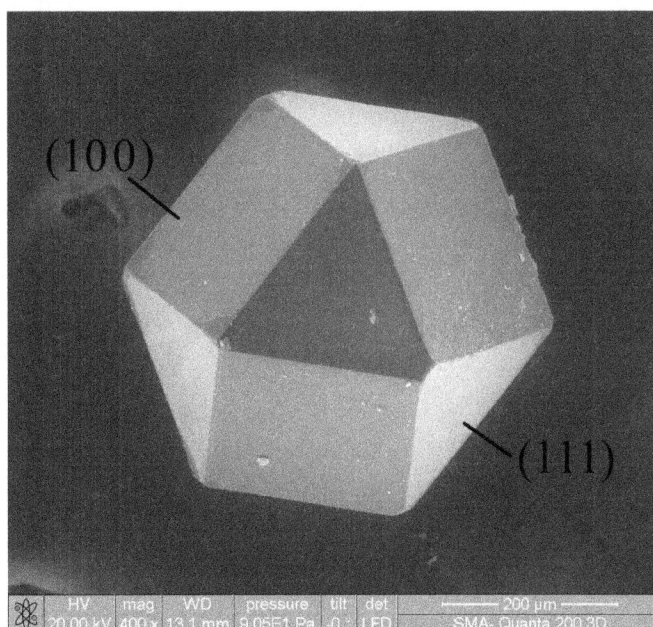

A diamond cuboctahedron showing seven crystallographic planes, imaged with scanning electron microscopy.

The interdisciplinarity field of materials science, also commonly termed materials science and engineering, involves the discovery and design of new materials, with an emphasis on solids. The intellectual origins of materials science stem from the Enlightenment, when researchers began to use analytical thinking from chemistry, physics, and engineering to understand ancient, phenomenological observations in metallurgy and mineralogy. Materials science still incorporates elements of physics, chemistry, and engineering. As such, the field was long considered by academic institutions as a sub-field of these related fields. Beginning in the 1940s, materials science began to be more widely recognized as a specific and distinct field of science and engineering, and major technical universities around the world created dedicated schools of the study.

> Materials science is a syncretic discipline hybridizing metallurgy, ceramics, solid-state physics, and chemistry. It is the first example of a new academic discipline emerging by fusion rather than fission.

Many of the most pressing scientific problems humans currently face are due to the limits of the materials that are available. Thus, breakthroughs in materials science are likely to affect the future of technology significantly.

Materials scientists emphasize understanding how the history of a material (its *processing*) influences its structure, and thus the material's properties and performance. The understanding of processing-structure-properties relationships is called the § materials paradigm. This paradigm is used to advance understanding in a variety of research areas, including nanotechnology, biomaterials, and metallurgy. Materials science is also an important part of forensic engineering and failure analysis - investigating materials, products, structures or components which fail or which do not operate or function as intended, causing personal injury or damage to property. Such investigations are key to understanding, for example, the causes of various aviation accidents and incidents.

History

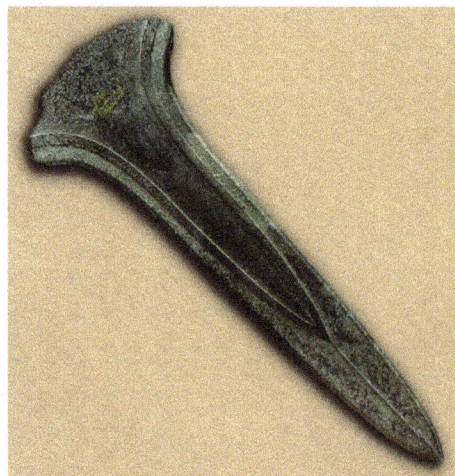

A late Bronze Age sword or dagger blade.

The material of choice of a given era is often a defining point. Phrases such as Stone Age, Bronze Age, Iron Age, and Steel Age are great examples. Originally deriving from the manufacture of ceramics and its putative derivative metallurgy, materials science is one of the oldest forms of engineering and applied science. Modern materials science evolved directly from metallurgy, which itself evolved from mining and (likely) ceramics and the use of fire. A major breakthrough in the understanding of materials occurred in the late 19th century, when the American scientist Josiah Willard Gibbs demonstrated that the thermodynamic properties related to atomic structure in various phases are related to the physical properties of a material. Important elements of modern materials science are a product of the space race: the understanding and engineering of the metallic alloys, and silica and carbon materials, used in building space vehicles enabling the exploration of space. Materials science has driven, and been driven by, the development of revolutionary technologies such as rubbers, plastics, semiconductors, and biomaterials.

Before the 1960s (and in some cases decades after), many *materials science* departments were named *metallurgy* departments, reflecting the 19th and early 20th century emphasis on metals. The growth of materials science in the United States was catalyzed in part by the Advanced Research Projects Agency, which funded a series of university-hosted laboratories in the early 1960s "to expand the national program of basic research and training in the materials sciences." The field has since broadened to include every class of materials, including ceramics, polymers, semiconductors, magnetic materials, medical implant materials, biological materials, and nanomaterials.

Fundamentals

The materials paradigm represented in the form of a tetrahedron.

A material is defined as a substance (most often a solid, but other condensed phases can be included) that is intended to be used for certain applications. There are a myriad of materials around us—they can be found in anything from buildings to spacecraft. Materials can generally be divided into two classes: crystalline and non-crystalline. The traditional examples of materials are metals, semiconductors, ceramics and polymers. New and advanced materials that are being developed include nanomaterials and biomaterials, etc.

The basis of materials science involves studying the structure of materials, and relating them to their properties. Once a materials scientist knows about this structure-property correlation, they can then go on to study the relative performance of a material in a given application. The major determinants of the structure of a material and thus of its properties are its constituent chemical elements and the way in which it has been processed into its final form. These characteristics, taken together and related through the laws of thermodynamics and kinetics, govern a material's microstructure, and thus its properties.

Structure

As mentioned above, structure is one of the most important components of the field of materials science. Materials science examines the structure of materials from the atomic scale, all the way up to the macro scale. Characterization is the way materials scientists examine the structure of a material. This involves methods such as diffraction with X-rays, electrons, or neutrons, and various forms of spectroscopy and chemical analysis such as Raman spectroscopy, energy-dispersive spectroscopy (EDS), chromatography, thermal analysis, electron microscope analysis, etc. Structure is studied at various levels, as detailed below.

Atomic Structure

This deals with the atoms of the materials, and how they are arranged to give molecules, crystals, etc. Much of the electrical, magnetic and chemical properties of materials arise from this level of structure. The length scales involved are in angstroms. The way in which the atoms and molecules are bonded and arranged is fundamental to studying the properties and behavior of any material.

Nanostructure

Nanostructure deals with objects and structures that are in the 1—100 nm range. In many materials, atoms or molecules agglomerate together to form objects at the nanoscale. This causes many interesting electrical, magnetic, optical, and mechanical properties.

In describing nanostructures it is necessary to differentiate between the number of dimensions on the nanoscale. Nanotextured surfaces have *one dimension* on the nanoscale, i.e., only the thickness of the surface of an object is between 0.1 and 100 nm. Nanotubes have *two dimensions* on the nanoscale, i.e., the diameter of the tube is between 0.1 and 100 nm; its length could be much greater. Finally, spherical nanoparticles have *three dimensions* on the nanoscale, i.e., the particle is between 0.1 and 100 nm in each spatial dimension. The terms nanoparticles and ultrafine particles (UFP) often are used synonymously although UFP can reach into the micrometre range. The term 'nanostructure' is often used when referring to magnetic technology. Nanoscale structure in biology is often called ultrastructure.

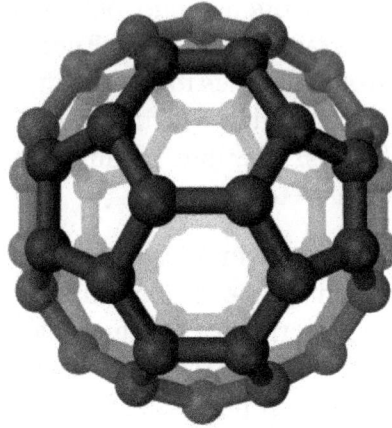

Buckminsterfullerene nanostructure.

Materials which atoms and molecules form constituents in the nanoscale (i.e., they form nanostructure) are called nanomaterials. Nanomaterials are subject of intense research in the materials science community due to the unique properties that they exhibit.

Microstructure

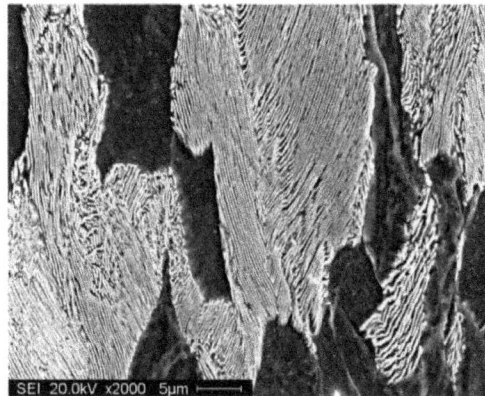

Microstructure of pearlite.

Microstructure is defined as the structure of a prepared surface or thin foil of material as revealed by a microscope above $25\times$ magnification. It deals with objects from 100 nm to a few cm. The microstructure of a material (which can be broadly classified into metallic, polymeric, ceramic and composite) can strongly influence physical properties such as strength, toughness, ductility, hardness, corrosion resistance, high/low temperature behavior, wear resistance, and so on. Most of the traditional materials (such as metals and ceramics) are microstructured.

The manufacture of a perfect crystal of a material is physically impossible. For example, a crystalline material will contain defects such as precipitates, grain boundaries (Hall–Petch relationship), interstitial atoms, vacancies or substitutional atoms. The microstructure of materials reveals these defects, so that they can be studied.

Macro Structure

Macro structure is the appearance of a material in the scale millimeters to meters—it is the structure of the material as seen with the naked eye.

Crystallography

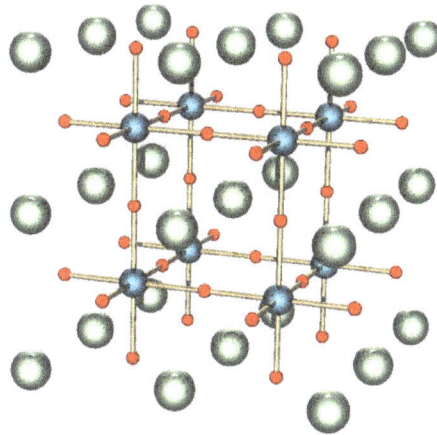

Crystal structure of a perovskite with a chemical formula ABX_3.

Crystallography is the science that examines the arrangement of atoms in crystalline solids. Crystallography is a useful tool for materials scientists. In single crystals, the effects of the crystalline arrangement of atoms is often easy to see macroscopically, because the natural shapes of crystals reflect the atomic structure. Further, physical properties are often controlled by crystalline defects. The understanding of crystal structures is an important prerequisite for understanding crystallographic defects. Mostly, materials do not occur as a single crystal, but in polycrystalline form, i.e., as an aggregate of small crystals with different orientations. Because of this, the powder diffraction method, which uses diffraction patterns of polycrystalline samples with a large number of crystals, plays an important role in structural determination. Most materials have a crystalline structure, but some important materials do not exhibit regular crystal structure. Polymers display varying degrees of crystallinity, and many are completely noncrystalline. Glass, some ceramics, and many natural materials are amorphous, not possessing any long-range order in their atomic arrangements. The study of polymers combines elements of chemical and statistical thermodynamics to give thermodynamic and mechanical, descriptions of physical properties.

Bonding

To obtain a full understanding of the material structure and how it relates to its properties, the materials scientist must study how the different atoms, ions and molecules are arranged and bonded to each other. This involves the study and use of quantum chemistry or quantum physics. Solid-state physics, solid-state chemistry and physical chemistry are also involved in the study of bonding and structure.

Properties

Materials exhibit myriad properties, including the following.

- Mechanical properties

- Chemical properties

- Electrical properties

- Thermal properties

- Optical properties

- Magnetic properties

The properties of a material determine its usability and hence its engineering application.

Synthesis and Processing

Synthesis and processing involves the creation of a material with the desired micro-nanostructure. From an engineering standpoint, a material cannot be used in industry if no economical production method for it has been developed. Thus, the processing of materials is vital to the field of materials science.

Different materials require different processing or synthesis methods. For example, the processing of metals has historically been very important and is studied under the branch of materials science named *physical metallurgy*. Also, chemical and physical methods are also used to synthesize other materials such as polymers, ceramics, thin films, etc. As of the early 21st century, new methods are being developed to synthesize nanomaterials such as graphene.

Thermodynamics

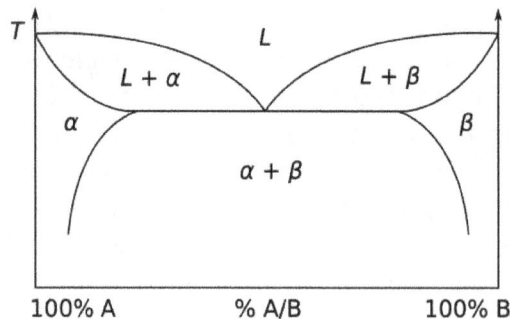

A phase diagram for a binary system displaying a eutectic point.

Thermodynamics is concerned with heat and temperature and their relation to energy and work. It defines macroscopic variables, such as internal energy, entropy, and pressure, that partly describe a body of matter or radiation. It states that the behavior of those variables is subject to general constraints, that are common to all materials, not the peculiar properties of particular materials. These general constraints are expressed in the four laws of thermodynamics. Thermodynamics describes the bulk behavior of the body, not the microscopic behaviors of the very large numbers of its microscopic constituents, such as molecules. The behavior of these microscopic particles is described by, and the laws of thermodynamics are derived from, statistical mechanics.

The study of thermodynamics is fundamental to materials science. It forms the foundation to treat general phenomena in materials science and engineering, including chemical reactions, magnetism, polarizability, and elasticity. It also helps in the understanding of phase diagrams and phase equilibrium.

Kinetics

Chemical kinetics is the study of the rates at which systems that are out of equilibrium change under the influence of various forces. When applied to materials science, it deals with how a material changes with time (moves from non-equilibrium to equilibrium state) due to application of a certain field. It details the rate of various processes evolving in materials including shape, size, composition and structure. Diffusion is important in the study of kinetics as this is the most common mechanism by which materials undergo change.

Kinetics is essential in processing of materials because, among other things, it details how the microstructure changes with application of heat.

In Research

Materials science has received much attention from researchers. In most universities, many departments ranging from physics to chemistry to chemical engineering, along with materials science departments, are involved in materials research. Research in materials science is vibrant and consists of many avenues. The following list is in no way exhaustive. It serves only to highlight certain important research areas.

Nanomaterials

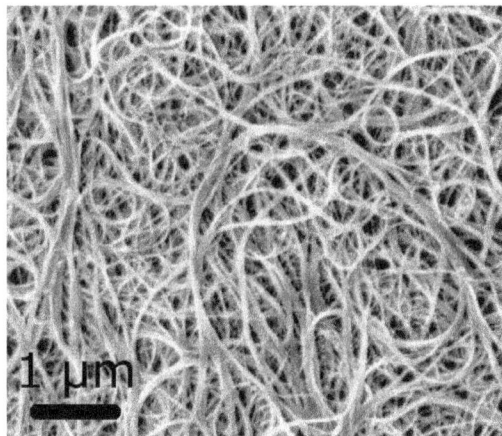

A scanning electron microscopy image of carbon nanotubes bundles

Nanomaterials describe, in principle, materials of which a single unit is sized (in at least one dimension) between 1 and 1000 nanometers (10^{-9} meter) but is usually 1–100 nm.

Nanomaterials research takes a materials science-based approach to nanotechnology, leveraging advances in materials metrology and synthesis which have been developed in support of microfabrication research. Materials with structure at the nanoscale often have unique optical, electronic, or mechanical properties.

The field of nanomaterials is loosely organized, like the traditional field of chemistry, into organic (carbon-based) nanomaterials such as fullerenes, and inorganic nanomaterials based on other elements, such as silicon. Examples of nanomaterials include fullerenes, carbon nanotubes, nanocrystals, etc.

Biomaterials

The iridescent nacre inside a nautilus shell.

A biomaterial is any matter, surface, or construct that interacts with biological systems. As a science, *bio materials* is about fifty years old. The study of biomaterials is called *bio materials science*. It has experienced steady and strong growth over its history, with many companies investing large amounts of money into developing new products. Biomaterials science encompasses elements of medicine, biology, chemistry, tissue engineering, and materials science.

Biomaterials can be derived either from nature or synthesized in a laboratory using a variety of chemical approaches using metallic components, polymers, ceramics, or composite materials. They are often used and/or adapted for a medical application, and thus comprises whole or part of a living structure or biomedical device which performs, augments, or replaces a natural function. Such functions may be benign, like being used for a heart valve, or may be bioactive with a more interactive functionality such as hydroxylapatite coated hip implants. Biomaterials are also used everyday in dental applications, surgery, and drug delivery. For example, a construct with impregnated pharmaceutical products can be placed into the body, which permits the prolonged release of a drug over an extended period of time. A biomaterial may also be an autograft, allograft or xenograft used as an organ transplant material.

Electronic, Optical, and Magnetic

Semiconductors, metals, and ceramics are used today to form highly complex systems, such as integrated electronic circuits, optoelectronic devices, and magnetic and optical mass storage media. These materials form the basis of our modern computing world, and hence research into these materials is of vital importance.

Semiconductors are a traditional example of these types of materials. They are materials that have properties that are intermediate between conductors and insulators. Their electrical conductivities are very sensitive to impurity concentrations, and this allows for the use of doping to achieve desirable electronic properties. Hence, semiconductors form the basis of the traditional computer.

This field also includes new areas of research such as superconducting materials, spintronics, metamaterials, etc. The study of these materials involves knowledge of materials science and solid-state physics or condensed matter physics.

Negative index metamaterial.

Computational Science and Theory

With the increase in computing power, simulating the behavior of materials has become possible. This enables materials scientists to discover properties of materials formerly unknown, as well as to design new materials. Up until now, new materials were found by time consuming trial and error processes. But, now it is hoped that computational methods could drastically reduce that time, and allow tailoring materials properties. This involves simulating materials at all length scales, using methods such as density functional theory, molecular dynamics, etc.

In Industry

Radical materials advances can drive the creation of new products or even new industries, but stable industries also employ materials scientists to make incremental improvements and troubleshoot issues with currently used materials. Industrial applications of materials science include materials design, cost-benefit tradeoffs in industrial production of materials, processing methods (casting, rolling, welding, ion implantation, crystal growth, thin-film deposition, sintering, glass-blowing, etc.), and analytic methods (characterization methods such as electron microscopy, X-ray diffraction, calorimetry, nuclear microscopy (HEFIB), Rutherford backscattering, neutron diffraction, small-angle X-ray scattering (SAXS), etc.).

Besides material characterization, the material scientist or engineer also deals with extracting materials and convering them into useful forms. Thus ingot casting, foundry methods, blast furnace extraction, and electrolytic extraction are all part of the required knowledge of a materials engineer. Often the presence, absence, or variation of minute quantities of secondary elements and compounds in a bulk material will greatly affect the final properties of the materials produced. For example, steels are classified based on 1/10 and 1/100 weight percentages of the carbon and other alloying elements they contain. Thus, the extracting and purifying methods used to extract iron in a blast furnace can affect the quality of steel that is produced.

Ceramics and Glasses

Another application of material science is the structures of ceramics and glass, typically associated with the most brittle materials. Bonding in ceramics and glasses uses covalent and ionic-covalent types with SiO_2 (silica or sand) as a fundamental building block. Ceramics are as soft as clay or as hard as stone and concrete. Usually, they are crystalline in form. Most glasses contain a metal oxide fused with silica. At high temperatures used to prepare glass, the material is a viscous liquid. The structure of glass forms into an amorphous state upon cooling. Windowpanes and eyeglasses are important examples. Fibers of glass are also available. Scratch resistant Corning Gorilla Glass is a well-known example of the application of materials science to drastically improve the properties of common components. Diamond and carbon in its graphite form are considered to be ceramics.

Si3N4 ceramic bearing parts

Engineering ceramics are known for their stiffness and stability under high temperatures, compression and electrical stress. Alumina, silicon carbide, and tungsten carbide are made from a fine powder of their constituents in a process of sintering with a binder. Hot pressing provides higher density material. Chemical vapor deposition can place a film of a ceramic on another material. Cermets are ceramic particles containing some metals. The wear resistance of tools is derived from cemented carbides with the metal phase of cobalt and nickel typically added to modify properties.

Composites

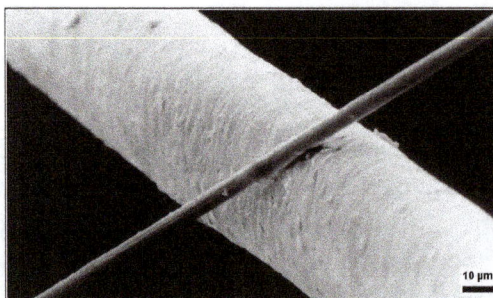

A 6 μm diameter carbon filament (running from bottom left to top right) siting atop the much larger human hair.

Filaments are commonly used for reinforcement in composite materials.

Another application of materials science in industry is making composite materials. These are structured materials composed of two or more macroscopic phases. Applications range from structural elements such as steel-reinforced concrete, to the thermal insulating tiles which play a key and integral role in NASA's Space Shuttle thermal protection system which is used to protect the surface of the shuttle from the heat of re-entry into the Earth's atmosphere. One example is reinforced Carbon-Carbon (RCC), the light gray material which withstands re-entry temperatures up to 1,510 °C (2,750 °F) and protects the Space Shuttle's wing leading edges and nose cap. RCC is a laminated composite material made from graphite rayon cloth and impregnated with a phenolic resin. After curing at high temperature in an autoclave, the laminate is pyrolized to convert the resin to carbon, impregnated with furfural alcohol in a vacuum chamber, and cured-pyrolized to convert the furfural alcohol to carbon. To provide oxidation resistance for reuse ability, the outer layers of the RCC are converted to silicon carbide.

Other examples can be seen in the "plastic" casings of television sets, cell-phones and so on. These plastic casings are usually a composite material made up of a thermoplastic matrix such as acrylonitrile butadiene styrene (ABS) in which calcium carbonate chalk, talc, glass fibers or carbon fibers have been added for added strength, bulk, or electrostatic dispersion. These additions may be termed reinforcing fibers, or dispersants, depending on their purpose.

Polymers

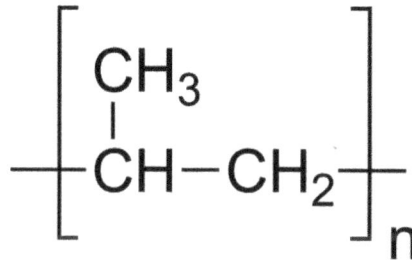

The repeating unit of the polymer polypropylene

Expanded polystyrene polymer packaging.

Polymers are chemical compounds made up of a large number of identical components linked together like chains. They are an important part of materials science. Polymers are the raw materials (the resins) used to make what are commonly called plastics and rubber. Plastics and rubber are really the final product, created after one or more polymers or additives have been added to a resin during processing, which is then shaped into a final form. Plastics which have been around, and which are in current widespread use, include polyethylene, polypropylene, polyvinyl chloride (PVC), polystyrene, nylons, polyesters, acrylics, polyurethanes, and polycarbonates and also rubbers which have been around are natural rubber, styrene butadiene rubber, chloroprene, and butadiene rubber. Plastics are generally classified as *commodity*, *specialty* and *engineering* plastics.

Polyvinyl chloride (PVC) is widely used, inexpensive, and annual production quantities are large. It lends itself to an vast array of applications, from artificial leather to electrical insulation and cabling, packaging, and containers. Its fabrication and processing are simple and well-established. The versatility of PVC is due to the wide range of plasticisers and other additives that it accepts. The term "additives" in polymer science refers to the chemicals and compounds added to the polymer base to modify its material properties.

Polycarbonate would be normally considered an engineering plastic (other examples include PEEK, ABS). Such plastics are valued for their superior strengths and other special material properties. They are usually not used for disposable applications, unlike commodity plastics.

Specialty plastics are materials with unique characteristics, such as ultra-high strength, electrical conductivity, electro-fluorescence, high thermal stability, etc.

The dividing lines between the various types of plastics is not based on material but rather on their properties and applications. For example, polyethylene (PE) is a cheap, low friction polymer commonly used to make disposable bags for shopping and trash, and is considered a commodity plastic, whereas medium-density polyethylene (MDPE) is used for underground gas and water pipes, and another variety called ultra-high-molecular-weight polyethylene (UHMWPE) is an engineering plastic which is used extensively as the glide rails for industrial equipment and the low-friction socket in implanted hip joints.

Metal Alloys

Wire rope made from steel alloy.

The study of metal alloys is a significant part of materials science. Of all the metallic alloys in use today, the alloys of iron (steel, stainless steel, cast iron, tool steel, alloy steels) make up the largest proportion both by quantity and commercial value. Iron alloyed with various proportions of carbon gives low, mid and high carbon steels. An iron carbon alloy is only considered steel if the carbon level is between 0.01% and 2.00%. For the steels, the hardness and tensile strength of the steel is related to the amount of carbon present, with increasing carbon levels also leading to lower ductility and toughness. Heat treatment processes such as quenching and tempering can significantly change these properties however. Cast Iron is defined as an iron–carbon alloy with more than 2.00% but less than 6.67% carbon. Stainless steel is defined as a regular steel alloy with greater than 10% by weight alloying content of Chromium. Nickel and Molybdenum are typically also found in stainless steels.

Other significant metallic alloys are those of aluminium, titanium, copper and magnesium. Copper alloys have been known for a long time (since the Bronze Age), while the alloys of the other three metals have been relatively recently developed. Due to the chemical reactivity of these metals, the electrolytic extraction processes required were only developed relatively recently. The alloys of aluminium, titanium and magnesium are also known and valued for their high strength-to-weight ratios and, in the case of magnesium, their ability to provide electromagnetic shielding. These materials are ideal for situations where high strength-to-weight ratios are more important than bulk cost, such as in the aerospace industry and certain automotive engineering applications.

Semiconductors

The study of semiconductors is a significant part of materials science. A semiconductor is a material that has a resistivity between a metal and insulator. It's electronic properties can be greatly altered through intentionally introducing impurities, or doping. From these semiconductor materials, things such as diodes, transistors, light-emitting diodes (LEDs), and analog and digital electric circuits can be built, making them materials of interest in industry. Semiconductor devices have replaced thermionic devices (vacuum tubes) in most applications. Semiconductor devices are manufactured both as single discrete devices and as integrated circuits (ICs), which consist of a number—from a few to millions—of devices manufactured and interconnected on a single semiconductor substrate.

Of all the semiconductors in use today, silicon makes up the largest portion both by quantity and commercial value. Monocrystalline silicon is used to produce wafers used in the semiconductor and electronics industry. Second to silicon, gallium arsenide (GaAs) is the second most popular semiconductor used. Due to its higher electron mobility and saturation velocity compared to silicon, its a material of choice for high speed electronics applications. These superior properties are compelling reasons to use GaAs circuitry in mobile phones, satellite communications, microwave point-to-point links and higher frequency radar systems. Other semiconductor materials include germanium, silicon carbide, and gallium nitride and have various applications.

Relation to Other Fields

Materials science evolved—starting from the 1960s—because it was recognized that to create, discover and design new materials, one had to approach it in a unified manner. Thus, materials science and engineering emerged at the intersection of various fields such as metallurgy, solid state physics, chemistry, chemical engineering, mechanical engineering and electrical engineering.

The field is inherently interdisciplinary, and the materials scientists/engineers must be aware and make use of the methods of the physicist, chemist and engineer. The field thus, maintains close relationships with these fields. Also, many physicists, chemists and engineers also find themselves working in materials science.

The overlap between physics and materials science has led to the offshoot field of *materials physics*, which is concerned with the physical properties of materials. The approach is generally more macroscopic and applied than in condensed matter physics.

The field of materials science and engineering is important both from a scientific perspective, as well as from an engineering one. When discovering new materials, one encounters new phenomena that may not have been observed before. Hence, there is lot of science to be discovered when working with materials. Materials science also provides test for theories in condensed matter physics.

Materials are of the utmost importance for engineers, as the usage of the appropriate materials is crucial when designing systems. As a result, materials science is an increasingly important part of an engineer's education.

Emerging Technologies in Materials Science

Emerging technology	Status	Potentially marginalized technologies	Potential applications	Related articles
Aerogel	Hypothetical, experiments, diffusion, early uses	Traditional insulation, glass	Improved insulation, insulative glass if it can be made clear, sleeves for oil pipelines, aerospace, high-heat & extreme cold applications	
Amorphous metal	Experiments	Kevlar	Armor	
Conductive polymers	Research, experiments, prototypes	Conductors	Lighter and cheaper wires, antistatic materials, organic solar cells	
Femtotechnology, picotechnology	Hypothetical	Present nuclear	New materials; nuclear weapons, power	
Fullerene	Experiments, diffusion	Synthetic diamond and carbon nanotubes (e.g., Buckypaper)	Programmable matter	
Graphene	Hypothetical, experiments, diffusion, early uses	Silicon-based integrated circuit	Components with higher strength to weight ratios, transistors that operate at higher frequency, lower cost of display screens in mobile devices, storing hydrogen for fuel cell powered cars, filtration systems, longer-lasting and faster-charging batteries, sensors to diagnose diseases	Potential applications of graphene

High-temperature superconductivity	Cryogenic receiver front-end (CRFE) RF and microwave filter systems for mobile phone base stations; prototypes in dry ice; Hypothetical and experiments for higher temperatures	Copper wire, semiconductor integral circuits	No loss conductors, frictionless bearings, magnetic levitation, lossless high-capacity accumulators, electric cars, heat-free integral circuits and processors	
LiTraCon	Experiments, already used to make Europe Gate	Glass	Building skyscrapers, towers, and sculptures like Europe Gate	
Metamaterials	Hypothetical, experiments, diffusion	Classical optics	Microscopes, cameras, metamaterial cloaking, cloaking devices	
Metal foam	Research, commercialization	Hulls	Space colonies, floating cities	
Multi-function structures	Hypothetical, experiments, some prototypes, few commercial	Composite materials mostly	Wide range, e.g., self health monitoring, self healing material, morphing, ...	
Nanomaterials: carbon nanotubes	Hypothetical, experiments, diffusion, early uses	Structural steel and aluminium	Stronger, lighter materials, space elevator	Potential applications of carbon nanotubes, carbon fiber
Programmable matter	Hypothetical, experiments	Coatings, catalysts	Wide range, e.g., claytronics, synthetic biology	
Quantum dots	Research, experiments, prototypes	LCD, LED	Quantum dot laser, future use as programmable matter in display technologies (TV, projection), optical data communications (high-speed data transmission), medicine (laser scalpel)	
Silicene	Hypothetical, research	Field-effect transistors		
Superalloy	Research, diffusion	Aluminum, titanium, composite materials	Aircraft jet engines	
Synthetic diamond	early uses (drill bits, jewelry)	Silicon transistors	Electronics	

Essential Concepts of Materials Science

The essential concepts of materials science are biomaterials, composite materials, acoustic emission, crystal growth, Hume-Rothery rules and hysteresis. Biomaterial is the substance that is made to interact with biological systems whereas a composite material is a material made from constituents that have different physical or chemical properties. The major components of materials science are discussed in this section.

Biomaterial

A biomaterial is any substance that has been engineered to interact with biological systems for a medical purpose - either a therapeutic (treat, augment, repair or replace a tissue function of the body) or a diagnostic one. As a science, biomaterials is about fifty years old. The study of biomaterials is called biomaterials science or biomaterials engineering. It has experienced steady and strong growth over its history, with many companies investing large amounts of money into the development of new products. Biomaterials science encompasses elements of medicine, biology, chemistry, tissue engineering and materials science.

Note that a biomaterial is different from a biological material, such as bone, that is produced by a biological system. Additionally, care should be exercised in defining a biomaterial as biocompatible, since it is application-specific. A biomaterial that is biocompatible or suitable for one application may not be biocompatible in another.

IUPAC Definition

Material exploited in contact with living tissues, organisms, or microorganisms.

Introduction

Biomaterials can be derived either from nature or synthesized in the laboratory using a variety of chemical approaches utilizing metallic components, polymers, ceramics or composite materials. They are often used and/or adapted for a medical application, and thus comprises whole or part of a living structure or biomedical device which performs, augments, or replaces a natural function. Such functions may be benign, like being used for a heart valve, or may be bioactive with a more interactive functionality such as hydroxy-apatite coated hip implants. Biomaterials are also used every day in dental applications, surgery, and drug delivery. For example, a construct with impregnated pharmaceutical products can be placed into the body, which permits the prolonged release of a drug over an extended period of time. A biomaterial may also be an autograft, allograft or xenograft used as a transplant material.

Biomineralization

Biomineralization is the process by which living organisms produce minerals, often to harden or stiffen existing tissues. Such tissues are called mineralized tissues. It is an extremely widespread phenomenon; all six taxonomic kingdoms contain members that are able to form minerals, and over 60 different minerals have been identified in organisms. Examples include silicates in algae and diatoms, carbonates in invertebrates, and calcium phosphates and carbonates in vertebrates. These minerals often form structural features such as sea shells and the bone in mammals and birds. Organisms have been producing mineralised skeletons for the past 550 million years. Other examples include copper, iron and gold deposits involving bacteria. Biologically-formed minerals often have special uses such as magnetic sensors in magnetotactic bacteria (Fe_3O_4), gravity sensing devices ($CaCO_3$, $CaSO_4$, $BaSO_4$) and iron storage and mobilization ($Fe_2O_3 \cdot H_2O$ in the protein ferritin).

Self-assembly

Self-assembly is the most common term in use in the modern scientific community to describe the spontaneous aggregation of particles (atoms, molecules, colloids, micelles, etc.) without the influence of any external forces. Large groups of such particles are known to assemble themselves into thermodynamically stable, structurally well-defined arrays, quite reminiscent of one of the 7 crystal systems found in metallurgy and mineralogy (e.g. face-centered cubic, body-centered cubic, etc.). The fundamental difference in equilibrium structure is in the spatial scale of the unit cell (or lattice parameter) in each particular case.

Molecular self-assembly is found widely in biological systems and provides the basis of a wide variety of complex biological structures. This includes an emerging class of mechanically superior biomaterials based on microstructural features and designs found in nature. Thus, self-assembly is also emerging as a new strategy in chemical synthesis and nanotechnology. Molecular crystals, liquid crystals, colloids, micelles, emulsions, phase-separated polymers, thin films and self-assembled monolayers all represent examples of the types of highly ordered structures which are obtained using these techniques. The distinguishing feature of these methods is self-organization.

Structural Hierarchy

Nearly all materials could be seen as hierarchically structured, especially since the changes in spatial scale bring about different mechanisms of deformation and damage. However, in biological materials this hierarchical organization is inherent to the microstructure. One of the first examples of this, in the history of structural biology, is the early X-ray scattering work on the hierarchical structure of hair and wool by Astbury and Woods. In bone, for example, collagen is the building block of the organic matrix — a triple helix with diameter of 1.5 nm. These tropocollagen molecules are intercalated with the mineral phase (hydroxyapatite, a calcium phosphate) forming fibrils that curl into helicoids of alternating directions. These "osteons" are the basic building blocks of bones, with the volume fraction distribution between organic and mineral phase being about 60/40.

In another level of complexity, the hydroxyapatite crystals are mineral platelets that have a diameter of approximately 70–100 nm and thickness of 1 nm. They originally nucleate at the gaps between collagen fibrils.

Similarly, the hierarchy of abalone shell begins at the nanolevel, with an organic layer having a thickness of 20–30 nm. This layer proceeds with single crystals of aragonite (a polymorph of $CaCO_3$) consisting of "bricks" with dimensions of 0.5 and finishing with layers approximately 0.3 mm (mesostructure).

Crabs are arthropods whose carapace is made of a mineralized hard component (which exhibits brittle fracture) and a softer organic component composed primarily of chitin. The brittle component is arranged in a helical pattern. Each of these mineral 'rods' (1 µm diameter) contains chitin–protein fibrils with approximately 60 nm diameter. These fibrils are made of 3 nm diameter canals which link the interior and exterior of the shell.

Applications

Biomaterials are used in:

- Joint replacements
- Bone plates
- Bone cement
- Artificial ligaments and tendons
- Dental implants for tooth fixation
- Blood vessel prostheses
- Heart valves
- Skin repair devices (artificial tissue)
- Cochlear replacements
- Contact lenses
- Breast implants
- Drug delivery mechanisms
- Sustainable materials
- Vascular grafts
- Stents
- Nerve conduits
- Surgical sutures, clips, and staples for wound closure
- Pins and screws for fracture stabilisation

- Surgical mesh

Biomaterials must be compatible with the body, and there are often issues of biocompatibility which must be resolved before a product can be placed on the market and used in a clinical setting. Because of this, biomaterials are usually subjected to the same requirements as those undergone by new drug therapies.

All manufacturing companies are also required to ensure traceability of all of their products so that if a defective product is discovered, others in the same batch may be traced.

Heart Valves

In the United States, 45% of the 250,000 valve replacement procedures performed annually involve a mechanical valve implant. The most widely used valve is a bileaflet disc heart valve, or St. Jude valve. The mechanics involve two semicircular discs moving back and forth, with both allowing the flow of blood as well as the ability to form a seal against backflow. The valve is coated with pyrolytic carbon, and secured to the surrounding tissue with a mesh of woven fabric called Dacron™ (du Pont's trade name for polyethylene terephthalate). The mesh allows for the body's tissue to grow while incorporating the valve.

Skin Repair

Most of the time, 'artificial' tissue is grown from the patient's own cells. However, when the damage is so extreme that it is impossible to use the patient's own cells, artificial tissue cells are grown. The difficulty is in finding a scaffold that the cells can grow and organize on. The characteristics of the scaffold must be that it is biocompatible, cells can adhere to the scaffold, mechanically strong and biodegradable. One successful scaffold is a copolymer of lactic acid and glycolic acid.

Compatibility

Biocompatibility is related to the behavior of biomaterials in various environments under various chemical and physical conditions. The term may refer to specific properties of a material without specifying where or how the material is to be used. For example, a material may elicit little or no immune response in a given organism, and may or may not able to integrate with a particular cell type or tissue. The ambiguity of the term reflects the ongoing development of insights into how biomaterials interact with the human body and eventually how those interactions determine the clinical success of a medical device (such as pacemaker or hip replacement). Modern medical devices and prostheses are often made of more than one material—so it might not always be sufficient to talk about the biocompatibility of a specific material.

Biopolymers

Biopolymers are polymers produced by living organisms. Cellulose and starch, proteins and peptides, and DNA and RNA are all examples of biopolymers, in which the monomeric units, respectively, are sugars, amino acids, and nucleotides. Cellulose is both the most common biopolymer and the most common organic compound on Earth. About 33% of all plant matter is cellulose.

Composite Material

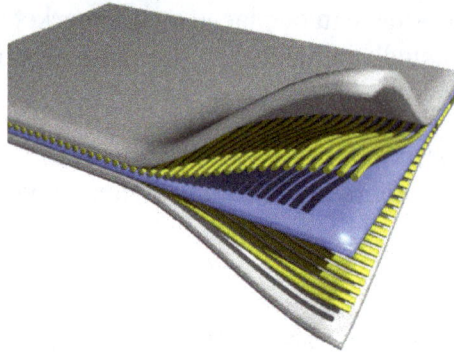

Composites are formed by combining materials together to form an overall structure with properties that differ from the sum of the individual components

A composite material (also called a composition material or shortened to composite which is the common name) is a material made from two or more constituent materials with significantly different physical or chemical properties that, when combined, produce a material with characteristics different from the individual components. The individual components remain separate and distinct within the finished structure. The new material may be preferred for many reasons: common examples include materials which are stronger, lighter, or less expensive when compared to traditional materials. More recently, researchers have also begun to actively include sensing, actuation, computation and communication into composites, which are known as Robotic Materials.

Typical engineered composite materials include:

- mortars, concrete
- Reinforced plastics, such as fiber-reinforced polymer
- Metal composites
- Ceramic composites (composite ceramic and metal matrices)

Composite materials are generally used for buildings, bridges, and structures such as boat hulls, swimming pool panels, race car bodies, shower stalls, bathtubs, storage tanks, imitation granite and cultured marble sinks and countertops. The most advanced examples perform routinely on spacecraft and aircraft in demanding environments.

History

The earliest man-made composite materials were straw and mud combined to form bricks for building construction. Ancient brick-making was documented by Egyptian tomb paintings .

Wattle and daub is one of the oldest man-made composite materials, at over 6000 years old. Concrete is also a composite material, and is used more than any other man-made material in the world. As of 2006, about 7.5 billion cubic metres of concrete are made each year—more than one cubic metre for every person on Earth.

- Woody plants, both true wood from trees and such plants as palms and bamboo, yield natural composites that were used prehistorically by mankind and are still used widely in construction and scaffolding.

- Plywood 3400 BC by the Ancient Mesopotamians; gluing wood at different angles gives better properties than natural wood

- Cartonnage layers of linen or papyrus soaked in plaster dates to the First Intermediate Period of Egypt c. 2181–2055 BC and was used for death masks

- Cob (material) Mud Bricks, or Mud Walls, (using mud (clay) with straw or gravel as a binder) have been used for thousands of years.

- Concrete was described by Vitruvius, writing around 25 BC in his *Ten Books on Architecture*, distinguished types of aggregate appropriate for the preparation of lime mortars. For structural mortars, he recommended *pozzolana*, which were volcanic sands from the sandlike beds of Pozzuoli brownish-yellow-gray in colour near Naples and reddish-brown at Rome. Vitruvius specifies a ratio of 1 part lime to 3 parts pozzolana for cements used in buildings and a 1:2 ratio of lime to pulvis Puteolanus for underwater work, essentially the same ratio mixed today for concrete used at sea. Natural cement-stones, after burning, produced cements used in concretes from post-Roman times into the 20th century, with some properties superior to manufactured Portland cement.

- Papier-mâché, a composite of paper and glue, has been used for hundreds of years

- The first artificial fibre reinforced plastic was bakelite which dates to 1907, although natural polymers such as shellac predate it

- One of the most common and familiar composite is fiberglass, in which small glass fiber are embedded within a polymeric material (normally an epoxy or polyester). The glass fiber is relatively strong and stiff (but also brittle), whereas the polymer is ductile (but also weak and flexible). Thus the resulting fiberglass is relatively stiff, strong, flexible, and ductile.

Examples

Materials

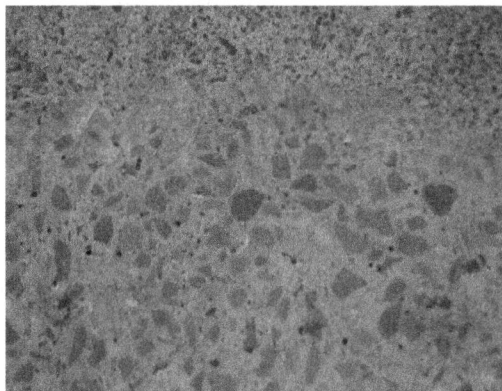

Concrete is a mixture of cement and aggregate, giving a robust, strong material that is very widely used.

Plywood is used widely in construction

Composite sandwich structure panel used for testing at NASA

A. composites reinforced by particles;
B. composites reinforced by chopped strands;
C. unidirectional composites;
D. laminates;
E. fabric reinforced plastics;
F. honeycomb composite structure;

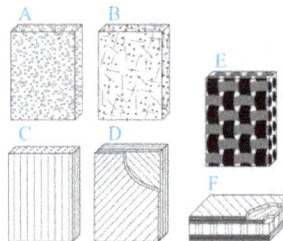

"Structural Integrity Analysis : Composites" (PDF).

Concrete is the most common artificial composite material of all and typically consists of loose stones (aggregate) held with a matrix of cement. Concrete is an inexpensive material, and will not compress or shatter even under quite a large compressive force. However, concrete cannot survive tensile loading (i.e., if stretched it will quickly break apart). Therefore, to give concrete the ability to resist being stretched, steel bars, which can resist high stretching forces, are often added to concrete to form reinforced concrete.

Fibre-reinforced polymers or FRPs include carbon-fiber-reinforced polymer or CFRP, and glass-reinforced plastic or GRP. If classified by matrix then there are thermoplastic composites, short fiber thermoplastics, long fibre thermoplastics or long fibre-reinforced thermoplastics. There are numerous thermoset composites, including paper composite panels. Many advanced systems usually incorporate aramid fibre and carbon fibre in an epoxy resin matrix.

Shape memory polymer composites are high-performance composites, formulated using fibre or fabric reinforcement and shape memory polymer resin as the matrix. Since a shape memory polymer resin is used as the matrix, these composites have the ability to be easily manipulated into various configurations when they are heated above their activation temperatures and will exhibit high strength and stiffness at lower temperatures. They can also be reheated and reshaped repeatedly without losing their material properties. These composites are ideal for applications such as lightweight, rigid, deployable structures; rapid manufacturing; and dynamic reinforcement.

High strain composites are another type of high-performance composites that are designed to perform in a high deformation setting and are often used in deployable systems where structural flexing is advantageous. Although high strain composites exhibit many similarities to shape memory polymers, their performance is generally dependent on the fiber layout as opposed to the resin content of the matrix.

Composites can also use metal fibres reinforcing other metals, as in metal matrix composites (MMC) or ceramic matrix composites (CMC), which includes bone (hydroxyapatite reinforced with collagen fibres), cermet (ceramic and metal) and concrete. Ceramic matrix composites are built primarily for fracture toughness, not for strength.

Organic matrix/ceramic aggregate composites include asphalt concrete, polymer concrete, mastic asphalt, mastic roller hybrid, dental composite, syntactic foam and mother of pearl. Chobham armour is a special type of composite armour used in military applications.

Additionally, thermoplastic composite materials can be formulated with specific metal powders resulting in materials with a density range from 2 g/cm^3 to 11 g/cm^3 (same density as lead). The most common name for this type of material is "high gravity compound" (HGC), although "lead replacement" is also used. These materials can be used in place of traditional materials such as aluminium, stainless steel, brass, bronze, copper, lead, and even tungsten in weighting, balancing (for example, modifying the centre of gravity of a tennis racquet), vibration damping, and radiation shielding applications. High density composites are an economically viable option when certain materials are deemed hazardous and are banned (such as lead) or when secondary operations costs (such as machining, finishing, or coating) are a factor.

A sandwich-structured composite is a special class of composite material that is fabricated by attaching two thin but stiff skins to a lightweight but thick core. The core material is normally low strength material, but its higher thickness provides the sandwich composite with high bending stiffness with overall low density.

Wood is a naturally occurring composite comprising cellulose fibres in a lignin and hemicellulose matrix. Engineered wood includes a wide variety of different products such as wood fibre board, plywood, oriented strand board, wood plastic composite (recycled wood fibre in polyethylene matrix), Pykrete (sawdust in ice matrix), Plastic-impregnated or laminated paper or textiles, Arborite, Formica (plastic) and Micarta. Other engineered laminate composites, such as Mallite, use a central core of end grain balsa wood, bonded to surface skins of light alloy or GRP. These generate low-weight, high rigidity materials.

Products

Fiber-reinforced composite materials have gained popularity (despite their generally high cost) in high-performance products that need to be lightweight, yet strong enough to take harsh loading conditions such as aerospace components (tails, wings, fuselages, propellers), boat and scull hulls, bicycle frames and racing car bodies. Other uses include fishing rods, storage tanks, swimming pool panels, and baseball bats. The new Boeing 787 structure including the wings and fuselage is composed largely of composites. Composite materials are also becoming more common in the realm of orthopedic surgery.And It is the most common hockey stick material.

Carbon composite is a key material in today's launch vehicles and heat shields for the re-entry phase of spacecraft. It is widely used in solar panel substrates, antenna reflectors and yokes of spacecraft. It is also used in payload adapters, inter-stage structures and heat shields of launch vehicles. Furthermore, disk brake systems of airplanes and racing cars are using carbon/carbon material, and the composite material with carbon fibers and silicon carbide matrix has been introduced in luxury vehicles and sports cars.

In 2006, a fiber-reinforced composite pool panel was introduced for in-ground swimming pools, residential as well as commercial, as a non-corrosive alternative to galvanized steel.

In 2007, an all-composite military Humvee was introduced by TPI Composites Inc and Armor Holdings Inc, the first all-composite military vehicle. By using composites the vehicle is lighter, allowing higher payloads. In 2008, carbon fiber and DuPont Kevlar (five times stronger than steel) were combined with enhanced thermoset resins to make military transit cases by ECS Composites creating 30-percent lighter cases with high strength.

Pipes and fittings for various purpose like transportation of potable water, fire-fighting, irrigation, seawater, desalinated water, chemical and industrial waste, and sewage are now manufactured in glass reinforced plastics.

Overview

Carbon fiber composite part.

Composites are made up of individual materials referred to as constituent materials. There are two main categories of constituent materials: matrix and reinforcement. At least one portion of each type is required. The matrix material surrounds and supports the reinforcement materials by maintaining their relative positions. The reinforcements impart their special mechanical and physical properties to enhance the matrix properties. A synergism produces material properties unavailable from the individual constituent materials, while the wide variety of matrix and strengthening materials allows the designer of the product or structure to choose an optimum combination.

Engineered composite materials must be formed to shape. The matrix material can be introduced to the reinforcement before or after the reinforcement material is placed into the mould cavity or onto the mould surface. The matrix material experiences a melding event, after which the part shape is essentially set. Depending upon the nature of the matrix material, this melding event can occur in various ways such as chemical polymerization or solidification from the melted state.

A variety of moulding methods can be used according to the end-item design requirements. The principal factors impacting the methodology are the natures of the chosen matrix and reinforcement materials. Another important factor is the gross quantity of material to be produced. Large quantities can be used to justify high capital expenditures for rapid and automated manufacturing technology. Small production quantities are accommodated with lower capital expenditures but higher labour and tooling costs at a correspondingly slower rate.

Many commercially produced composites use a polymer matrix material often called a resin solution. There are many different polymers available depending upon the starting raw ingredients. There are several broad categories, each with numerous variations. The most common are known as polyester, vinyl ester, epoxy, phenolic, polyimide, polyamide, polypropylene, PEEK, and others. The reinforcement materials are often fibres but also commonly ground minerals. The various methods described below have been developed to reduce the resin content of the final product, or the fibre content is increased. As a rule of thumb, lay up results in a product containing 60% resin and 40% fibre, whereas vacuum infusion gives a final product with 40% resin and 60% fiber content. The strength of the product is greatly dependent on this ratio.

Martin Hubbe and Lucian A Lucia consider wood to be a natural composite of cellulose fibres in a matrix of lignin.

Constituents

Matrices

Organic

Polymers are common matrices (especially used for fiber reinforced plastics). Road surfaces are often made from asphalt concrete which uses bitumen as a matrix. Mud (wattle and daub) has seen extensive use. Typically, most common polymer-based composite materials, including fiberglass, carbon fiber, and Kevlar, include at least two parts, the substrate and the resin.

Polyester resin tends to have yellowish tint, and is suitable for most backyard projects. Its weaknesses are that it is UV sensitive and can tend to degrade over time, and thus generally is also coated to help preserve it. It is often used in the making of surfboards and for marine applications. Its hardener is a peroxide, often MEKP (methyl ethyl ketone peroxide). When the peroxide is mixed with the resin, it decomposes to generate free radicals, which initiate the curing reaction. Hardeners in these systems are commonly called catalysts, but since they do not re-appear unchanged at the end of the reaction, they do not fit the strictest chemical definition of a catalyst.

Vinylester resin tends to have a purplish to bluish to greenish tint. This resin has lower viscosity than polyester resin, and is more transparent. This resin is often billed as being fuel resistant, but will melt in contact with gasoline. This resin tends to be more resistant over time to degradation than polyester resin, and is more flexible. It uses the same hardeners as polyester resin (at a similar mix ratio) and the cost is approximately the same.

Epoxy resin is almost totally transparent when cured. In the aerospace industry, epoxy is used as a structural matrix material or as a structural glue.

Shape memory polymer (SMP) resins have varying visual characteristics depending on their formulation. These resins may be epoxy-based, which can be used for auto body and outdoor equipment repairs; cyanate-ester-based, which are used in space applications; and acrylate-based, which can be used in very cold temperature applications, such as for sensors that indicate whether perishable goods have warmed above a certain maximum temperature. These resins are unique in that their shape can be repeatedly changed by heating above their glass transition temperature (T_g). When heated, they become flexible and elastic, allowing for easy configuration. Once they are cooled, they will maintain their new shape. The resins will return to their original shapes when they are reheated above their T_g. The advantage of shape memory polymer resins is that they can be shaped and reshaped repeatedly without losing their material properties. These resins can be used in fabricating shape memory composites.

Inorganic

Cement (concrete), metals, ceramics, and sometimes glasses are employed. Unusual matrices such as ice are sometime proposed as in pykecrete.

Reinforcements

Fiber

Differences in the way the fibers are laid out give different strengths and ease of manufacture

Reinforcement usually adds rigidity and greatly impedes crack propagation. Thin fibers can have very high strength, and provided they are mechanically well attached to the matrix they can greatly improve the composite's overall properties.

Fiber-reinforced composite materials can be divided into two main categories normally referred to as short fiber-reinforced materials and continuous fiber-reinforced materials. Continuous reinforced materials will often constitute a layered or laminated structure. The woven and continuous fibre styles are typically available in a variety of forms, being pre-impregnated with the given matrix (resin), dry, uni-directional tapes of various widths, plain weave, harness satins, braided, and stitched.

The short and long fibers are typically employed in compression moulding and sheet moulding operations. These come in the form of flakes, chips, and random mate (which can also be made from a continuous fibre laid in random fashion until the desired thickness of the ply / laminate is achieved).

Common fibers used for reinforcement include glass fibers, carbon fibers, cellulose (wood/paper fiber and straw) and high strength polymers for example aramid. Silicon carbide fibers are used for some high temperature applications.

Other Reinforcement

Concrete uses aggregate, and reinforced concrete additionally uses steel bars (rebar) to tension the concrete. Steel mesh or wires are also used in some glass and plastic products.

Cores

Many composite layup designs also include a co-curing or post-curing of the prepreg with various other media, such as honeycomb or foam. This is commonly called a sandwich structure. This is a more common layup for the manufacture of radomes, doors, cowlings, or non-structural parts.

Open- and closed-cell-structured foams like polyvinylchloride, polyurethane, polyethylene or polystyrene foams, balsa wood, syntactic foams, and honeycombs are commonly used core materials. Open- and closed-cell metal foam can also be used as core materials. Recently, 3D graphene structures (also called graphene foam) have also been employed as core structures.A recent review by Khurram and Xu et al., have provided the summary of the state-of-the-art techniques for fabrication of the 3D structure of graphene, and the examples of the use of these foam like structures as a core for their respective polymer composites.

Fabrication Methods

Fabrication of composite materials is accomplished by a wide variety of techniques, including:

- Advanced fiber placement (Automated fiber placement)
- Tailored fiber placement
- Fiberglass spray lay-up process
- Filament winding

- Lanxide process

- Tufting

- Z-pinning

Composite fabrication usually involves wetting, mixing or saturating the reinforcement with the matrix, and then causing the matrix to bind together (with heat or a chemical reaction) into a rigid structure. The operation is usually done in an open or closed forming mold, but the order and ways of introducing the ingredients varies considerably.

Mold Overview

Within a mold, the reinforcing and matrix materials are combined, compacted, and cured (processed) to undergo a melding event. After the melding event, the part shape is essentially set, although it can deform under certain process conditions. For a thermoset polymeric matrix material, the melding event is a curing reaction that is initiated by the application of additional heat or chemical reactivity such as an organic peroxide. For a thermoplastic polymeric matrix material, the melding event is a solidification from the melted state. For a metal matrix material such as titanium foil, the melding event is a fusing at high pressure and a temperature near the melting point.

For many moulding methods, it is convenient to refer to one mould piece as a "lower" mould and another mould piece as an "upper" mould. Lower and upper refer to the different faces of the moulded panel, not the mould's configuration in space. In this convention, there is always a lower mould, and sometimes an upper mould. Part construction begins by applying materials to the lower mould. Lower mould and upper mould are more generalized descriptors than more common and specific terms such as male side, female side, a-side, b-side, tool side, bowl, hat, mandrel, etc. Continuous manufacturing uses a different nomenclature.

The moulded product is often referred to as a panel. For certain geometries and material combinations, it can be referred to as a casting. For certain continuous processes, it can be referred to as a profile.

Vacuum Bag Moulding

Vacuum bag moulding uses a flexible film to enclose the part and seal it from outside air. Vacuum bag material is available in a tube shape or a sheet of material. A vacuum is then drawn on the vacuum bag and atmospheric pressure compresses the part during the cure. When a tube shaped bag is used, the entire part can be enclosed within the bag. When using sheet bagging materials, the edges of the vacuum bag are sealed against the edges of the mould surface to enclose the part against an air-tight mould. When bagged in this way, the lower mould is a rigid structure and the upper surface of the part is formed by the flexible membrane vacuum bag. The flexible membrane can be a reusable silicone material or an extruded polymer film. After sealing the part inside the vacuum bag, a vacuum is drawn on the part (and held) during cure. This process can be performed at either ambient or elevated temperature with ambient atmospheric pressure acting upon the vacuum bag. A vacuum pump is typically used to draw a vacuum. An economical method of drawing a vacuum is with a venturi vacuum and air compressor.

A vacuum bag is a bag made of strong rubber-coated fabric or a polymer film used to compress the part during cure or hardening. In some applications the bag encloses the entire material, or in other applications a mold is used to form one face of the laminate with the bag being a single layer to seal to the outer edge of the mold face. When using a tube shaped bag, the ends of the bag are sealed and the air is drawn out of the bag through a nipple using a vacuum pump. As a result, uniform pressure approaching one atmosphere is applied to the surfaces of the object inside the bag, holding parts together while the adhesive cures. The entire bag may be placed in a temperature-controlled oven, oil bath or water bath and gently heated to accelerate curing.

Vacuum bagging is widely used in the composites industry as well. Carbon fiber fabric and fiberglass, along with resins and epoxies are common materials laminated together with a vacuum bag operation.

Woodworking applications

In commercial woodworking facilities, vacuum bags are used to laminate curved and irregular shaped workpieces.

Typically, polyurethane or vinyl materials are used to make the bag. A tube shaped bag is open at both ends. The piece, or pieces to be glued are placed into the bag and the ends sealed. One method of sealing the open ends of the bag is by placing a clamp on each end of the bag. A plastic rod is laid across the end of the bag, the bag is then folded over the rod. A plastic sleeve with an opening in it, is then snapped over the rod. This procedure forms a seal at both ends of the bag, when the vacuum is ready to be drawn.

A "platen" is sometimes used inside the bag for the piece being glued to lie on. The platen has a series of small slots cut into it, to allow the air under it to be evacuated. The platen must have rounded edges and corners to prevent the vacuum from tearing the bag.

When a curved part is to be glued in a vacuum bag, it is important that the pieces being glued be placed over a solidly built form, or have an air bladder placed under the form. This air bladder has access to "free air" outside the bag. It is used to create an equal pressure under the form, preventing it from being crushed.

Pressure Bag Moulding

This process is related to vacuum bag molding in exactly the same way as it sounds. A solid female mold is used along with a flexible male mold. The reinforcement is placed inside the female mold with just enough resin to allow the fabric to stick in place (wet lay up). A measured amount of resin is then liberally brushed indiscriminately into the mold and the mold is then clamped to a machine that contains the male flexible mold. The flexible male membrane is then inflated with heated compressed air or possibly steam. The female mold can also be heated. Excess resin is forced out along with trapped air. This process is extensively used in the production of composite helmets due to the lower cost of unskilled labor. Cycle times for a helmet bag moulding machine vary from 20 to 45 minutes, but the finished shells require no further curing if the molds are heated.

Autoclave Moulding

A process using a two-sided mould set that forms both surfaces of the panel. On the lower side is a rigid mould and on the upper side is a flexible membrane made from silicone or an extruded polymer film such as nylon. Reinforcement materials can be placed manually or robotically. They include continuous fibre forms fashioned into textile constructions. Most often, they are pre-impregnated with the resin in the form of prepreg fabrics or unidirectional tapes. In some instances, a resin film is placed upon the lower mould and dry reinforcement is placed above. The upper mould is installed and vacuum is applied to the mould cavity. The assembly is placed into an autoclave. This process is generally performed at both elevated pressure and elevated temperature. The use of elevated pressure facilitates a high fibre volume fraction and low void content for maximum structural efficiency.

Resin Transfer Moulding (RTM)

RTM is a process using a rigid two-sided mould set that forms both surfaces of the panel. The mould is typically constructed from aluminum or steel, but composite molds are sometimes used. The two sides fit together to produce a mould cavity. The distinguishing feature of resin transfer moulding is that the reinforcement materials are placed into this cavity and the mould set is closed prior to the introduction of matrix material. Resin transfer moulding includes numerous varieties which differ in the mechanics of how the resin is introduced to the reinforcement in the mould cavity. These variations include everything from the RTM methods used in out of autoclave composite manufacturing for high-tech aerospace components to vacuum infusion to vacuum assisted resin transfer moulding (VARTM). This process can be performed at either ambient or elevated temperature.

Other Fabrication Methods

Other types of fabrication include press moulding, transfer moulding, pultrusion moulding, filament winding, casting, centrifugal casting, continuous casting and slip forming. There are also forming capabilities including CNC filament winding, vacuum infusion, wet lay-up, compression moulding, and thermoplastic moulding, to name a few. The use of curing ovens and paint booths is also needed for some projects.

Finishing Methods

The finishing of the composite parts is also critical in the final design. Many of these finishes will include rain-erosion coatings or polyurethane coatings.

Tooling

The mold and mold inserts are referred to as "tooling." The mold/tooling can be constructed from a variety of materials. Tooling materials include invar, steel, aluminium, reinforced silicone rubber, nickel, and carbon fiber. Selection of the tooling material is typically based on, but not limited to, the coefficient of thermal expansion, expected number of cycles, end item tolerance, desired or required surface condition, method of cure, glass transition temperature of the material being moulded, moulding method, matrix, cost and a variety of other considerations.

Physical Properties

The physical properties of composite materials are generally not isotropic (independent of direction of applied force) in nature, but rather are typically anisotropic (different depending on the direction of the applied force or load). For instance, the stiffness of a composite panel will often depend upon the orientation of the applied forces and/or moments. Panel stiffness is also dependent on the design of the panel. For instance, the fibre reinforcement and matrix used, the method of panel build, thermoset versus thermoplastic, type of weave, and orientation of fibre axis to the primary force.

In contrast, isotropic materials (for example, aluminium or steel), in standard wrought forms, typically have the same stiffness regardless of the directional orientation of the applied forces and/ or moments.

The relationship between forces/moments and strains/curvatures for an isotropic material can be described with the following material properties: Young's Modulus, the shear Modulus and the Poisson's ratio, in relatively simple mathematical relationships. For the anisotropic material, it requires the mathematics of a second order tensor and up to 21 material property constants. For the special case of orthogonal isotropy, there are three different material property constants for each of Young's Modulus, Shear Modulus and Poisson's ratio—a total of 9 constants to describe the relationship between forces/moments and strains/curvatures.

Techniques that take advantage of the anisotropic properties of the materials include mortise and tenon joints (in natural composites such as wood) and Pi Joints in synthetic composites.

Failure

Shock, impact, or repeated cyclic stresses can cause the laminate to separate at the interface between two layers, a condition known as delamination. Individual fibres can separate from the matrix e.g. fibre pull-out.

Composites can fail on the microscopic or macroscopic scale. Compression failures can occur at both the macro scale or at each individual reinforcing fiber in compression buckling. Tension failures can be net section failures of the part or degradation of the composite at a microscopic scale where one or more of the layers in the composite fail in tension of the matrix or failure of the bond between the matrix and fibers.

Some composites are brittle and have little reserve strength beyond the initial onset of failure while others may have large deformations and have reserve energy absorbing capacity past the onset of damage. The variations in fibers and matrices that are available and the mixtures that can be made with blends leave a very broad range of properties that can be designed into a composite structure. The best known failure of a brittle ceramic matrix composite occurred when the carbon-carbon composite tile on the leading edge of the wing of the Space Shuttle Columbia fractured when impacted during take-off. It led to catastrophic break-up of the vehicle when it re-entered the Earth's atmosphere on 1 February 2003.

Compared to metals, composites have relatively poor bearing strength.

Testing

To aid in predicting and preventing failures, composites are tested before and after construction. Pre-construction testing may use finite element analysis (FEA) for ply-by-ply analysis of curved surfaces and predicting wrinkling, crimping and dimpling of composites. Materials may be tested during manufacturing and after construction through several nondestructive methods including ultrasonics, thermography, shearography and X-ray radiography, and laser bond inspection for NDT of relative bond strength integrity in a localized area.

Various Engineered Composite Materials

Mortar (Masonry)

Mortar holding weathered bricks

Mortar is a workable paste used to bind building blocks such as stones, bricks, and concrete masonry units together, fill and seal the irregular gaps between them, and sometimes add decorative colors or patterns in masonry walls. In its broadest sense mortar includes pitch, asphalt, and soft mud or clay, such as used between mud bricks. *Mortar* comes from Latin *mortarium* meaning crushed.

Cement mortar becomes hard when it cures, resulting in a rigid aggregate structure; however the mortar is intended to be weaker than the building blocks and the sacrificial element in the masonry, because the mortar is easier and less expensive to repair than the building blocks. Mortars are typically made from a mixture of sand, a binder, and water. The most common binder since the early 20th century is Portland cement but the ancient binder lime mortar is still used in some new construction. Lime and gypsum in the form of plaster of Paris are used particularly in the repair and repointing of buildings and structures because it is important the repair materials are similar to the original materials: The type and ratio of the repair mortar is determined by a *mortar analysis*. There are several types of cement mortars and additives.

Ancient Mortar

The first mortars were made of mud and clay. Because of a lack of stone and an abundance of clay, Babylonian constructions were of baked brick, using lime or pitch for mortar. According to Roman Ghirshman, the first evidence of humans using a form of mortar was at the Mehrgarh of Baluchistan in Pakistan, built of sun-dried bricks in 6500 BCE. The ancient sites of Harappan civilization of third millennium BCE are built with kiln-fired bricks and a *gypsum mortar*. Gypsum mortar, also called plaster of Paris, was used in the construction of the Egyptian pyramids and many other ancient structures. It is made from gypsum which requires a lower firing temperature so is easier to make than lime mortar and sets up much faster which may be a reason it was used as the typical mortar in ancient, brick arch and vault construction. Gypsum mortar is not as durable as other mortars in damp conditions.

Roman mortar on display at Chetham's School of Music.

In early Egyptian pyramids constructed about 2600–2500 BCE, the limestone blocks were bound by mortar of mud and clay, or clay and sand. In later Egyptian pyramids, the mortar was made of either gypsum or lime. Gypsum mortar was essentially a mixture of plaster and sand and was quite soft.

In the Indian subcontinent, multiple cement types have been observed in the sites of the Indus Valley Civilization, such as the Mohenjo-daro city-settlement that dates to earlier than 2600 BCE. Gypsum cement that was "*light grey and contained sand, clay, traces of calcium carbonate, and a high percentage of lime*" was used in the construction of wells, drains and on the exteriors of "*important looking buildings.*" Bitumen mortar was also used at a lower-frequency, including in the Great Bath at Mohenjo-daro.

Historically, building with concrete and mortar next appeared in Greece. The excavation of the underground aqueduct of Megara revealed that a reservoir was coated with a pozzolanic mortar 12 mm thick. This aqueduct dates back to c. 500 BCE. Pozzolanic mortar is a lime based mortar, but is made with an additive of volcanic ash that allows it to be hardened underwater; thus it is known as hydraulic cement. The Greeks obtained the volcanic ash from the Greek islands Thira and Nisiros, or from the then Greek colony of Dicaearchia (Pozzuoli) near Naples, Italy. The Romans later improved the use and methods of making what became known as pozzolanic mortar and cement. Even later, the Romans used a mortar without pozzolana using crushed terra cotta, introducing aluminum oxide and silicon dioxide into the mix. This mortar was not as strong as pozzolanic mortar, but, because it was denser, it better resisted penetration by water.

Hydraulic mortar was not available in ancient China, possibly due to a lack of volcanic ash. Around 500 CE, sticky rice soup was mixed with slaked lime to make an inorganic–organic composite mortar that had more strength and water resistance than lime mortar.

It is not understood how the art of making hydraulic mortar and cement, which was perfected and in such widespread use by both the Greeks and Romans, was then lost for almost two millennia. During the Middle Ages when the Gothic cathedrals were being built, the only active ingredient in the mortar was lime. Since cured lime mortar can be degraded by contact with water, many structures suffered from wind blown rain over the centuries.

Laying bricks with Portland cement mortar

Ordinary Portland Cement Mortar

Ordinary Portland cement mortar, commonly known as OPC mortar or just cement mortar, is created by mixing powdered Ordinary Portland Cement, aggregate and water.

It was invented in 1794 by Joseph Aspdin and patented on 18 December 1824, largely as a result of efforts to develop stronger mortars. It was made popular during the late nineteenth century, and had by 1930 became more popular than lime mortar as construction material. The advantages of Portland cement is that it sets hard and quickly, allowing a faster pace of construction. Furthermore, fewer skilled workers are required in building a structure with Portland cement.

As a general rule, however, Portland cement should not be used for the repair or repointing of older buildings built in lime mortar, which require the flexibility, softness and breathability of lime if they are to function correctly.

In the United States and other countries, five standard types of mortar (available as dry pre-mixed products) are generally used for both new construction and repair. Strengths of mortar change based on the ratio of cement, lime, and sand used in mortar. The ingredients and the mix ratio for each type of mortars are specified under the ASTM standards. These premixed mortar products are designated by one of the five letters, M, S, N, O, and K. Type M mortar is the strongest, and Type K the weakest. These type letters are taken from the alternate letters of the words "MaSoN wOrK".

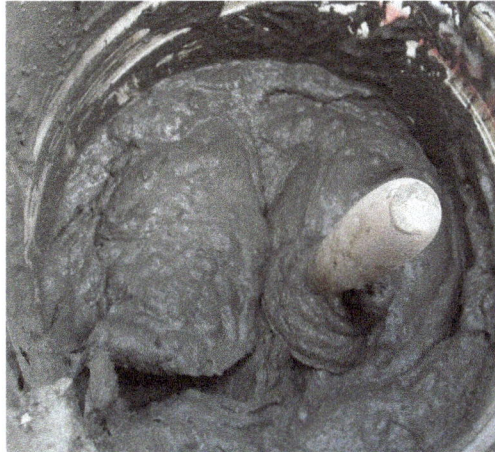

Mortar mixed inside a 5-gallon bucket using clean water and mortar from a bag. When it's the right consistency as in the photo (trowel stands up) it's ready to apply.

Polymer Cement Mortar

Polymer cement mortars (PCM) are the materials which are made by partially replacing the cement hydrate binders of conventional cement mortar with polymers. The polymeric admixtures include latexes or emulsions, redispersible polymer powders, water-soluble polymers, liquid resins and monomers. It has low permeability, and it reduces the incidence of drying shrinkage cracking, mainly designed for repairing concrete structures.

Lime Mortar

The setting speed can be increased by using impure limestone in the kiln, to form a hydraulic lime that will set on contact with water. Such a lime must be stored as a dry powder. Alternatively, a pozzolanic material such as calcined clay or brick dust may be added to the mortar mix. Addition of a pozzolanic material will make the mortar set reasonably quickly by reaction with the water.

It would be problematic to use Portland cement mortars to repair older buildings originally constructed using lime mortar. Lime mortar is softer than cement mortar, allowing brickwork a certain degree of flexibility to adapt to shifting ground or other changing conditions. Cement mortar is harder and allows little flexibility. The contrast can cause brickwork to crack where the two mortars are present in a single wall.

Lime mortar is considered breathable in that it will allow moisture to freely move through and evaporate from the surface. In old buildings with walls that shift over time, cracks can be found which allow rain water into the structure. The lime mortar allows this moisture to escape through evaporation and keeps the wall dry. Re–pointing or rendering an old wall with cement mortar stops the evaporation and can cause problems associated with moisture behind the cement.

Pozzolanic Mortar

Pozzolana is a fine, sandy volcanic ash. It was originally discovered and dug at Pozzuoli, nearby Mount Vesuvius in Italy, and was subsequently mined at other sites, too. The Romans learned that pozzolana added to lime mortar allowed the lime to set relatively quickly and even under water.

Vitruvius, the Roman architect, spoke of four types of pozzolana. It is found in all the volcanic areas of Italy in various colours: black, white, grey and red. Pozzolana has since become a generic term for any siliceous and/or aluminous additive to slaked lime to create hydraulic cement.

Finely ground and mixed with lime it is a hydraulic cement, like Portland cement, and makes a strong mortar that will also set under water.

Firestop Mortar

Firestop mortars are mortars most typically used to firestop large openings in walls and floors required to have a fire-resistance rating. They are passive fire protection items. Firestop mortars differ in formula and properties from most other cementitious substances and cannot be substituted with generic mortars without violating the listing and approval use and compliance.

Firestop mortar is usually a combination of powder mixed with water, forming a cementatious stone which dries hard. It is sometimes mixed with lightweight aggregates, such as perlite or vermiculite. It is sometimes pigmented to distinguish it from generic materials in an effort to prevent unlawful substitution and to enable verification of the certification listing.

Mortar constituents.

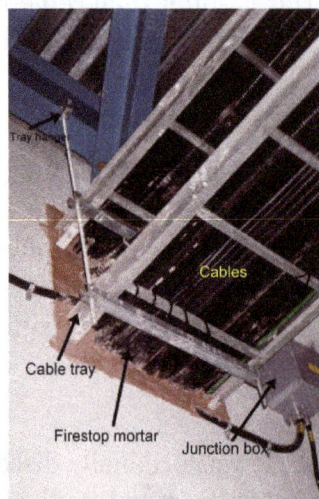

Firestopped cable tray penetration. The cables and the tray are penetrants.

Cable tray cross barrier firestop test, full scale wall

Radiocarbon Dating

An international team headed by Åbo Akademi University has developed a method of determining the age of mortar using radiocarbon dating. As the mortar hardens, the current atmosphere is encased in the mortar and thus provides a sample for analysis. Various factors affect the sample and raise the margin of error for the analysis.

Concrete

Exterior of the Roman Pantheon, finished 128 AD, still the largest unreinforced solid concrete dome.

Interior of the Pantheon dome, seen from beneath. The concrete for the coffered dome was laid on moulds, probably mounted on temporary scaffolding.

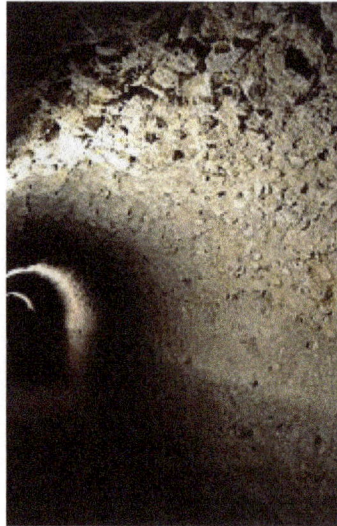

Opus caementicium exposed in a characteristic Roman arch. In contrast to modern concrete structures, the concrete used in Roman buildings was usually covered with brick or stone.

Concrete is a composite material composed of coarse aggregate bonded together with a fluid cement which hardens over time. Most concretes used are lime-based concretes such as Portland cement concrete or concretes made with other hydraulic cements, such as ciment fondu. However, asphalt concrete which is very frequently used for road surfaces is also a type of concrete, where the cement material is bitumen, and polymer concretes are sometimes used where the cementing material is a polymer.

In Portland cement concrete (and other hydraulic cement concretes), when the aggregate is mixed together with the dry cement and water, they form a fluid mass that is easily molded into shape. The cement reacts chemically with the water and other ingredients to form a hard matrix which binds all the materials together into a durable stone-like material that has many uses. Often, additives (such as pozzolans or superplasticizers) are included in the mixture to improve the physical properties of the wet mix or the finished material. Most concrete is poured with reinforcing materials (such as rebar) embedded to provide tensile strength, yielding reinforced concrete.

Famous concrete structures include the Hoover Dam, the Panama Canal and the Roman Pantheon. The earliest large-scale users of concrete technology were the ancient Romans, and concrete was widely used in the Roman Empire. The Colosseum in Rome was built largely of concrete, and the concrete dome of the Pantheon is the world's largest unreinforced concrete dome. Today, large concrete structures (for example, dams and multi-storey car parks) are usually made with reinforced concrete.

After the Roman Empire collapsed, use of concrete became rare until the technology was redeveloped in the mid-18th century. Today, concrete is the most widely used man-made material (measured by tonnage).

History

The word concrete comes from the Latin word *"concretus"* (meaning compact or condensed), the perfect passive participle of *"concrescere"*, from *"con-"* (together) and *"crescere"* (to grow).

Prehistory

Perhaps the earliest known occurrence of cement was twelve million years ago. A deposit of cement was formed after an occurrence of oil shale located adjacent to a bed of limestone burned due to natural causes. These ancient deposits were investigated in the 1960s and 1970s.

On a human timescale, small usages of concrete go back for thousands of years. Concrete-like materials were used since 6500 BC by the Nabataea traders or Bedouins who occupied and controlled a series of oases and developed a small empire in the regions of southern Syria and northern Jordan. They discovered the advantages of hydraulic lime, with some self-cementing properties, by 700 BC. They built kilns to supply mortar for the construction of rubble-wall houses, concrete floors, and underground waterproof cisterns. The cisterns were kept secret and were one of the reasons the Nabataea were able to thrive in the desert. Some of these structures survive to this day.

Classical Era

In both Roman and Egyptian times, it was re-discovered that adding volcanic ash to the mix allowed it to set underwater. Similarly, the Romans knew that adding horse hair made concrete less liable to crack while it hardened, and adding blood made it more frost-resistant. Crystallization of strätlingite and the introduction of pyroclastic clays creates further fracture resistance.

German archaeologist Heinrich Schliemann found concrete floors, which were made of lime and pebbles, in the royal palace of Tiryns, Greece, which dates roughly to 1400–1200 BC. Lime mortars were used in Greece, Crete, and Cyprus in 800 BC. The Assyrian Jerwan Aqueduct (688 BC) made use of waterproof concrete. Concrete was used for construction in many ancient structures.

The Romans used concrete extensively from 300 BC to 476 AD, a span of more than seven hundred years. During the Roman Empire, Roman concrete (or *opus caementicium*) was made from quicklime, pozzolana and an aggregate of pumice. Its widespread use in many Roman structures, a key event in the history of architecture termed the Roman Architectural Revolution, freed Roman construction from the restrictions of stone and brick material and allowed for revolutionary new designs in terms of both structural complexity and dimension.

Concrete, as the Romans knew it, was a new and revolutionary material. Laid in the shape of arches, vaults and domes, it quickly hardened into a rigid mass, free from many of the internal thrusts and strains that troubled the builders of similar structures in stone or brick.

Modern tests show that *opus caementicium* had as much compressive strength as modern Portland-cement concrete (ca. 200 kg/cm^2 [20 MPa; 2,800 psi]). However, due to the absence of reinforcement, its tensile strength was far lower than modern reinforced concrete, and its mode of application was also different:

Modern structural concrete differs from Roman concrete in two important details. First, its mix consistency is fluid and homogeneous, allowing it to be poured into forms rather than requiring hand-layering together with the placement of aggregate, which, in Roman practice, often consisted of rubble. Second, integral reinforcing steel gives modern concrete assemblies great strength in tension, whereas Roman concrete could depend only upon the strength of the concrete bonding to resist tension.

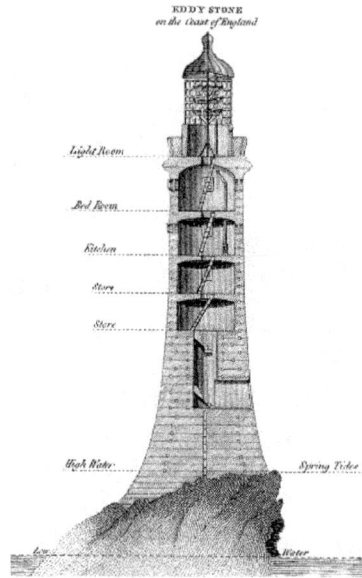

Smeaton's Tower

The widespread use of concrete in many Roman structures ensured that many survive to the present day. The Baths of Caracalla in Rome are just one example. Many Roman aqueducts and bridges such as the magnificent Pont du Gard have masonry cladding on a concrete core, as does the dome of the Pantheon.

Middle Ages

After the Roman Empire, the use of burned lime and pozzolana was greatly reduced until the technique was all but forgotten between 500 and the 14th century. From the 14th century to the mid-18th century, the use of cement gradually returned. The *Canal du Midi* was built using concrete in 1670, and there are concrete structures in Finland that date from the 16th century.

Industrial Era

Perhaps the greatest driver behind the modern usage of concrete was Smeaton's Tower, the third Eddystone Lighthouse in Devon, England. To create this structure, between 1756 and 1759, British engineer John Smeaton pioneered the use of hydraulic lime in concrete, using pebbles and powdered brick as aggregate.

A method for producing Portland cement was patented by Joseph Aspdin in 1824.

Reinforced concrete was invented in 1849 by Joseph Monier. In 1889 the first concrete reinforced bridge was built, and the first large concrete dams were built in 1936, Hoover Dam and Grand Coulee Dam.

Composition of Concrete

There are many types of concrete available, created by varying the proportions of the main ingredients below. In this way or by substitution for the cementitious and aggregate phases, the finished product can be tailored to its application with varying strength, density, or chemical and thermal resistance properties.

Aggregate consists of large chunks of material in a concrete mix, generally a coarse gravel or crushed rocks such as limestone, or granite, along with finer materials such as sand.

Cement, most commonly Portland cement, is associated with the general term "concrete." A range of materials can be used as the cement in concrete. One of the most familiar of these alternative cements is asphalt concrete. Other cementitious materials such as fly ash and slag cement, are sometimes added as mineral admixtures - either pre-blended with the cement or directly as a concrete component - and become a part of the binder for the aggregate.

To produce concrete from most cements (excluding asphalt), water is mixed with the dry powder and aggregate, which produces a semi-liquid that workers can shape, typically by pouring it into a form. The concrete solidifies and hardens through a chemical process called hydration. The water reacts with the cement, which bonds the other components together, creating a robust stone-like material.

Chemical admixtures are added to achieve varied properties. These ingredients may accelerate or slow down the rate at which the concrete hardens, and impart many other useful properties including increased tensile strength, entrainment of air, and/or water resistance.

Reinforcement is often included in concrete. Concrete can be formulated with high compressive strength, but always has lower tensile strength. For this reason it is usually reinforced with materials that are strong in tension, often steel.

Mineral admixtures are becoming more popular in recent decades. The use of recycled materials as concrete ingredients has been gaining popularity because of increasingly stringent environmental legislation, and the discovery that such materials often have complementary and valuable properties. The most conspicuous of these are fly ash, a by-product of coal-fired power plants, ground granulated blast furnace slag, a byproduct of steelmaking, and silica fume, a byproduct of industrial electric arc furnaces. The use of these materials in concrete reduces the amount of resources required, as the mineral admixtures act as a partial cement replacement. This displaces some cement production, an energetically expensive and environmentally problematic process, while reducing the amount of industrial waste that must be disposed of. Mineral admixtures can be pre-blended with the cement during its production for sale and use as a blended cement, or mixed directly with other components when the concrete is produced.

The *mix design* depends on the type of structure being built, how the concrete is mixed and delivered, and how it is placed to form the structure.

Cement

Portland cement is the most common type of cement in general usage. It is a basic ingredient of concrete, mortar and many plasters. English masonry worker Joseph Aspdin patented Portland cement in 1824. It was named because of the similarity of its color to Portland limestone, quarried from the English Isle of Portland and used extensively in London architecture. It consists of a mixture of calcium silicates (alite, belite), aluminates and ferrites - compounds which combine calcium, silicon, aluminium and iron in forms which will react with water. Portland cement and similar materials are made by heating limestone (a source of calcium) with clay and/or shale (a source of silicon, aluminium and iron) and grinding this product (called *clinker*) with a source of sulfate (most commonly gypsum).

A few tons of bagged cement. This amount represents about two minutes of output from a 10,000 ton per day cement kiln.

In modern cement kilns many advanced features are used to lower the fuel consumption per ton of clinker produced. Cement kilns are extremely large, complex, and inherently dusty industrial installations, and have emissions which must be controlled. Of the various ingredients used to produce a given quantity of concrete, the cement is the most energetically expensive. Even complex and efficient kilns require 3.3 to 3.6 gigajoules of energy to produce a ton of clinker and then grind it into cement. Many kilns can be fueled with difficult-to-dispose-of wastes, the most common being used tires. The extremely high temperatures and long periods of time at those temperatures allows cement kilns to efficiently and completely burn even difficult-to-use fuels.

Water

Combining water with a cementitious material forms a cement paste by the process of hydration. The cement paste glues the aggregate together, fills voids within it, and makes it flow more freely.

A lower water-to-cement ratio yields a stronger, more durable concrete, whereas more water gives a freer-flowing concrete with a higher slump. Impure water used to make concrete can cause problems when setting or in causing premature failure of the structure.

Hydration involves many different reactions, often occurring at the same time. As the reactions proceed, the products of the cement hydration process gradually bond together the individual sand and gravel particles and other components of the concrete to form a solid mass.

Reaction:

Cement chemist notation: $C_3S + H \rightarrow C\text{-}S\text{-}H + CH$

Standard notation: $Ca_3SiO_5 + H_2O \rightarrow (CaO)\cdot(SiO_2)\cdot(H_2O)(gel) + Ca(OH)_2$

Balanced: $2Ca_3SiO_5 + 7H_2O \rightarrow 3(CaO)\cdot2(SiO_2)\cdot4(H_2O)(gel) + 3Ca(OH)_2$ (approximately; the exact ratios of the CaO, SiO_2 and H_2O in C-S-H can vary)

Aggregates

Crushed stone aggregate

Fine and coarse aggregates make up the bulk of a concrete mixture. Sand, natural gravel, and crushed stone are used mainly for this purpose. Recycled aggregates (from construction, demolition, and excavation waste) are increasingly used as partial replacements for natural aggregates, while a number of manufactured aggregates, including air-cooled blast furnace slag and bottom ash are also permitted.

The size distribution of the aggregate determines how much binder is required. Aggregate with a very even size distribution has the biggest gaps whereas adding aggregate with smaller particles tends to fill these gaps. The binder must fill the gaps between the aggregate as well as pasting the surfaces of the aggregate together, and is typically the most expensive component. Thus variation in sizes of the aggregate reduces the cost of concrete. The aggregate is nearly always stronger than the binder, so its use does not negatively affect the strength of the concrete.

Redistribution of aggregates after compaction often creates inhomogeneity due to the influence of vibration. This can lead to strength gradients.

Decorative stones such as quartzite, small river stones or crushed glass are sometimes added to the surface of concrete for a decorative "exposed aggregate" finish, popular among landscape designers.

In addition to being decorative, exposed aggregate may add robustness to a concrete.

Reinforcement

Concrete is strong in compression, as the aggregate efficiently carries the compression load. However, it is weak in tension as the cement holding the aggregate in place can crack, allowing the structure to fail. Reinforced concrete adds either steel reinforcing bars, steel fibers, glass fibers, or plastic fibers to carry tensile loads.

Constructing a rebar cage. This cage will be permanently embedded in poured concrete to create a reinforced concrete structure.

Chemical Admixtures

Chemical admixtures are materials in the form of powder or fluids that are added to the concrete to give it certain characteristics not obtainable with plain concrete mixes. In normal use, admixture dosages are less than 5% by mass of cement and are added to the concrete at the time of batching/mixing. The common types of admixtures are as follows:

- Accelerators speed up the hydration (hardening) of the concrete. Typical materials used are $CaCl2$, $Ca(NO_3)_2$ and $NaNO_3$. However, use of chlorides may cause corrosion in steel reinforcing and is prohibited in some countries, so that nitrates may be favored. Accelerating admixtures are especially useful for modifying the properties of concrete in cold weather.

- Retarders slow the hydration of concrete and are used in large or difficult pours where partial setting before the pour is complete is undesirable. Typical polyol retarders are sugar, sucrose, sodium gluconate, glucose, citric acid, and tartaric acid.

- Air entraining agents add and entrain tiny air bubbles in the concrete, which reduces damage during freeze-thaw cycles, increasing durability. However, entrained air entails a trade off with strength, as each 1% of air may decrease compressive strength 5%. If too much air becomes trapped in the concrete as a result of the mixing process, Defoamers can be used to encourage the air bubble to agglomerate, rise to the surface of the wet concrete and then disperse.

- Plasticizers increase the workability of plastic or "fresh" concrete, allowing it be placed more easily, with less consolidating effort. A typical plasticizer is lignosulfonate. Plasticizers can be used to reduce the water content of a concrete while maintaining workability and are sometimes called *water-reducers* due to this use. Such treatment improves its strength and durability characteristics. Superplasticizers (also called *high-range water-reducers*) are a class of plasticizers that have fewer deleterious effects and can be used to increase workability more than is practical with traditional plasticizers. Compounds used as superplasticizers include sulfonated naphthalene formaldehyde condensate, sulfonated melamine formaldehyde condensate, acetone formaldehyde condensate and polycarboxylate ethers.

- Pigments can be used to change the color of concrete, for aesthetics.

- Corrosion inhibitors are used to minimize the corrosion of steel and steel bars in concrete.

- Bonding agents are used to create a bond between old and new concrete (typically a type of polymer) with wide temperature tolerance and corrosion resistance.

- Pumping aids improve pumpability, thicken the paste and reduce separation and bleeding.

Mineral Admixtures and Blended Cements

	Components of Cement Comparison of Chemical and Physical Characteristics[a]				
Property	Portland Cement	Siliceous (ASTM C618 Class F) Fly Ash	Calcareous (ASTM C618 Class C) Fly Ash	Slag Cement	Silica Fume
SiO_2 content (%)	21.9	52	35	35	85–97
Al_2O_3 content (%)	6.9	23	18	12	—
Fe_2O_3 content (%)	3	11	6	1	—
CaO content (%)	63	5	21	40	< 1
MgO content (%)	2.5	—	—	—	—
SO_3 content (%)	1.7	—	—	—	—
Specific surface[b] (m^2/kg)	370	420	420	400	15,000–30,000
Specific gravity	3.15	2.38	2.65	2.94	2.22
General use in concrete	Primary binder	Cement replacement	Cement replacement	Cement replacement	Property enhancer
[a]Values shown are approximate: those of a specific material may vary.					
[b]Specific surface measurements for silica fume by nitrogen adsorption (BET) method, others by air permeability method (Blaine).					

Inorganic materials that have pozzolanic or latent hydraulic properties, these very fine-grained materials are added to the concrete mix to improve the properties of concrete (mineral admixtures), or as a replacement for Portland cement (blended cements). Products which incorporate limestone, fly ash, blast furnace slag, and other useful materials with pozzolanic properties into the mix, are being tested and used. This development is due to cement production being one of the largest producers (at about 5 to 10%) of global greenhouse gas emissions, as well as lowering costs, improving concrete properties, and recycling wastes.

- Fly ash: A by-product of coal-fired electric generating plants, it is used to partially replace Portland cement (by up to 60% by mass). The properties of fly ash depend on the type of coal burnt. In general, siliceous fly ash is pozzolanic, while calcareous fly ash has latent hydraulic properties.

- Ground granulated blast furnace slag (GGBFS or GGBS): A by-product of steel production is used to partially replace Portland cement (by up to 80% by mass). It has latent hydraulic properties.

- Silica fume: A byproduct of the production of silicon and ferrosilicon alloys. Silica fume is similar to fly ash, but has a particle size 100 times smaller. This results in a higher surface-to-volume ratio and a much faster pozzolanic reaction. Silica fume is used to increase strength and durability of concrete, but generally requires the use of superplasticizers for workability.

- High reactivity Metakaolin (HRM): Metakaolin produces concrete with strength and durability similar to concrete made with silica fume. While silica fume is usually dark gray or black in color, high-reactivity metakaolin is usually bright white in color, making it the preferred choice for architectural concrete where appearance is important.

- Carbon nanofibres can be added to concrete to enhance compressive strength and gain a higher Young's modulus, and also to improve the electrical properties required for strain monitoring, damage evaluation and self-health monitoring of concrete. Carbon fiber has many advantages in terms of mechanical and electrical properties (e.g. higher strength) and self-monitoring behavior due to the high tensile strength and high conductivity.

- Carbon products have been added to make concrete electrically conductive, for deicing purposes.

Concrete Production

Concrete production is the process of mixing together the various ingredients—water, aggregate, cement, and any additives—to produce concrete. Concrete production is time-sensitive. Once the ingredients are mixed, workers must put the concrete in place before it hardens. In modern usage, most concrete production takes place in a large type of industrial facility called a concrete plant, or often a batch plant.

In general usage, concrete plants come in two main types, ready mix plants and central mix plants. A ready mix plant mixes all the ingredients except water, while a central mix plant mixes all the ingredients including water. A central mix plant offers more accurate control of the concrete quality through better measurements of the amount of water added, but must be placed closer to the work site where the concrete will be used, since hydration begins at the plant.

Concrete plant facility showing a Concrete mixer being filled from the ingredient silos.

A concrete plant consists of large storage hoppers for various reactive ingredients like cement, storage for bulk ingredients like aggregate and water, mechanisms for the addition of various additives and amendments, machinery to accurately weigh, move, and mix some or all of those ingredients, and facilities to dispense the mixed concrete, often to a concrete mixer truck.

Modern concrete is usually prepared as a viscous fluid, so that it may be poured into forms, which are containers erected in the field to give the concrete its desired shape. There are many different ways in which concrete formwork can be prepared, such as Slip forming and Steel plate construction. Alternatively, concrete can be mixed into dryer, non-fluid forms and used in factory settings to manufacture Precast concrete products.

There is a wide variety of equipment for processing concrete, from hand tools to heavy industrial machinery. Whichever equipment builders use, however, the objective is to produce the desired building material; ingredients must be properly mixed, placed, shaped, and retained within time constraints. Any interruption in pouring the concrete can cause the initially placed material to begin to set before the next batch is added on top. This creates a horizontal plane of weakness called a *cold joint* between the two batches. Once the mix is where it should be, the curing process must be controlled to ensure that the concrete attains the desired attributes. During concrete preparation, various technical details may affect the quality and nature of the product.

When initially mixed, Portland cement and water rapidly form a gel of tangled chains of interlocking crystals, and components of the gel continue to react over time. Initially the gel is fluid, which improves workability and aids in placement of the material, but as the concrete sets, the chains of crystals join into a rigid structure, counteracting the fluidity of the gel and fixing the particles of aggregate in place. During curing, the cement continues to react with the residual water in a process of hydration. In properly formulated concrete, once this curing process has terminated the product has the desired physical and chemical properties. Among the qualities typically desired, are mechanical strength, low moisture permeability, and chemical and volumetric stability.

Mixing Concrete

Thorough mixing is essential for the production of uniform, high-quality concrete. For this reason equipment and methods should be capable of effectively mixing concrete materials containing the largest specified aggregate to produce *uniform mixtures* of the lowest slump practical for the work.

Separate paste mixing has shown that the mixing of cement and water into a paste before combining these materials with aggregates can increase the compressive strength of the resulting concrete. The paste is generally mixed in a *high-speed*, shear-type mixer at a w/cm (water to cement ratio) of 0.30 to 0.45 by mass. The cement paste premix may include admixtures such as accelerators or retarders, superplasticizers, pigments, or silica fume. The premixed paste is then blended with aggregates and any remaining batch water and final mixing is completed in conventional concrete mixing equipment.

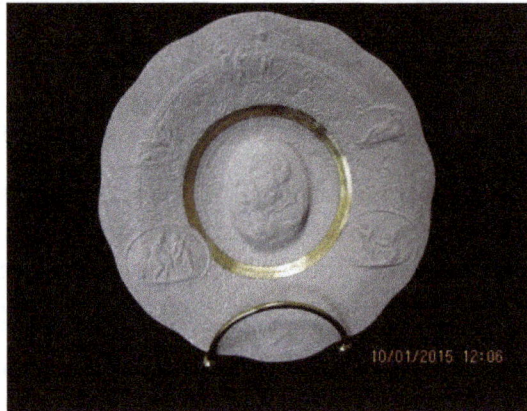

Decorative plate made of Nano concrete with High-Energy Mixing (HEM)

Workability

Pouring and smoothing out concrete at Palisades Park in Washington DC.

Workability is the ability of a fresh (plastic) concrete mix to fill the form/mold properly with the desired work (vibration) and without reducing the concrete's quality. Workability depends on water content, aggregate (shape and size distribution), cementitious content and age (level of hydration) and can be modified by adding chemical admixtures, like superplasticizer. Raising the water content or adding chemical admixtures increases concrete workability. Excessive water leads to increased bleeding and/or segregation of aggregates (when the cement and aggregates start to separate), with the resulting concrete having reduced quality. The use of an aggregate with an undesirable gradation can result in a very harsh mix design with a very low slump, which cannot readily be made more workable by addition of reasonable amounts of water.

Workability can be measured by the concrete slump test, a simplistic measure of the plasticity of a fresh batch of concrete following the ASTM C 143 or EN 12350-2 test standards. Slump is normally measured by filling an "Abrams cone" with a sample from a fresh batch of concrete. The cone is placed with the wide end down onto a level, non-absorptive surface. It is then filled in three layers of equal volume, with each layer being tamped with a steel rod to consolidate the layer. When the cone is carefully lifted off, the enclosed material slumps a certain amount, owing to gravity. A relatively dry sample slumps very little, having a slump value of one or two inches (25 or 50 mm) out of one foot (305 mm). A relatively wet concrete sample may slump as much as eight inches. Workability can also be measured by the flow table test.

Slump can be increased by addition of chemical admixtures such as plasticizer or superplasticizer without changing the water-cement ratio. Some other admixtures, especially air-entraining admixture, can increase the slump of a mix.

High-flow concrete, like self-consolidating concrete, is tested by other flow-measuring methods. One of these methods includes placing the cone on the narrow end and observing how the mix flows through the cone while it is gradually lifted.

After mixing, concrete is a fluid and can be pumped to the location where needed.

Curing

A concrete slab ponded while curing.

A common misconception is that concrete dries as it sets, but the opposite is true - damp concrete sets better than dry concrete. In other words, cement is "hydraulic": water allows it to gain strength. Too much water is counterproductive, but too little water is deleterious. Curing allows concrete to achieve optimal strength and hardness. Curing is the hydration process that occurs after the concrete has been placed. In chemical terms, curing allows calcium-silicate hydrate (C-S-H) to form. To gain strength and harden fully, concrete curing requires time. In around 4 weeks, typically over 90% of the final strength is reached, although strengthening may continue for decades. The conversion of calcium hydroxide in the concrete into calcium carbonate from absorption of CO_2 over several decades further strengthens the concrete and makes it more resistant to damage. This carbonation reaction, however, lowers the pH of the cement pore solution and can corrode the reinforcement bars.

Hydration and hardening of concrete during the first three days is critical. Abnormally fast drying and shrinkage due to factors such as evaporation from wind during placement may lead to increased tensile stresses at a time when it has not yet gained sufficient strength, resulting in greater shrinkage cracking. The early strength of the concrete can be increased if it is kept damp during the curing process. Minimizing stress prior to curing minimizes cracking. High-early-strength concrete is designed to hydrate faster, often by increased use of cement that increases shrinkage and cracking. The strength of concrete changes (increases) for up to three years. It depends on cross-section dimension of elements and conditions of structure exploitation. Addition of short-cut polymer fibers can improve (reduce) shrinkage-induced stresses during curing and increase early and ultimate compression strength.

Properly curing concrete leads to increased strength and lower permeability and avoids cracking where the surface dries out prematurely. Care must also be taken to avoid freezing or overheating due to the exothermic setting of cement. Improper curing can cause scaling, reduced strength, poor abrasion resistance and cracking.

Curing Techniques

During the curing period, concrete is ideally maintained at controlled temperature and humidity. To ensure full hydration during curing, concrete slabs are often sprayed with "curing compounds" that create a water-retaining film over the concrete. Typical films are made of wax or related hydrophobic compounds. After the concrete is sufficiently cured, the film is allowed to abrade from the concrete through normal use.

Traditional conditions for curing involve by spraying or ponding the concrete surface with water. The picture to the right shows one of many ways to achieve this, ponding – submerging setting concrete in water and wrapping in plastic to prevent dehydration. Additional common curing methods include wet burlap and/or plastic sheeting covering the fresh concrete.

For higher-strength applications, accelerated curing techniques may be applied to the concrete. One common technique involves heating the poured concrete with steam, which serves to both keep it damp and raise the temperature, so that the hydration process proceeds more quickly and more thoroughly.

Specialty Concretes

Pervious Concrete

Pervious concrete is a mix of specially graded coarse aggregate, cement, water and little-to-no fine aggregates. This concrete is also known as "no-fines" or porous concrete. Mixing the ingredients in a carefully controlled process creates a paste that coats and bonds the aggregate particles. The hardened concrete contains interconnected air voids totalling approximately 15 to 25 percent. Water runs through the voids in the pavement to the soil underneath. Air entrainment admixtures are often used in freeze–thaw climates to minimize the possibility of frost damage.

Nanoconcrete

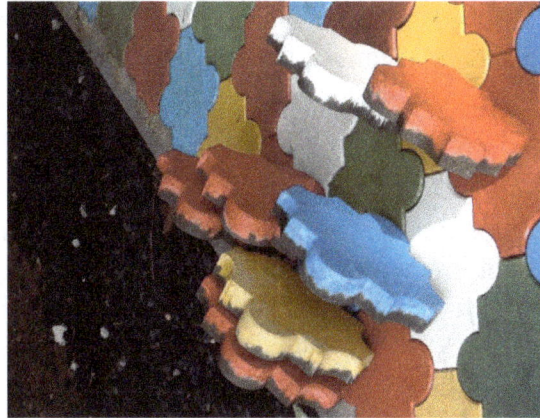

Two-layered pavers, top layer made of pigmented HEM Nanoconcrete.

Nanoconcrete is created by high-energy mixing (HEM) of cement, sand and water. To ensure the mixing is thorough enough to create nano-concrete, the mixer must apply a total mixing power to the mixture of 30 - 600 watts per kilogram of the mix. This mixing must continue long enough to yield a net specific energy expended upon the mix of at least 5000 joules per kilogram of the mix. A plasticizer or a superplasticizer is then added to the activated mixture which can later be mixed with aggregates in a conventional concrete mixer. In the HEM process, the intense mixing of cement and water with sand provides dissipation of energy and increases shear stresses on the surface of cement particles. This intense mixing serves to divide the cement particles into extremely fine nanometer scale sizes, which provides for extremely thorough mixing. This results in the increased volume of water interacting with cement and acceleration of Calcium Silicate Hydrate (C-S-H) colloid creation.

The initial natural process of cement hydration with formation of colloidal globules about 5 nm in diameter spreads into the entire volume of cement – water matrix as the energy expended upon the mix approaches and exceeds 5000 joules per kilogram.

The liquid activated high-energy mixture can be used by itself for casting small architectural details and decorative items, or foamed (expanded) for lightweight concrete. HEM Nanoconcrete hardens in low and subzero temperature conditions and possesses an increased volume of gel, which reduces capillarity in solid and porous materials.

Microbial Concrete

Bacteria such as *Bacillus pasteurii, Bacillus pseudofirmus, Bacillus cohnii, Sporosarcina pasteurii*, and *Arthrobacter crystallopoietes* increase the compression strength of concrete through their biomass. Not all bacteria increase the strength of concrete significantly with their biomass. Bacillus sp. CT-5. can reduce corrosion of reinforcement in reinforced concrete by up to four times.

Sporosarcina pasteurii reduces water and chloride permeability. *B. pasteurii* increases resistance to acid. *Bacillus pasteurii* and *B. sphaericuscan* induce calcium carbonate precipitation in the surface of cracks, adding compression strength.

Safety

Concrete, when ground, can result in the creation of hazardous dust. The National Institute for Occupational Safety and Health in the United States recommends attaching local exhaust ventilation shrouds to electric concrete grinders to control the spread of this dust.

Properties

Concrete has relatively high compressive strength, but much lower tensile strength. For this reason it is usually reinforced with materials that are strong in tension (often steel). The elasticity of concrete is relatively constant at low stress levels but starts decreasing at higher stress levels as matrix cracking develops. Concrete has a very low coefficient of thermal expansion and shrinks as it matures. All concrete structures crack to some extent, due to shrinkage and tension. Concrete that is subjected to long-duration forces is prone to creep.

Tests can be performed to ensure that the properties of concrete correspond to specifications for the application.

Compression testing of a concrete cylinder

Different mixes of concrete ingredients produce different strengths. Concrete strength values are usually specified as the compressive strength of either a cylindrical or cubic specimen, where these values usually differ by around 20% for the same concrete mix.

Different strengths of concrete are used for different purposes. Very low-strength - 14 MPa (2,000 psi) or less - concrete may be used when the concrete must be lightweight. Lightweight concrete is often achieved by adding air, foams, or lightweight aggregates, with the side effect that the strength is reduced. For most routine uses, 20 MPa (2,900 psi) to 32 MPa (4,600 psi) concrete is often used. 40 MPa (5,800 psi) concrete is readily commercially available as a more durable, although more expensive, option. Higher-strength concrete is often used for larger civil projects. Strengths above 40 MPa (5,800 psi) are often used for specific building elements. For example, the lower floor columns of high-rise concrete buildings may use concrete of 80 MPa (11,600 psi) or more, to keep the size of the columns small. Bridges may use long beams of high-strength concrete to lower the number of spans required. Occasionally, other structural needs may require high-strength concrete. If a structure must be very rigid, concrete of very high strength may be specified, even much stronger than is required to bear the service loads. Strengths as high as 130 MPa (18,900 psi) have been used commercially for these reasons.

Building With Concrete

The Buffalo City Court Building in Buffalo, NY.

Concrete is one of the most durable building materials. It provides superior fire resistance compared with wooden construction and gains strength over time. Structures made of concrete can have a long service life. Concrete is used more than any other manmade material in the world. As of 2006, about 7.5 billion cubic meters of concrete are made each year, more than one cubic meter for every person on Earth.

Mass Concrete Structures

Aerial photo of reconstruction at Taum Sauk (Missouri) pumped storage facility in late November, 2009. After the original reservoir failed, the new reservoir was made of roller-compacted concrete.

Large concrete structures such as dams, navigation locks, large mat foundations, and large break-waters generate excessive heat during cement hydration and associated expansion. To mitigate these effects post-cooling is commonly applied during construction. An early example at Hoover Dam, installed a network of pipes between vertical concrete placements to circulate cooling water during the curing process to avoid damaging overheating. Similar systems are still used; depending on volume of the pour, the concrete mix used, and ambient air temperature, the cooling process may last for many months after the concrete is placed. Various methods also are used to pre-cool the concrete mix in mass concrete structures.

Another approach to mass concrete structures that is becoming more widespread is the use of roller-compacted concrete, which uses much lower amounts of cement and water than conventional concrete mixtures and is generally not poured into place. Instead it is placed in thick layers as a semi-dry material and compacted into a dense, strong mass with rolling compactors. Because it uses less cementitious material, roller-compacted concrete has a much lower cooling requirement than conventional concrete.

Surface Finishes

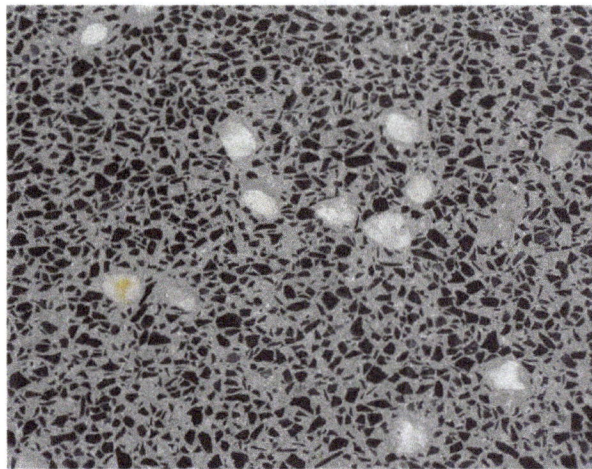

Black basalt polished concrete floor

Raw concrete surfaces tend to be porous, and have a relatively uninteresting appearance. Many different finishes can be applied to improve the appearance and preserve the surface against staining, water penetration, and freezing.

Examples of improved appearance include stamped concrete where the wet concrete has a pattern impressed on the surface, to give a paved, cobbled or brick-like effect, and may be accompanied with coloration. Another popular effect for flooring and table tops is polished concrete where the concrete is polished optically flat with diamond abrasives and sealed with polymers or other sealants.

Other finishes can be achieved with chiselling, or more conventional techniques such as painting or covering it with other materials.

The proper treatment of the surface of concrete, and therefore its characteristics, is an important stage in the construction and renovation of architectural structures.

Prestressed Concrete Structures

Prestressed concrete is a form of reinforced concrete that builds in compressive stresses during construction to oppose those experienced in use. This can greatly reduce the weight of beams or slabs, by better distributing the stresses in the structure to make optimal use of the reinforcement. For example, a horizontal beam tends to sag. Prestressed reinforcement along the bottom of the beam counteracts this. In pre-tensioned concrete, the prestressing is achieved by using steel or polymer tendons or bars that are subjected to a tensile force prior to casting, or for post-tensioned concrete, after casting.

40-foot cacti decorate a sound/retaining wall in Scottsdale, Arizona

More than 55,000 miles (89,000 km) of highways in the United States are paved with this material. Reinforced concrete, prestressed concrete and precast concrete are the most widely used types of concrete functional extensions in modern days.

Concrete Roads

Concrete roads are more fuel efficient to drive on, more reflective and last significantly longer than other paving surfaces, yet have a much smaller market share than other paving solutions. Modern paving methods and design practices have changed the economics of concrete paving, so that a well designed and placed concrete pavement will be less expensive on initial costs and significantly less expensive over the life cycle. Another major benefit is that pervious concrete can be used, which eliminates the need to place storm drains near the road, and reducing the need for slightly sloped roadway to help rainwater to run off. No longer requiring to discard the rainwater using drains also means that less electricity is needed (more pumping is otherwise needed in the water distribution system), and no rainwater gets polluted as it no longer mixes with polluted water; rather it is immediately absorbed by the ground.

Energy Efficiency

Energy requirements for transportation of concrete are low because it is produced locally from local resources, typically manufactured within 100 kilometers of the job site. Similarly, relatively little energy is used in producing and combining the raw materials (although large amounts of CO_2 are produced by the chemical reactions in cement manufacture). The overall embodied energy of concrete is therefore lower than for most structural materials other than wood.

Once in place, concrete offers great energy efficiency over the lifetime of a building. Concrete walls leak air far less than those made of wood frames. Air leakage accounts for a large percentage of energy loss from a home. The thermal mass properties of concrete increase the efficiency of both residential and commercial buildings. By storing and releasing the energy needed for heating or cooling, concrete's thermal mass delivers year-round benefits by reducing temperature swings inside and minimizing heating and cooling costs. While insulation reduces energy loss through the building envelope, thermal mass uses walls to store and release energy. Modern concrete wall systems use both external insulation and thermal mass to create an energy-efficient building. Insulating concrete forms (ICFs) are hollow blocks or panels made of either insulating foam or rastra that are stacked to form the shape of the walls of a building and then filled with reinforced concrete to create the structure.

Fire Safety

A modern building: Boston City Hall (completed 1968) is constructed largely of concrete, both precast and poured in place. Of Brutalist architecture, it was voted "The World's Ugliest Building" in 2008.

Concrete buildings are more resistant to fire than those constructed using steel frames, since concrete has lower heat conductivity than steel and can thus last longer under the same fire conditions. Concrete is sometimes used as a fire protection for steel frames, for the same effect as above. Concrete as a fire shield, for example Fondu fyre, can also be used in extreme environments like a missile launch pad.

Options for non-combustible construction include floors, ceilings and roofs made of cast-in-place and hollow-core precast concrete. For walls, concrete masonry technology and Insulating Concrete Forms (ICFs) are additional options. ICFs are hollow blocks or panels made of fireproof insulating foam that are stacked to form the shape of the walls of a building and then filled with reinforced concrete to create the structure.

Concrete also provides good resistance against externally applied forces such as high winds, hurricanes, and tornadoes owing to its lateral stiffness, which results in minimal horizontal movement. However this stiffness can work against certain types of concrete structures, particularly where a relatively higher flexing structure is required to resist more extreme forces.

Earthquake Safety

As discussed above, concrete is very strong in compression, but weak in tension. Larger earthquakes can generate very large shear loads on structures. These shear loads subject the structure to both tensile and compressional loads. Concrete structures without reinforcement, like other unreinforced masonry structures, can fail during severe earthquake shaking. Unreinforced masonry structures constitute one of the largest earthquake risks globally. These risks can be reduced through seismic retrofitting of at-risk buildings, (e.g. school buildings in Istanbul, Turkey).

Concrete Degradation

Concrete spalling caused by the corrosion of rebar

Concrete can be damaged by many processes, such as the expansion of corrosion products of the steel reinforcement bars, freezing of trapped water, fire or radiant heat, aggregate expansion, sea water effects, bacterial corrosion, leaching, erosion by fast-flowing water, physical damage and chemical damage (from carbonatation, chlorides, sulfates and distillate water). The micro fungi Aspergillus Alternaria and Cladosporium were able to grow on samples of concrete used as a radioactive waste barrier in the Chernobyl reactor; leaching aluminium, iron, calcium and silicon.

Useful Life

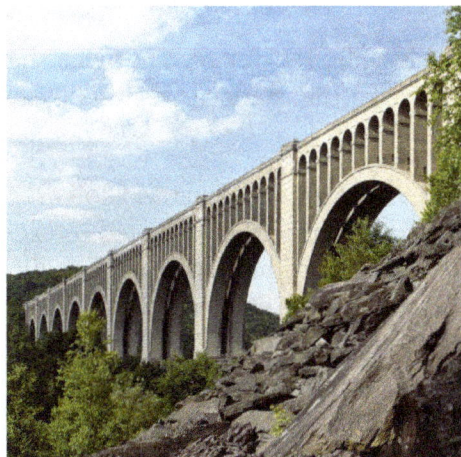

The Tunkhannock Viaduct was begun in 1912 and is still in regular service as of 2016.

Concrete can be viewed as a form of artificial sedimentary rock. As a type of mineral, the compounds of which it is composed are extremely stable. Many concrete structures are built with an expected lifetime of approximately 100 years, but researchers have suggested that adding silica fume could extend the useful life of bridges and other concrete uses to as long as 16,000 years.

Coatings are also available to protect concrete from damage, and extend the useful life. Epoxy coatings may be applied only to interior surfaces, though, as they would otherwise trap moisture in the concrete.

A self-healing concrete has been developed that can also last longer than conventional concrete.

Effect of Modern Concrete Use

Concrete mixing plant in Birmingham, Alabama in 1936

Concrete is widely used for making architectural structures, foundations, brick/block walls, pavements, bridges/overpasses, highways, runways, parking structures, dams, pools/reservoirs, pipes, footings for gates, fences and poles and even boats. Concrete is used in large quantities almost everywhere mankind has a need for infrastructure. Concrete is one of the most frequently used building materials in animal houses and for manure and silage storage structures in agriculture.

The amount of concrete used worldwide, ton for ton, is twice that of steel, wood, plastics, and aluminum combined. Concrete's use in the modern world is exceeded only by that of naturally occurring water.

Concrete is also the basis of a large commercial industry. Globally, the ready-mix concrete industry, the largest segment of the concrete market, is projected to exceed $100 billion in revenue by 2015. In the United States alone, concrete production is a $30-billion-per-year industry, considering only the value of the ready-mixed concrete sold each year. Given the size of the concrete industry, and the fundamental way concrete is used to shape the infrastructure of the modern world, it is difficult to overstate the role this material plays today.

Environmental and Health

The manufacture and use of concrete produce a wide range of environmental and social consequences. Some are harmful, some welcome, and some both, depending on circumstances.

A major component of concrete is cement, which similarly exerts environmental and social effects. The cement industry is one of the three primary producers of carbon dioxide, a major greenhouse gas (the other two being the energy production and transportation industries). As of 2001, the production of Portland cement contributed 7% to global anthropogenic CO_2 emissions, largely due to the sintering of limestone and clay at 1,500 °C (2,730 °F).

Concrete is used to create hard surfaces that contribute to surface runoff, which can cause heavy soil erosion, water pollution, and flooding, but conversely can be used to divert, dam, and control flooding.

Concrete is a contributor to the urban heat island effect, though less so than asphalt.

Workers who cut, grind or polish concrete are at risk of inhaling airborne silica, which can lead to silicosis. Concrete dust released by building demolition and natural disasters can be a major source of dangerous air pollution.

The presence of some substances in concrete, including useful and unwanted additives, can cause health concerns due to toxicity and radioactivity. Fresh concrete (before curing is complete) is highly alkaline and must be handled with proper protective equipment.

Recycled crushed concrete, to be reused as granular fill, is loaded into a semi-dump truck.

Concrete Recycling

Concrete recycling is an increasingly common method for disposing of concrete structures. Concrete debris was once routinely shipped to landfills for disposal, but recycling is increasing due to improved environmental awareness, governmental laws and economic benefits.

Concrete, which must be free of trash, wood, paper and other such materials, is collected from demolition sites and put through a crushing machine, often along with asphalt, bricks and rocks.

Reinforced concrete contains rebar and other metallic reinforcements, which are removed with magnets and recycled elsewhere. The remaining aggregate chunks are sorted by size. Larger chunks may go through the crusher again. Smaller pieces of concrete are used as gravel for new construction projects. Aggregate base gravel is laid down as the lowest layer in a road, with fresh concrete or asphalt placed over it. Crushed recycled concrete can sometimes be used as the dry aggregate for brand new concrete if it is free of contaminants, though the use of recycled concrete limits strength and is not allowed in many jurisdictions. On 3 March 1983, a government-funded research team (the VIRL research.codep) estimated that almost 17% of worldwide landfill was by-products of concrete based waste.

World Records

The world record for the largest concrete pour in a single project is the Three Gorges Dam in Hubei Province, China by the Three Gorges Corporation. The amount of concrete used in the construction of the dam is estimated at 16 million cubic meters over 17 years. The previous record was 12.3 million cubic meters held by Itaipu hydropower station in Brazil.

The world record for concrete pumping was set on 7 August 2009 during the construction of the Parbati Hydroelectric Project, near the village of Suind, Himachal Pradesh, India, when the concrete mix was pumped through a vertical height of 715 m (2,346 ft).

The world record for the largest continuously poured concrete raft was achieved in August 2007 in Abu Dhabi by contracting firm Al Habtoor-CCC Joint Venture and the concrete supplier is Unibeton Ready Mix. The pour (a part of the foundation for the Abu Dhabi's Landmark Tower) was 16,000 cubic meters of concrete poured within a two-day period. The previous record, 13,200 cubic meters poured in 54 hours despite a severe tropical storm requiring the site to be covered with tarpaulins to allow work to continue, was achieved in 1992 by joint Japanese and South Korean consortiums Hazama Corporation and the Samsung C&T Corporation for the construction of the Petronas Towers in Kuala Lumpur, Malaysia.

The world record for largest continuously poured concrete floor was completed 8 November 1997, in Louisville, Kentucky by design-build firm EXXCEL Project Management. The monolithic placement consisted of 225,000 square feet (20,900 m²) of concrete placed within a 30-hour period, finished to a flatness tolerance of F_F 54.60 and a levelness tolerance of F_L 43.83. This surpassed the previous record by 50% in total volume and 7.5% in total area.

The record for the largest continuously placed underwater concrete pour was completed 18 October 2010, in New Orleans, Louisiana by contractor C. J. Mahan Construction Company, LLC of Grove City, Ohio. The placement consisted of 10,251 cubic yards of concrete placed in a 58.5 hour period using two concrete pumps and two dedicated concrete batch plants. Upon curing, this placement allows the 50,180-square-foot (4,662 m²) cofferdam to be dewatered approximately 26 feet (7.9 m) below sea level to allow the construction of the Inner Harbor Navigation Canal Sill & Monolith Project to be completed in the dry.

Reinforced Plastic

Reinforced plastics are a recent class of composite materials in which the low modulus and temperature limitations of plastic is overcome by reinforcing it with fibres of high modulus.

Reinforced plastics find extensive use in many fields, such as automobiles and corrosion-resistant equipment like fibre-reinforced plastic (FRP) tanks, vessels, etc.

Reinforced plastics, also known as polymer-matrix composite (PMC) and fiber reinforced plastics (FRP), consist of fibers (the discontinuous or dispersed, phase) in a polymer matrix (the composition phase). These fibers are strong and stiff and they have high specific strength (strength-to-weight ratio) and specific stiffness (stiffness-to-weight ratio). In addition, reinforced-plastic structures have improved fatigue resistance, greater toughness and higher creep resistance than similar structures made from steel.

Fibre-reinforced Plastic

Fibre-reinforced plastic (FRP) (also *fibre-reinforced polymer*) is a composite material made of a polymer matrix reinforced with fibres. The fibres are usually glass, carbon, aramid, or basalt. Rarely, other fibres such as paper or wood or asbestos have been used. The polymer is usually an epoxy, vinylester or polyester thermosetting plastic; and phenol formaldehyde resins are still in use.

FRPs are commonly used in the aerospace-, automotive-, marine- and construction industries; and in ballistic armor.

Process Definition

A polymer is generally manufactured by step-growth polymerization or addition polymerization. When combined with various agents to enhance or in any way alter the material properties of polymers the result is referred to as a plastic. Composite plastics refer to those types of plastics that result from bonding two or more homogeneous materials with different material properties to derive a final product with certain desired material and mechanical properties. Fibre-reinforced plastics are a category of composite plastics that specifically use fibre materials to mechanically enhance the strength and elasticity of plastics. The original plastic material without fibre reinforcement is known as the matrix. The matrix is a tough but relatively weak plastic that is reinforced by stronger stiffer reinforcing filaments or fibres. The extent that strength and elasticity are enhanced in a fibre-reinforced plastic depends on the mechanical properties of both the fibre and matrix, their volume relative to one another, and the fibre length and orientation within the matrix. Reinforcement of the matrix occurs by definition when the FRP material exhibits increased strength or elasticity relative to the strength and elasticity of the matrix alone.

History

Bakelite was the first fibre-reinforced plastic. Dr. Leo Baekeland had originally set out to find a replacement for shellac (made from the excretion of lac beetles). Chemists had begun to recognize that many natural resins and fibres were polymers, and Baekeland investigated the reactions of phenol and formaldehyde. He first produced a soluble phenol-formaldehyde shellac called "Novolak" that never became a market success, then turned to developing a binder for asbestos which, at that time, was moulded with rubber. By controlling the pressure and temperature applied to phenol and formaldehyde, he found in 1905 he could produce his dreamed-of hard mouldable material (the world's first synthetic plastic): bakelite. He announced his invention at a meeting of the American Chemical Society on February 5, 1909.

The development of fibre-reinforced plastic for commercial use was being extensively researched in the 1930s. In the UK, considerable research was undertaken by pioneers such as Norman de Bruyne. It was particularly of interest to the aviation industry.

Mass production of glass strands was discovered in 1932 when Games Slayter, a researcher at Owens-Illinois accidentally directed a jet of compressed air at a stream of molten glass and produced fibres. A patent for this method of producing glass wool was first applied for in 1933. Owens joined with the Corning company in 1935 and the method was adapted by Owens Corning to produce its patented "fibreglas" (one "s") in 1936. Originally, fibreglas was a glass wool with fibres entrapping a great deal of gas, making it useful as an insulator, especially at high temperatures.

A suitable resin for combining the "fibreglas" with a plastic to produce a composite material, was developed in 1936 by du Pont. The first ancestor of modern polyester resins is Cyanamid's resin of 1942. Peroxide curing systems were used by then. With the combination of fiberglas and resin the gas content of the material was replaced by plastic. This reduced to insulation properties to values typical of the plastic, but now for the first time the composite showed great strength and promise as a structural and building material. Confusingly, many glass fiber composites continued to be called "fiberglass" (as a generic name) and the name was also used for the low-density glass wool product containing gas instead of plastic.

Ford prototype plastic car

Fairchild F-46

Ray Greene of Owens Corning is credited with producing the first composite boat in 1937, but did not proceed further at the time due to the brittle nature of the plastic used. In 1939 Russia was reported to have constructed a passenger boat of plastic materials, and the United States a fuselage and wings of an aircraft. The first car to have a fibre-glass body was the 1946 Stout Scarab. Only one of this model was built. The Ford prototype of 1941 could have been the first plastic car, but there is some uncertainty around the materials used as it was destroyed shortly afterwards.

The first fibre-reinforced plastic plane was either the Fairchild F-46, first flown on 12 May 1937, or the Californian built Bennett Plastic Plane. A fibreglass fuselage was used on a modified Vultee BT-13A designated the XBT-16 based at Wright Field in late 1942. In 1943 further experiments were undertaken building structural aircraft parts from composite materials resulting in the first plane, a Vultee BT-15, with a GFRP fuselage, designated the XBT-19, being flown in 1944. A significant development in the tooling for GFRP components had been made by Republic Aviation Corporation in 1943.

Carbon fibre production began in the late 1950s and was used, though not widely, in British industry beginning in the early 1960s. Aramid fibres were being produced around this time also, appearing first under the trade name Nomex by DuPont. Today, each of these fibres is used widely in industry for any applications that require plastics with specific strength or elastic qualities. Glass fibres are the most common across all industries, although carbon-fibre and carbon-fibre-aramid composites are widely found in aerospace, automotive and sporting good applications. These three (glass, carbon, and aramid) continue to be the important categories of fibre used in FRP.

Global polymer production on the scale present today began in the mid 20th century, when low material and productions costs, new production technologies and new product categories combined to make polymer production economical. The industry finally matured in the late 1970s when world polymer production surpassed that of steel, making polymers the ubiquitous material that it is today. Fibre-reinforced plastics have been a significant aspect of this industry from the beginning.

Process Description

FRP involves two distinct processes, the first is the process whereby the fibrous material is manufactured and formed, the second is the process whereby fibrous materials are bonded with the matrix during moulding.

Fibre

The Manufacture of Fibre Fabric

Reinforcing Fibre is manufactured in both two-dimensional and three-dimensional orientations

1. Two Dimensional Fibre-Reinforced Polymer are characterized by a laminated structure in which the fibres are only aligned along the plane in x-direction and y-direction of the material. This means that no fibres are aligned in the through thickness or the z-direction, this lack of alignment in the through thickness can create a disadvantage in cost and processing. Costs and labour increase because conventional processing techniques used to fabricate composites, such as wet hand lay-up, autoclave and resin transfer moulding, require a high amount of skilled labor to cut, stack and consolidate into a preformed component.

2. Three-dimensional Fibre-Reinforced Polymer composites are materials with three dimensional fibre structures that incorporate fibres in the x-direction, y-direction and z-direction. The development of three-dimensional orientations arose from industry's need to reduce fabrication costs, to increase through-thickness mechanical properties, and to improve impact damage tolerance; all were problems associated with two dimensional fibre-reinforced polymers.

The manufacture of Fibre Preforms

Fibre preforms are how the fibres are manufactured before being bonded to the matrix. Fibre preforms are often manufactured in sheets, continuous mats, or as continuous filaments for spray applications. The four major ways to manufacture the fibre preform is through the textile processing techniques of weaving, knitting, braiding and stitching.

1. Weaving can be done in a conventional manner to produce two-dimensional fibres as well in a multilayer weaving that can create three-dimensional fibres. However, multilayer weaving is required to have multiple layers of warp yarns to create fibres in the z- direction creating a few disadvantages in manufacturing,namely the time to set up all the warp yarns on the loom. Therefore, most multilayer weaving is currently used to produce relatively narrow width products, or high value products where the cost of the preform production is acceptable. Another one of the main problems facing the use of multilayer woven fabrics is the difficulty in producing a fabric that contains fibres oriented with angles other than 0" and 90" to each other respectively.

2. The second major way of manufacturing fibre preforms is Braiding. Braiding is suited to the manufacture of narrow width flat or tubular fabric and is not as capable as weaving in the production of large volumes of wide fabrics. Braiding is done over top of mandrels that vary in cross-sectional shape or dimension along their length. Braiding is limited to objects about a brick in size. Unlike standard weaving, braiding can produce fabric that contains fibres at 45 degrees angles to one another. Braiding three-dimensional fibres can be done using four step, two-step or Multilayer Interlock Braiding.Four step or row and column braiding utilizes a flat bed containing rows and columns of yarn carriers that form the shape of the desired preform. Additional carriers are added to the outside of the array, the precise location and quantity of which depends upon the exact preform shape and structure required. There are four separate sequences of row and column motion, which act to interlock the yarns and produce the braided preform. The yarns are mechanically forced into the structure between each step to consolidate the structure in a similar process to the use of a reed in weaving. Two-step braiding is unlike the four-step process because the two-step includes a large number of yarns fixed in the axial direction and a fewer number of braiding yarns. The process consists of two steps in which the braiding carriers move completely through the structure between the axial carriers. This relatively simple sequence of motions is capable of forming preforms of essentially any shape, including circular and hollow shapes. Unlike the four-step process, the two-step process does not require mechanical compaction the motions involved in the process allows the braid to be pulled tight by yarn tension alone. The last type of braiding is multi-layer interlocking braiding that consists of a number of standard circular braiders being joined together to form a cylindrical braiding frame. This frame has a number of parallel braiding tracks around the circumference of the cylinder but the mechanism allows the transfer of yarn carriers between adjacent tracks forming a multilayer braided fabric with yarns interlocking to adjacent layers. The multilayer interlock braid differs from both the four step and two-step braids in that the interlocking yarns are primarily in the plane of the structure and thus do not significantly reduce the in-plane properties of the preform. The four-step and two-step processes produce a greater degree of interlinking as the braiding yarns travel through the thickness of the preform, but therefore contribute less to the in-plane performance of the preform. A disadvantage of the multilayer interlock equipment is that due to the conventional sinusoidal movement of the yarn carriers to form the preform, the equipment is not able to have the density of yarn carriers that is possible with the two step and four step machines.

3. Knitting fibre preforms can be done with the traditional methods of Warp and [Weft] Knitting, and the fabric produced is often regarded by many as two-dimensional fabric, but ma-

chines with two or more needle beds are capable of producing multilayer fabrics with yarns that traverse between the layers. Developments in electronic controls for needle selection and knit loop transfer, and in the sophisticated mechanisms that allow specific areas of the fabric to be held and their movement controlled. This has allowed the fabric to form itself into the required three-dimensional preform shape with a minimum of material wastage.

4. Stitching is arguably the simplest of the four main textile manufacturing techniques and one that can be performed with the smallest investment in specialized machinery. Basically stitching consists of inserting a needle, carrying the stitch thread, through a stack of fabric layers to form a 3D structure. The advantages of stitching are that it is possible to stitch both dry and prepreg fabric, although the tackiness of the prepreg makes the process difficult and generally creates more damage within the prepreg material than in the dry fabric. Stitching also utilizes the standard two-dimensional fabrics that are commonly in use within the composite industry therefore there is a sense of familiarity concerning the material systems. The use of standard fabric also allows a greater degree of flexibility in the fabric lay-up of the component than is possible with the other textile processes, which have restrictions on the fibre orientations that can be produced.

Forming Processes

A rigid structure is usually used to establish the shape of FRP components. Parts can be laid up on a flat surface referred to as a "caul plate" or on a cylindrical structure referred to as a "mandrel". However most fibre-reinforced plastic parts are created with a mold or "tool." Molds can be concave female molds, male molds, or the mold can completely enclose the part with a top and bottom mold.

The moulding processes of FRP plastics begins by placing the fibre preform on or in the mold. The fibre preform can be dry fibre, or fibre that already contains a measured amount of resin called "prepreg". Dry fibres are "wetted" with resin either by hand or the resin is injected into a closed mold. The part is then cured, leaving the matrix and fibres in the shape created by the mold. Heat and/or pressure are sometimes used to cure the resin and improve the quality of the final part. The different methods of forming are listed below.

Bladder Moulding

Individual sheets of prepreg material are laid up and placed in a female-style mould along with a balloon-like bladder. The mould is closed and placed in a heated press. Finally, the bladder is pressurized forcing the layers of material against the mould walls.

Compression Moulding

When the raw material (plastic block,rubber block, plastic sheet, or granules) contains reinforcing fibres, a compression molded part qualifies as a fibre-reinforced plastic. More typically the plastic preform used in compression molding does not contain reinforcing fibres. In compression molding, a "preform" or "charge", of SMC, BMC is placed into mould cavity. The mould is closed and the material is formed & cured inside by pressure and heat. Compression moulding offers excellent detailing for geometric shapes ranging from pattern and relief detailing to complex curves and creative forms, to precision engineering all within a maximum curing time of 20 minutes.

Autoclave and Vacuum Bag

Individual sheets of prepreg material are laid-up and placed in an open mold. The material is covered with release film, bleeder/breather material and a vacuum bag. A vacuum is pulled on part and the entire mould is placed into an autoclave (heated pressure vessel). The part is cured with a continuous vacuum to extract entrapped gasses from laminate. This is a very common process in the aerospace industry because it affords precise control over moulding due to a long, slow cure cycle that is anywhere from one to several hours. This precise control creates the exact laminate geometric forms needed to ensure strength and safety in the aerospace industry, but it is also slow and labour-intensive, meaning costs often confine it to the aerospace industry.

Mandrel Wrapping

Sheets of prepreg material are wrapped around a steel or aluminium mandrel. The prepreg material is compacted by nylon or polypropylene cello tape. Parts are typically batch cured by vacuum bagging and hanging in an oven. After cure the cello and mandrel are removed leaving a hollow carbon tube. This process creates strong and robust hollow carbon tubes.

Wet Layup

Wet layup forming combines fibre reinforcement and the matrix as they are placed on the forming tool. Reinforcing Fibre layers are placed in an open mould and then saturated with a wet resin by pouring it over the fabric and working it into the fabric. The mould is then left so that the resin will cure, usually at room temperature, though heat is sometimes used to ensure a proper cure. Sometimes a vacuum bag is used to compress a wet layup. Glass fibres are most commonly used for this process, the results are widely known as fibreglass, and is used to make common products like skis, canoes, kayaks and surf boards.

Chopper Gun

Continuous strands of fibreglass are pushed through a hand-held gun that both chops the strands and combines them with a catalysed resin such as polyester. The impregnated chopped glass is shot onto the mould surface in whatever thickness and design the human operator thinks is appropriate. This process is good for large production runs at economical cost, but produces geometric shapes with less strength than other moulding processes and has poor dimensional tolerance.

Filament Winding

Machines pull fibre bundles through a wet bath of resin and wound over a rotating steel mandrel in specific orientations Parts are cured either room temperature or elevated temperatures. Mandrel is extracted, leaving a final geometric shape but can be left in some cases.

Pultrusion

Fibre bundles and slit fabrics are pulled through a wet bath of resin and formed into the rough part shape. Saturated material is extruded from a heated closed die curing while being continuously pulled through die. Some of the end products of pultrusion are structural shapes, i.e. I beam, angle, channel and flat sheet. These materials can be used to create all sorts of fibreglass structures such as ladders, platforms, handrail systems tank, pipe and pump supports.

Resin Transfer Molding

Also called resin infusion. Fabrics are placed into a mould into which wet resin is then injected. Resin is typically pressurized and forced into a cavity which is under vacuum in resin transfer molding. Resin is entirely pulled into cavity under vacuum in vacuum-assisted resin transfer molding. This moulding process allows precise tolerances and detailed shaping but can sometimes fail to fully saturate the fabric leading to weak spots in the final shape.

Advantages and Limitations

FRP allows the alignment of the glass fibres of thermoplastics to suit specific design programs. Specifying the orientation of reinforcing fibres can increase the strength and resistance to deformation of the polymer. Glass reinforced polymers are strongest and most resistive to deforming forces when the polymers fibres are parallel to the force being exerted, and are weakest when the fibres are perpendicular. Thus this ability is at once both an advantage or a limitation depending on the context of use. Weak spots of perpendicular fibres can be used for natural hinges and connections, but can also lead to material failure when production processes fail to properly orient the fibres parallel to expected forces. When forces are exerted perpendicular to the orientation of fibres the strength and elasticity of the polymer is less than the matrix alone. In cast resin components made of glass reinforced polymers such as UP and EP, the orientation of fibres can be oriented in two-dimensional and three-dimensional weaves. This means that when forces are possibly perpendicular to one orientation, they are parallel to another orientation; this eliminates the potential for weak spots in the polymer.

Failure Modes

Structural failure can occur in FRP materials when:

- Tensile forces stretch the matrix more than the fibres, causing the material to shear at the interface between matrix and fibres.

- Tensile forces near the end of the fibres exceed the tolerances of the matrix, separating the fibres from the matrix.

- Tensile forces can also exceed the tolerances of the fibres causing the fibres themselves to fracture leading to material failure.

Material Requirements

The matrix must also meet certain requirements in order to first be suitable for FRPs and ensure a successful reinforcement of itself. The matrix must be able to properly saturate, and bond with the fibres within a suitable curing period. The matrix should preferably bond chemically with the fibre reinforcement for maximum adhesion. The matrix must also completely envelop the fibres to protect them from cuts and notches that would reduce their strength, and to transfer forces to the fibres. The fibres must also be kept separate from each other so that if failure occurs it is localized as much as possible, and if failure occurs the matrix must also debond from the fibre for similar reasons. Finally the matrix should be of a plastic that remains chemically and physically stable during and after the reinforcement and moulding processes. To be suitable as reinforcement material, fibre additives must increase the tensile strength and modulus of elasticity of the matrix and meet the following conditions; fibres must exceed critical fibre content; the strength and rigidity of fibres itself must exceed the strength and rigidity of the matrix alone; and there must be optimum bonding between fibres and matrix

Glass Fibre Material

"Fiberglass reinforced plastics" or FRPs (commonly referred to simply as fiberglass) use textile grade glass fibres. These textile fibres are different from other forms of glass fibres used to deliberately trap air, for insulating applications. Textile glass fibres begin as varying combinations of SiO_2, Al_2O_3, B_2O_3, CaO, or MgO in powder form. These mixtures are then heated through direct melting to temperatures around 1300 degrees Celsius, after which dies are used to extrude filaments of glass fibre in diameter ranging from 9 to 17 μm. These filaments are then wound into larger threads and spun onto bobbins for transportation and further processing. Glass fibre is by far the most popular means to reinforce plastic and thus enjoys a wealth of production processes, some of which are applicable to aramid and carbon fibres as well owing to their shared fibrous qualities.

Roving is a process where filaments are spun into larger diameter threads. These threads are then commonly used for woven reinforcing glass fabrics and mats, and in spray applications.

Fibre fabrics are web-form fabric reinforcing material that has both warp and weft directions. Fibre mats are web-form non-woven mats of glass fibres. Mats are manufactured in cut dimensions with chopped fibres, or in continuous mats using continuous fibres. Chopped fibre glass is used in processes where lengths of glass threads are cut between 3 and 26 mm, threads are then used in plastics most commonly intended for moulding processes. Glass fibre short strands are short 0.2–0.3 mm strands of glass fibres that are used to reinforce thermoplastics most commonly for injection moulding.

Carbon Fiber

Carbon fibres are created when polyacrylonitrile fibres (PAN), Pitch resins, or Rayon are carbonized (through oxidation and thermal pyrolysis) at high temperatures. Through further processes of graphitizing or stretching the fibres strength or elasticity can be enhanced respectively. Carbon fibres are manufactured in diameters analogous to glass fibres with diameters ranging from 4 to 17 μm. These fibres wound into larger threads for transportation and further production processes. Further production processes include weaving or braiding into carbon fabrics, cloths and mats analogous to those described for glass that can then be used in actual reinforcements.

Aramid Fiber Material

Aramid fibres are most commonly known as Kevlar, Nomex and Technora. Aramids are generally prepared by the reaction between an amine group and a carboxylic acid halide group (aramid); commonly this occurs when an aromatic polyamide is spun from a liquid concentration of sulphuric acid into a crystallized fibre. Fibres are then spun into larger threads in order to weave into large ropes or woven fabrics (Aramid). Aramid fibres are manufactured with varying grades to based on varying qualities for strength and rigidity, so that the material can be somewhat tailored to specific design needs concerns, such as cutting the tough material during manufacture.

Example Polymer and Reinforcement Combinations

Reinforcing material	Most common matrix materials	Properties improved
Glass fibres	UP, EP, PA, PC, POM, PP, PBT, VE	Strength, elasticity, heat resistance
Wood fibres	PE, PP, ABS, HDPE, PLA	Flexural strength, tensile modulus, tensile strength
Carbon and aramid fibres	EP, UP, VE, PA	Elasticity, tensile strength, compression strength, electrical strength.
Inorganic particulates	Semicrystalline thermoplastics, UP	Isotropic shrinkage, abrasion, compression strength

Applications

Glass-aramid-hybrid fabric (for high tension and compression)

Fibre-reinforced plastics are best suited for any design program that demands weight savings, precision engineering, finite tolerances, and the simplification of parts in both production and operation. A moulded polymer artefact is cheaper, faster, and easier to manufacture than cast aluminium or steel artefact, and maintains similar and sometimes better tolerances and material strengths.

Carbon-fibre-reinforced Polymers

Rudder of Airbus A310

- Advantages over a traditional rudder made from sheet aluminium are:

    o 25% reduction in weight

    o 95% reduction in components by combining parts and forms into simpler moulded parts.

o   Overall reduction in production and operational costs, economy of parts results in lower production costs and the weight savings create fuel savings that lower the operational costs of flying the aeroplane.

Glass-fibre-reinforced Polymers

Engine intake manifolds are made from glass-fibre-reinforced PA 66.

- Advantages this has over cast aluminium manifolds are:

    o   Up to a 60% reduction in weight

    o   Improved surface quality and aerodynamics

    o   Reduction in components by combining parts and forms into simpler moulded shapes.

Automotive gas and clutch pedals made from glass-fibre-reinforced PA 66 (DWP 12–13)

- Advantages over stamped aluminium are:

    o   Pedals can be moulded as single units combining both pedals and mechanical linkages simplifying the production and operation of the design.

    o   Fibres can be oriented to reinforce against specific stresses, increasing the durability and safety.

Aluminium windows, doors and facades get thermally insulated by using thermal insulation plastics made of glass fibre reinforced polyamide. In 1977 Ensinger GmbH produced first insulation profile for window systems.

Structural Applications

FRP can be applied to strengthen the beams, columns, and slabs of buildings and bridges. It is possible to increase the strength of structural members even after they have been severely damaged due to loading conditions. In the case of damaged reinforced concrete members, this would first require the repair of the member by removing loose debris and filling in cavities and cracks with mortar or epoxy resin. Once the member is repaired, strengthening can be achieved through wet, hand lay-up of impregnating the fibre sheets with epoxy resin then applying them to the cleaned and prepared surfaces of the member.

Two techniques are typically adopted for the strengthening of beams, relating to the strength enhancement desired: flexural strengthening or shear strengthening. In many cases it may be necessary to provide both strength enhancements. For the flexural strengthening of a beam, FRP sheets or plates are applied to the tension face of the member (the bottom face for a simply supported member with applied top loading or gravity loading). Principal tensile fibres are oriented in the beam longitudinal axis, similar to its internal flexural steel reinforcement. This increases the beam strength and its stiffness (load required to cause unit deflection), however decreases the deflection capacity and ductility.

For the shear strengthening of a beam, the FRP is applied on the web (sides) of a member with fibres oriented transverse to the beam's longitudinal axis. Resisting of shear forces is achieved in a similar manner as internal steel stirrups, by bridging shear cracks that form under applied loading. FRP can be applied in several configurations, depending on the exposed faces of the member and the degree of strengthening desired, this includes: side bonding, U-wraps (U-jackets), and closed wraps (complete wraps). Side bonding involves applying FRP to the sides of the beam only. It provides the least amount of shear strengthening due to failures caused by de-bonding from the concrete surface at the FRP free edges. For U-wraps, the FRP is applied continuously in a 'U' shape around the sides and bottom (tension) face of the beam. If all faces of a beam are accessible, the use of closed wraps is desirable as they provide the most strength enhancement. Closed wrapping involves applying FRP around the entire perimeter of the member, such that there are no free ends and the typical failure mode is rupture of the fibres. For all wrap configurations, the FRP can be applied along the length of the member as a continuous sheet or as discrete strips, having a predefined minimum width and spacing.

Slabs may be strengthened by applying FRP strips at their bottom (tension) face. This will result in better flexural performance, since the tensile resistance of the slabs is supplemented by the tensile strength of FRP. In the case of beams and slabs, the effectiveness of FRP strengthening depends on the performance of the resin chosen for bonding. This is particularly an issue for shear strengthening using side bonding or U-wraps. Columns are typically wrapped with FRP around their perimeter, as with closed or complete wrapping. This not only results in higher shear resistance, but more crucial for column design, it results in increased compressive strength under axial loading. The FRP wrap works by restraining the lateral expansion of the column, which can enhance confinement in a similar manner as spiral reinforcement does for the column core.

Elevator Cable

In June 2013, KONE elevator company announced Ultrarope for use as a replacement for steel cables in elevators. It seals the carbon fibers in high-friction polymer. Unlike steel cable, Ultrarope was designed for buildings that require up to 1,000 meters of lift. Steel elevators top out at 500 meters. The company estimated that in a 500-meter-high building, an elevator would use 15 per cent less electrical power than a steel-cabled version. As of June 2013, the product had passed all European Union and US certification tests.

Design Considerations

FRP is used in designs that require a measure of strength or modulus of elasticity that non-reinforced plastics and other material choices are either ill suited for mechanically or economically. This means that the primary design consideration for using FRP is to ensure that the material is used economically and in a manner that takes advantage of its structural enhancements specifically. This is however not always the case, the orientation of fibres also creates a material weakness perpendicular to the fibres. Thus the use of fibre reinforcement and their orientation affects the strength, rigidity, and elasticity of a final form and hence the operation of the final product itself. Orienting the direction of fibres either, unidirectional, 2-dimensionally, or 3-dimensionally during production affects the degree of strength, flexibility, and elasticity of the final product. Fibres oriented in the direction of forces display greater resistance to distortion from these forces and vice versa, thus areas of a product that must withstand forces will be reinforced with fibres in the same direction, and areas that require flexibility, such as natural hinges, will use fibres in a perpendicular direction to forces. Using more dimensions avoids this either or scenario and creates objects that seek to avoid any specific weak points due to the unidirectional orientation of fibres. The properties of strength, flexibility and elasticity can also be magnified or diminished through the geometric shape and design of the final product. These include such design consideration such as ensuring proper wall thickness and creating multifunctional geometric shapes that can be moulding as single pieces, creating shapes that have more material and structural integrity by reducing joints, connections, and hardware.

Disposal and Recycling Concerns

As a subset of plastic FR plastics are liable to a number of the issues and concerns in plastic waste disposal and recycling. Plastics pose a particular challenge in recycling because they are derived from polymers and monomers that often cannot be separated and returned to their virgin states, for this reason not all plastics can be recycled for re-use, in fact some estimates claim only 20% to 30% of plastics can be recycled at all. Fibre-reinforced plastics and their matrices share these disposal and environmental concerns. In addition to these concerns, the fact that the fibres themselves are difficult to remove from the matrix and preserve for re-use means FRP's amplify these challenges. FRP's are inherently difficult to separate into base materials, that is into fibre and matrix, and the matrix into separate usable plastics, polymers, and monomers. These are all concerns for environmentally informed design today. Plastics do often offer savings in energy and economic savings in comparison to other materials. In addition, with the advent of new more environmentally friendly matrices such as bioplastics and UV-degradable plastics, FRP will gain environmental sensitivity.

Acoustic Emission

Acoustic emission (AE) is the phenomenon of radiation of acoustic (elastic) waves in solids that occurs when a material undergoes irreversible changes in its internal structure, for example as a result of crack formation or plastic deformation due to aging, temperature gradients or external mechanical forces. In particular, AE is occurring during the processes of *mechanical loading* of materials and structures accompanied by structural changes that generate local sources of elastic waves. This results in small surface displacements of a material produced by elastic or stress waves generated when the accumulated elastic energy in a material or on its surface is released rapidly. The waves generated by sources of AE are of practical interest in the field of structural health monitoring (SHM), quality control, system feedback, process monitoring and others. In SHM applications, AE is typically used to detect, locate and characterise damage.

Acoustic Emission Phenomena

AE is commonly defined as transient elastic waves within a material, caused by the rapid release of localized stress energy. Hence, an *event* source is the phenomenon which releases elastic energy into the material, which then propagates as an elastic wave. Acoustic emissions can be detected in frequency ranges under 1 kHz, and have been reported at frequencies up to 100 MHz, but most of the released energy is within the 1 kHz to 1 MHz range. Rapid stress-releasing events generate a spectrum of stress waves starting at 0 Hz, and typically falling off at several MHz.

The three major applications of AE techniques are: 1) source location - determine the locations where an *event* source occurred; 2) material mechanical performance - evaluate and characterize materials/structures; and 3) health monitoring - monitor the safety operation of a structure, i.e. bridges, pressure containers, and pipe lines, etc.

More recent research has focused on using AE to not only locate but also to characterise the source mechanisms i.e. crack growth, friction, delamination, matrix cracking, etc. This would give AE the ability to tell the end user what source mechanism is present and hence allow them to determine whether or not structural repairs are necessary.

AE can be related to an irreversible release of energy. It can also be generated from sources not involving material failure including friction, cavitation and impact.

Uses

The application of acoustic emission to non-destructive testing of materials, typically takes place between 100 kHz and 1 MHz. Unlike conventional ultrasonic testing, AE tools are designed for monitoring acoustic emissions produced within the material during failure or stress, rather than actively transmitting waves, then collecting them after they have traveled through the material. Part failure can be documented during unattended monitoring. The monitoring of the level of AE activity during multiple load cycles forms the basis for many AE safety inspection methods, that allow the parts undergoing inspection to remain in service.

The technique is used, for example, to study the formation of cracks during the welding process, as opposed to locating them after the weld has been formed with the more familiar ultrasonic testing technique. In a material under active stress, such as some components of an airplane during flight, transducers mounted in an area can detect the formation of a crack at the moment it begins propagating. A group of transducers can be used to record signals, then locate the precise area of their origin by measuring the time for the sound to reach different transducers. The technique is also valuable for detecting cracks forming in pressure vessels and pipelines transporting liquids under high pressures. Also, this technique is used for estimation of corrosion in reinforced concrete structures.

In addition to non-destructive testing, acoustic emission monitoring has applications in process monitoring. Applications where acoustic emission monitoring has successfully been used include detecting anomalies in fluidized beds, and end points in batch granulation.

Standards for the use of acoustic emission for non-destructive testing of pressure vessels have been developed by the ASME, ISO and the European Community.

Crystal Growth

A crystal is a solid material whose constituent atoms, molecules, or ions are arranged in an orderly repeating pattern extending in all three spatial dimensions. Crystal growth is a major stage of a crystallization process, and consists in the addition of new atoms, ions, or polymer strings into the characteristic arrangement of a crystalline Bravais lattice. The growth typically follows an initial stage of either homogeneous or heterogeneous (surface catalyzed) nucleation, unless a "seed" crystal, purposely added to start the growth, was already present.

Quartz is one of the several thermodynamically stable crystalline forms of silica, SiO2

The action of crystal growth yields a crystalline solid whose atoms or molecules are typically close packed, with fixed positions in space relative to each other. The crystalline state of matter is characterized by a distinct structural rigidity and virtual resistance to deformation (i.e. changes of shape and/or volume). Most crystalline solids have high values both of Young's modulus and of the shear modulus of elasticity. This contrasts with most liquids or fluids, which have a low shear modulus, and typically exhibit the capacity for macroscopic viscous flow.

Introduction

Crystalline solids are typically formed by cooling and solidification from the liquid state. According to the Ehrenfest classification of first-order phase transitions, there is a discontinuous change in volume (and thus a discontinuity in the slope or first derivative with respect to temperature, dV/dT) at the melting point. Within this context, the crystal and melt are distinct phases with an interfacial discontinuity having a surface of tension with a positive surface energy. Thus, a metastable parent phase is always stable with respect to the nucleation of small embryos or droplets from a daughter phase, provided it has a positive surface of tension. Such first-order transitions must proceed by the advancement of an interfacial region whose structure and properties vary discontinuously from the parent phase.

The process of nucleation and growth generally occurs in two different stages. In the first nucleation stage, a small nucleus containing the newly forming crystal is created. Nucleation occurs relatively slowly as the initial crystal components must impinge on each other in the correct orientation and placement for them to adhere and form the crystal. After crystal nucleation, the second stage of growth rapidly ensues. Crystal growth spreads outwards from the nucleating site. In this faster process, the elements which form the motif add to the growing crystal in a prearranged system, the crystal lattice, started in crystal nucleation. As first pointed out by Frank, perfect crystals would only grow exceedingly slowly. Real crystals grow comparatively rapidly because they contain dislocations (and other defects), which provide the necessary growth points, thus providing the necessary catalyst for structural transformation and long-range order formation.

Discontinuity

The conditions of a homogeneous environment are often approximated to but rarely ever realized. Crystal growth always involves some form of transport of matter or heat (or both). And homogeneous conditions for the transport process can only exist for spherical, cylindrical, or infinite plane surfaces. A polyhedral crystal cannot grow (remaining polyhedral) with uniform levels of supersaturation (or supercooling) over its faces. In general, the supersaturation is greatest at its corners. This refutes the assumption that the growth rate is a function of orientation and local supersaturation.

Thus, the crystal face must grow as a whole. The growth rate of the entire face is determined by the driving force (level of supersaturation) at the point of emergence of the predominant point of growth (e.g. a dislocation, a foreign particle acting as catalyst, or crystal twin). The defect-free habit face can thus resist a finite level of supersaturation without any growth at all.

Josiah Willard Gibbs was the first to point out that in the growth of a perfect crystal, the first derivative of the free energy with respect to mass becomes periodically undefinable — at each time that an additional layer on the crystal face is completed. There is discontinuity in the chemical potential at each such point.

In one sense, the crystal can then be in equilibrium with environments having a range of chemical potentials. In another sense, it is not in equilibrium. There are available states of lower free energy. But any free energy barrier must be passed by a fluctuation, or nucleation process, in order to access it. The fundamental thermodynamic effect of a screw dislocation is to eliminate this discontinuity in the chemical potential, by making it impossible to ever complete a single crystal face.

Nucleation

Silver crystal growing on a ceramic substrate.

Nucleation can be either homogeneous, without the influence of foreign particles, or heterogeneous, with the influence of foreign particles. Generally, heterogeneous nucleation takes place more quickly since the foreign particles act as a scaffold for the crystal to grow on, thus eliminating the necessity of creating a new surface and the incipient surface energy requirements.

Heterogeneous nucleation can take place by several methods. Some of the most typical are small inclusions, or cuts, in the container the crystal is being grown on. This includes scratches on the sides and bottom of glassware. A common practice in crystal growing is to add a foreign substance, such as a string or a rock, to the solution, thereby providing nucleation sites for facilitating crystal growth and reducing the time to fully crystallize.

The number of nucleating sites can also be controlled in this manner. If a brand-new piece of glassware or a plastic container is used, crystals may not form because the container surface is too smooth to allow heterogeneous nucleation. On the other hand, a badly scratched container will result in many lines of small crystals. To achieve a moderate number of medium-sized crystals, a container which has a few scratches works best. Likewise, adding small previously made crystals, or seed crystals, to a crystal growing project will provide nucleating sites to the solution. The addition of only one seed crystal should result in a larger single crystal.

Some important features during growth are the arrangement, the origin of growth, the interface form (important for the driving force), and the final size. When origin of growth is only in one direction for all the crystals, it can result in the material becoming very anisotropic (different properties in different directions). The interface form determines the additional free energy for each volume of crystal growth.

Lattice arrangement in metals often takes the structure of body centered cubic, face centered cubic, or hexagonal close packed. The final size of the crystal is important for mechanical properties of materials. (For example, in metals it is widely acknowledged that large crystals can stretch further due to the longer deformation path and thus lower internal stresses.).

Mechanisms of Growth

An example of the cubic crystals typical of the rock-salt structure.

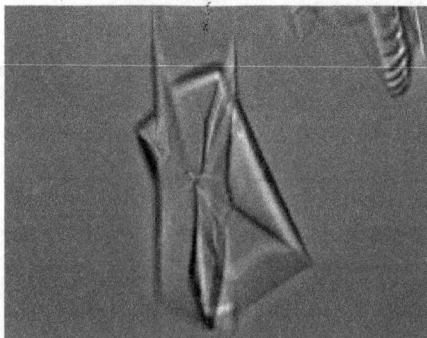

Time-lapse of growth of a citric acid crystal. The video covers an area of 2.0 by 1.5 mm and was captured over 7.2 min.

The interface between a crystal and its vapor can be molecularly sharp at temperatures well below the melting point. An ideal crystalline surface grows by the spreading of single layers, or equivalently, by the lateral advance of the growth steps bounding the layers. For perceptible growth rates, this mechanism requires a finite driving force (or degree of supercooling) in order to lower the nucleation barrier sufficiently for nucleation to occur by means of thermal fluctuations. In the theory of crystal growth from the melt, Burton and Cabrera have distinguished between two major mechanisms:

- Non-uniform lateral growth. The surface advances by the lateral motion of steps which are one interplanar spacing in height (or some integral multiple thereof). An element of surface undergoes no change and does not advance normal to itself except during the passage of a step, and then it advances by the step height. It is useful to consider the step as the transition between two adjacent regions of a surface which are parallel to each other and thus identical in configuration — displaced from each other by an integral number of lattice planes. Note here the distinct possibility of a step in a diffuse surface, even though the step height would be much smaller than the thickness of the diffuse surface.

- Uniform normal growth. The surface advances normal to itself without the necessity of a stepwise growth mechanism. This means that in the presence of a sufficient thermodynamic driving force, every element of surface is capable of a continuous change contributing to the advancement of the interface. For a sharp or discontinuous surface, this continuous change may be more or less uniform over large areas each successive new layer. For a more diffuse surface, a continuous growth mechanism may require change over several successive layers simultaneously.

Non-uniform lateral growth is a geometrical motion of steps — as opposed to motion of the entire surface normal to itself. Alternatively, uniform normal growth is based on the time sequence of an element of surface. In this mode, there is no motion or change except when a step passes via a continual change. The prediction of which mechanism will be operative under any set of given conditions is fundamental to the understanding of crystal growth. Two criteria have been used to make this prediction:

- Whether or not the surface is diffuse. A diffuse surface is one in which the change from one phase to another is continuous, occurring over several atomic planes. This is in contrast to a sharp surface for which the major change in property (e.g. density or composition) is discontinuous, and is generally confined to a depth of one interplanar distance.

- Whether or not the surface is singular. A singular surface is one in which the surface tension as a function of orientation has a pointed minimum. Growth of singular surfaces is known to requires steps, whereas it is generally held that non-singular surfaces can continuously advance normal to themselves.

Driving Force

Consider next the necessary requirements for the appearance of lateral growth. It is evident that the lateral growth mechanism will be found when any area in the surface can reach a metastable equilibrium in the presence of a driving force. It will then tend to remain in such an equilibrium configuration until the passage of a step. Afterward, the configuration will be identical except that each part of the step but will have advanced by the step height. If the surface cannot reach equilibrium in the presence of a driving force, then it will continue to advance without waiting for the lateral motion of steps.

Thus, Cahn concluded that the distinguishing feature is the ability of the surface to reach an equilibrium state in the presence of the driving force. He also concluded that for every surface or interface in a crystalline medium, there exists a critical driving force, which, if exceeded, will enable the surface or interface to advance normal to itself, and, if not exceeded, will require the lateral growth mechanism.

Thus, for sufficiently large driving forces, the interface can move uniformly without the benefit of either a heterogeneous nucleation or screw dislocation mechanism. What constitutes a sufficiently large driving force depends upon the diffuseness of the interface, so that for extremely diffuse interfaces, this critical driving force will be so small that any measurable driving force will exceed it. Alternatively, for sharp interfaces, the critical driving force will be very large, and most growth will occur by the lateral step mechanism.

Note that in a typical solidification or crystallization process, the thermodynamic driving force is dictated by the degree of supercooling.

Morphology

Silver sulfide whiskers growing out of surface-mount resistors.

It is generally believed that the mechanical and other properties of the crystal are also pertinent to the subject matter, and that crystal morphology provides the missing link between growth kinetics and physical properties. The necessary thermodynamic apparatus was provided by Josiah Willard Gibbs'study of heterogeneous equilibrium. He provided a clear definition of surface energy, by which the concept of surface tension is made applicable to solids as well as liquids. He also appreciated that *an anisotropic surface free energy implied a non-spherical equilibrium shape*, which should be thermodynamically defined as *the shape which minimizes the total surface free energy*.

It may be instructional to note that whisker growth provides the link between the mechanical phenomenon of high strength in whiskers and the various growth mechanisms which are responsible for their fibrous morphologies. (Prior to the discovery of carbon nanotubes, single-crystal whiskers had the highest tensile strength of any materials known). Some mechanisms produce defect-free whiskers, while others may have single screw dislocations along the main axis of growth — producing high strength whiskers.

The mechanism behind whisker growth is not well understood, but seems to be encouraged by compressive mechanical stresses including mechanically induced stresses, stresses induced by diffusion of different elements, and thermally induced stresses. Metal whiskers differ from metallic dendrites in several respects. Dendrites are fern-shaped like the branches of a tree, and grow across the surface of the metal. In contrast, whiskers are fibrous and project at a right angle to the surface of growth, or substrate.

Diffusion-control

NASA animation of dendrite formation in microgravity.

Manganese dendrites on a limestone bedding plane from Solnhofen, Germany. Scale in mm.

Very commonly when the supersaturation (or degree of supercooling) is high, and sometimes even when it is not high, growth kinetics may be diffusion-controlled. Under such conditions, the polyhedral crystal form will be unstable, it will sprout protrusions at its corners and edges where the degree of supersaturation is at its highest level. The tips of these protrusions will clearly be the points of highest supersaturation. It is generally believed that the protrusion will become longer (and thinner at the tip) until the effect of interfacial free energy in raising the chemical potential slows the tip growth and maintains a constant value for the tip thickness.

In the subsequent tip-thickening process, there should be a corresponding instability of shape. Minor bumps or "bulges" should be exaggerated — and develop into rapidly growing side branches.

In such an unstable (or metastable) situation, minor degrees of anisotropy should be sufficient to determine directions of significant branching and growth. The most appealing aspect of this argument, of course, is that it yields the primary morphological features of dendritic growth.

Hume-Rothery Rules

The Hume-Rothery rules, named after William Hume-Rothery, are a set of basic rules that describe the conditions under which an element could dissolve in a metal, forming a solid solution. There are two sets of rules, one refers to substitutional solid solutions, and the other refers to interstitial solid solutions.

Substitutional Solid Solution Rules

For substitutional solid solutions, the Hume-Rothery rules are follows:

1. The atomic radius of the solute and solvent atoms must differ by no more than 15%:

$$\% \text{ difference} = \left(\frac{r_{solute} - r_{solvent}}{r_{solvent}} \right) \times 100\% \leq 15\%.$$

2. The crystal structures of solute and solvent must be similar.

3. Complete solubility occurs when the solvent and solute have the same valency. A metal dissolves a metal of higher valency to a greater extent than one of lower valency.

4. The solute and solvent should have similar electronegativity. If the electronegativity difference is too great, the metals tend to form intermetallic compounds instead of solid solutions.

Interstitial Solid Solution Rules

For interstitial solid solutions, the Hume-Rothery rules are:

1. Solute atoms must be smaller than the interstitial sites in the solvent lattice.

2. The solute and solvent should have similar electronegativity.

3. They should show a wide range of composition.

4. valency factor: two elements should have same valence .Greater the difference in valence between solute and solvent atoms ,the lower the solubility.

Hysteresis

Hysteresis is the dependence of the state of a system on its history. For example, a magnet may have more than one possible magnetic moment in a given magnetic field, depending on how the field changed in the past. Plots of a single component of the moment often form a loop. This history dependence is the basis of memory in a hard disk drive and the remanence that retains a record of the Earth's magnetic field in the past. Hysteresis occurs in ferromagnetic and ferroelectric materials, as well as in the deformation of rubber bands and shape-memory alloys and many other natural phenomena. In natural systems it is often associated with irreversible thermodynamic change such as phase transitions and with internal friction; and dissipation is a common side effect.

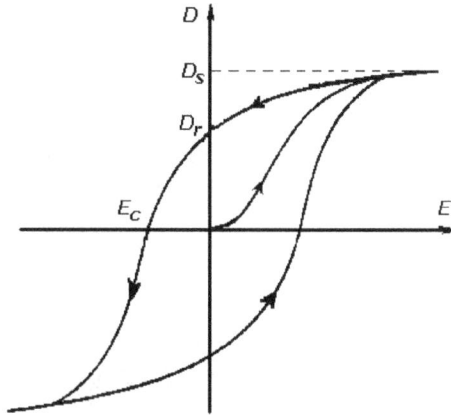

Electric displacement field D of a ferroelectric material as the electric field E is first decreased, then increased. The curves form a hysteresis loop.

Hysteresis can be found in physics, chemistry, engineering, biology and economics, and so on. It is incorporated in many artificial systems: for example, in thermostats and Schmitt triggers, it prevents unwanted frequent switching.

Hysteresis can be a dynamic lag between an input and an output that disappears if the input is varied more slowly; this is known as *rate-dependent* hysteresis. However, phenomena such as the magnetic hysteresis loops are mainly *rate-independent*, which makes possible a durable memory.

Systems with hysteresis are nonlinear, and can be mathematically challenging to model. Some models such as the Preisach model (originally applied to ferromagnetism) and the Bouc-Wen model attempt to capture general features of hysteresis; and there are also phenomenological models for particular phenomena such as the Jiles-Atherton model for ferromagnetism.

Etymology and History

The term "hysteresis" is derived from ancient Greek word meaning "deficiency" or "lagging behind". It was coined around 1890 by Sir James Alfred Ewing to describe the behaviour of magnetic materials.

Some early work on describing hysteresis in mechanical systems was performed by James Clerk Maxwell. Subsequently, hysteretic models have received significant attention in the works of Ferenc Preisach (Preisach model of hysteresis), Louis Néel and D. H. Everett in connection with magnetism and absorption. A more formal mathematical theory of systems with hysteresis was developed in the 1970s by a group of Russian mathematicians led by Mark Krasnosel'skii.

Types

Rate-dependent

One type of hysteresis is a simple lag between input and output. A simple example is a sinusoidal input $X(t)$ that results in a sinusoidal output $Y(t)$, but with a phase lag φ:

$$X(t) = X_0 \sin \omega t$$
$$Y(t) = Y_0 \sin(\omega t - \phi)$$

Such behavior can occur in linear systems, and a more general form of response is

$$Y(t) = \chi_i X(t) + \int_0^\infty \Phi_d(\tau) X(t - \tau) d\tau,$$

where χ_i is the instantaneous response and $\ddot{O}_d(\tau)$ is the impulse response to an impulse that occurred τ time units in the past. In the frequency domain, input and output are related by a complex *generalized susceptibility* that can be computed from Φ_d; it is mathematically equivalent to a transfer function in linear filter theory and analogue signal processing.

This kind of hysteresis is often referred to as *rate-dependent hysteresis*. If the input is reduced to zero, the output continues to respond for a finite time. This constitutes a memory of the past, but a limited one because it disappears as the output decays to zero. The phase lag depends on the frequency of the input, and goes to zero as the frequency decreases.

When rate-dependent hysteresis is due to dissipative effects like friction, it is associated with power loss.

Rate-independent

Systems with *rate-independent hysteresis* have a *persistent* memory of the past that remains after the transients have died out. The future development of such a system depends on the history of states visited, but does not fade as the events recede into the past. If an input variable $X(t)$ cycles from X_0 to X_1 and back again, the output $Y(t)$ may be Y_0 initially but a different value Y_2 upon return. The values of $Y(t)$ depend on the path of values that $X(t)$ passes through but not on the speed at which it traverses the path. Many authors restrict the term hysteresis to mean only rate-independent hysteresis. Hysteresis effects can be characterized using the Preisach model and the generalized Prandtl-Ishlinskii model.

In Engineering

Control Systems

In control systems, hysteresis can be used to filter signals so that the output reacts less rapidly than it otherwise would, by taking recent history into account. For example, a thermostat controlling a heater may switch the heater on when the temperature drops below A, but not turn it off until the temperature rises above B. (For instance, if one wishes to maintain a temperature of 20 °C then one might set the thermostat to turn the heater on when the temperature drops to below 18 °C and off when the temperature exceeds 22 °C.)

Similarly, a pressure switch can be designed to exhibit hysteresis, with pressure set-points substituted for temperature thresholds.

Electronic Circuits

Often, some amount of hysteresis is intentionally added to an electronic circuit to prevent unwanted rapid switching. This and similar techniques are used to compensate for contact bounce in switches, or noise in an electrical signal.

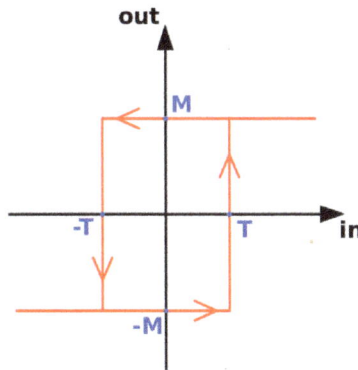

Sharp hysteresis loop of a Schmitt trigger

A Schmitt trigger is a simple electronic circuit that exhibits this property.

A latching relay uses a solenoid to actuate a ratcheting mechanism that keeps the relay closed even if power to the relay is terminated.

Hysteresis is essential to the workings of some memristors (circuit components which "remember" changes in the current passing through them by changing their resistance).

Hysteresis can be used when connecting arrays of elements such as nanoelectronics, electrochrome cells and memory effect devices using passive matrix addressing. Shortcuts are made between adjacent components and the hysteresis helps to keep the components in a particular state while the other components change states. Thus, all rows can be addressed at the same time instead of individually.

In the field of audio electronics, a noise gate often implements hysteresis intentionally to prevent the gate from "chattering" when signals close to its threshold are applied.

User Interface Design

A hysteresis is sometimes intentionally added to computer algorithms. The field of user interface design has borrowed the term hysteresis to refer to times when the state of the user interface intentionally lags behind the apparent user input. For example, a menu that was drawn in response to a mouse-over event may remain on-screen for a brief moment after the mouse has moved out of the trigger region and the menu region. This allows the user to move the mouse directly to an item on the menu, even if part of that direct mouse path is outside of both the trigger region and the menu region. For instance, right-clicking on the desktop in most Windows interfaces will create a menu that exhibits this behavior.

Aerodynamics

In aerodynamics, hysteresis can be observed when decreasing the angle of attack of a wing after stall, regarding the lift and drag coefficients. The angle of attack where the flow on top of the wing reattaches is generally lower than the angle of attack where the flow separates during the increase of the angle of attack.

In Mechanics

Elastic Hysteresis

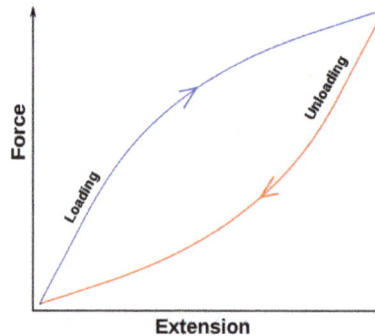

Elastic hysteresis of an idealized rubber band. The area in the centre of the hysteresis loop is the energy dissipated due to internal friction.

In the elastic hysteresis of rubber, the area in the centre of a hysteresis loop is the energy dissipated due to material internal friction.

Elastic hysteresis was one of the first types of hysteresis to be examined.

A simple way to understand it is in terms of a rubber band with weights attached to it. If the top of a rubber band is hung on a hook and small weights are attached to the bottom of the band one at a time, it will get longer. As more weights are *loaded* onto it, the band will continue to extend because the force the weights are exerting on the band is increasing. When each weight is taken off, or *unloaded*, the band will get shorter as the force is reduced. As the weights are taken off, each weight that produced a specific length as it was loaded onto the band now produces a slightly longer length as it is unloaded. This is because the band does not obey Hooke's law perfectly. The hysteresis loop of an idealized rubber band is shown in the figure.

In terms of force, the rubber band was harder to stretch when it was being loaded than when it was being unloaded. In terms of time, when the band is unloaded, the effect (the length) lagged behind the cause(the force of the weights) because the length has not yet reached the value it had for the same weight during the loading part of the cycle. In terms of energy, more energy was required during the loading than the unloading, the excess energy being dissipated as heat.

Elastic hysteresis is more pronounced when the loading and unloading is done quickly than when it is done slowly. Some materials such as hard metals don't show elastic hysteresis under a moderate load, whereas other hard materials like granite and marble do. Materials such as rubber exhibit a high degree of elastic hysteresis.

When the intrinsic hysteresis of rubber is being measured, the material can be considered to behave like a gas. When a rubber band is stretched it heats up, and if it is suddenly released, it cools down perceptibly. These effects correspond to a large hysteresis from the thermal exchange with the environment and a smaller hysteresis due to internal friction within the rubber. This proper, intrinsic hysteresis can be measured only if the rubber band is adiabatically isolated.

Small vehicle suspensions using rubber (or other elastomers) can achieve the dual function of springing and damping because rubber, unlike metal springs, has pronounced hysteresis and does not return all the absorbed compression energy on the rebound. Mountain bikes have made use of elastomer suspension, as did the original Mini car.

The primary cause of rolling resistance when a body (such as a ball, tire, or wheel) rolls on a surface is hysteresis. This is attributed to the viscoelastic characteristics of the material of the rolling body.

Contact Angle Hysteresis

The contact angle formed between a liquid and solid phase will exhibit a range of contact angles that are possible. There are two common methods for measuring this range of contact angles. The first method is referred to as the tilting base method. Once a drop is dispensed on the surface with the surface level, the surface is then tilted from 0° to 90°. As the drop is tilted, the downhill side will be in a state of imminent wetting while the uphill side will be in a state of imminent dewetting. As the tilt increases the downhill contact angle will increase and represents the advancing contact angle while the uphill side will decrease; this is the receding contact angle. The values for these angles just prior to the drop releasing will typically represent the advancing and receding contact angles. The difference between these two angles is the contact angle hysteresis.

The second method is often referred to as the add/remove volume method. When the maximum liquid volume is removed from the drop without the interfacial area decreasing the receding contact angle is thus measured. When volume is added to the maximum before the interfacial area increases, this is the advancing contact angle. As with the tilt method, the difference between the advancing and receding contact angles is the contact angle hysteresis. Most researchers prefer the tilt method; the add/remove method requires that a tip or needle stay embedded in the drop which can affect the accuracy of the values, especially the receding contact angle.

Adsorption Hysteresis

Hysteresis can also occur during physical adsorption processes. In this type of hysteresis, the quantity adsorbed is different when gas is being added than it is when being removed. The specific causes of adsorption hysteresis are still an active area of research, but it is linked to differences in the nucleation and evaporation mechanisms inside mesopores. These mechanisms are further complicated by effects such as cavitation and pore blocking.

In physical adsorption, hysteresis is evidence of mesoporosity-indeed, the definition of mesopores (2–50 nm) is associated with the appearance (50 nm) and disappearance (2 nm) of mesoporosity in nitrogen adsorption isotherms as a function of Kelvin radius. An adsorption isotherm showing hysteresis is said to be of Type IV (for a wetting adsorbate) or Type V (for a non-wetting adsorbate), and hysteresis loops themselves are classified according to how symmetric the loop is. Adsorption hysteresis loops also have the unusual property that it is possible to scan within a hysteresis loop by reversing the direction of adsorption while on a point on the loop. The resulting scans are called "crossing," "converging," or "returning," depending on the shape of the isotherm at this point.

Matric Potential Hysteresis

The relationship between matric water potential and water content is the basis of the water retention curve. Matric potential measurements (Ψ_m) are converted to volumetric water content (θ) measurements based on a site or soil specific calibration curve. Hysteresis is a source of water content measurement error. Matric potential hysteresis arises from differences in wetting behaviour causing dry medium to re-wet; that is, it depends on the saturation history of the porous medium. Hysteretic behaviour means that, for example, at a matric potential (Ψ_m) of 5 kPa, the volumetric water content (θ) of a fine sandy soil matrix could be anything between 8% to 25%.

Tensiometers are directly influenced by this type of hysteresis. Two other types of sensors used to measure soil water matric potential are also influenced by hysteresis effects within the sensor itself. Resistance blocks, both nylon and gypsum based, measure matric potential as a function of electrical resistance. The relation between the sensor's electrical resistance and sensor matric potential is hysteretic. Thermocouples measure matric potential as a function of heat dissipation. Hysteresis occurs because measured heat dissipation depends on sensor water content, and the sensor water content–matric potential relationship is hysteretic. As of 2002, only desorption curves are usually measured during calibration of soil moisture sensors. Despite the fact that it can be a source of significant error, the sensor specific effect of hysteresis is generally ignored.

In Materials

Magnetic Hysteresis

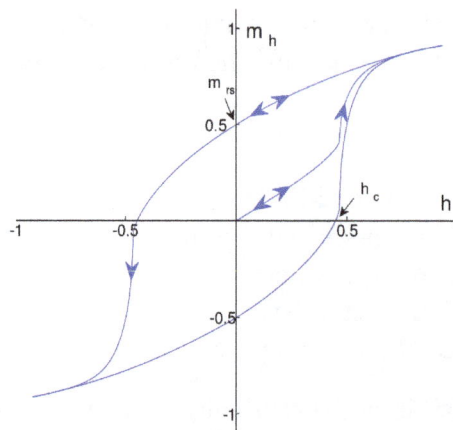

Theoretical model of magnetization m against magnetic field h. Starting at the origin, the upward curve is the *initial magnetization curve*. The downward curve after saturation, along with the lower return curve, form the *main loop*. The intercepts h_c and m_{rs} are the *coercivity* and *saturation remanence*.

When an external magnetic field is applied to a ferromagnetic material such as iron, the atomic dipoles align themselves with it. Even when the field is removed, part of the alignment will be retained: the material has become *magnetized*. Once magnetized, the magnet will stay magnetized indefinitely. To demagnetize it requires heat or a magnetic field in the opposite direction. This is the effect that provides the element of memory in a hard disk drive.

The relationship between field strength H and magnetization M is not linear in such materials. If a magnet is demagnetized ($H=M=0$) and the relationship between H and M is plotted for increasing levels of field strength, M follows the *initial magnetization curve*. This curve increases rapidly at first and then approaches an asymptote called magnetic saturation. If the magnetic field is now reduced monotonically, M follows a different curve. At zero field strength, the magnetization is offset from the origin by an amount called the remanence. If the H-M relationship is plotted for all strengths of applied magnetic field the result is a hysteresis loop called the *main loop*. The width of the middle section is twice the coercivity of the material.

A closer look at a magnetization curve generally reveals a series of small, random jumps in magnetization called Barkhausen jumps. This effect is due to crystallographic defects such as dislocations.

Magnetic hysteresis loops are not exclusive to materials with ferromagnetic ordering. Other magnetic orderings, such as spin glass ordering, also exhibit this phenomena.

Physical Origin

The phenomenon of hysteresis in ferromagnetic materials is the result of two effects: rotation of magnetization and changes in size or number of magnetic domains. In general, the magnetization varies (in direction but not magnitude) across a magnet, but in sufficiently small magnets, it does not. In these single-domain magnets, the magnetization responds to a magnetic field by rotating. Single-domain magnets are used wherever a strong, stable magnetization is needed (for example, magnetic recording).

Larger magnets are divided into regions called *domains*. Across each domain, the magnetization does not vary; but between domains are relatively thin *domain walls* in which the direction of magnetization rotates from the direction of one domain to another. If the magnetic field changes, the walls move, changing the relative sizes of the domains. Because the domains are not magnetized in the same direction, the magnetic moment per unit volume is smaller than it would be in a single-domain magnet; but domain walls involve rotation of only a small part of the magnetization, so it is much easier to change the magnetic moment. The magnetization can also change by addition or subtraction of domains (called *nucleation* and *denucleation*).

Magnetic Hysteresis Models

The most known empirical models in hysteresis are Preisach and Jiles-Atherton models. These models allow an accurate modeling of the hysteresis loop and are widely used in the industry. However, these models lose the connection with thermodynamics and the energy consistency is not ensured. Last models rely on a consistent thermodynamic formulation. VINCH model is inspired by the kinematic hardening laws and by the thermodynamics of irreversible processes.

In particular, in addition to provide an accurate modeling, the stored magnetic energy and the dissipated energy are known at all times. The obtained incremental formulation is variationally consistent, i.e., all internal variables follow from the minimization of a thermodynamic potential. That allows to obtain easily a vectorial model while Preisach and Jiles-Atherton are fundamentally scalar models.

Applications

There are a great variety of applications of the hysteresis in ferromagnets. Many of these make use of their ability to retain a memory, for example magnetic tape, hard disks, and credit cards. In these applications, *hard* magnets (high coercivity) like iron are desirable so the memory is not easily erased.

Soft magnets (low coercivity) are used as cores in electromagnets. The nonlinear response of the magnetic moment to a magnetic field boosts the response of the coil wrapped around it. The low coercivity reduces that energy loss associated with hysteresis.

Electrical Hysteresis

Electrical hysteresis typically occurs in ferroelectric material, where domains of polarization contribute to the total polarization. Polarization is the electrical dipole moment (either $C \cdot m^{-2}$ or $C \cdot m$). The mechanism, an organization of the polarization into domains, is similar to that of magnetic hysteresis.

Liquid–solid-phase Transitions

Hysteresis manifests itself in state transitions when melting temperature and freezing temperature do not agree. For example, agar melts at 85 °C and solidifies from 32 to 40 °C. This is to say that once agar is melted at 85 °C, it retains a liquid state until cooled to 40 °C. Therefore, from the temperatures of 40 to 85 °C, agar can be either solid or liquid, depending on which state it was before.

In Biology

Cell Biology and Genetics

Cells undergoing cell division exhibit hysteresis in that it takes a higher concentration of cyclins to switch them from G2 phase into mitosis than to stay in mitosis once begun.

Darlington in his classic works on genetics discussed hysteresis of the chromosomes, by which he meant "failure of the external form of the chromosomes to respond immediately to the internal stresses due to changes in their molecular spiral", as they lie in a somewhat rigid medium in the limited space of the cell nucleus.

In developmental biology, cell type diversity is regulated by long range-acting signaling molecules called morphogens that pattern uniform pools of cells in a concentration- and time-dependent manner. The morphogen Sonic Hedgehog (Shh), for example, acts on limb bud and neural progenitors to induce expression of a set of homeodomain-containing transcription factors to subdivide these tissues into distinct domains. It has been shown that these tissues have a 'memory' of previous exposure to Shh. In neural tissue, this hysteresis is regulated by a homeodomain (HD) feedback circuit that amplifies Shh signaling. In this circuit, expression of Gli transcription factors, the executors of the Shh pathway, is suppressed. Glis are processed to repressor forms (GliR) in the absence of Shh, but in the presence of Shh, a proportion of Glis are maintained as full-length proteins allowed to translocate to the nucleus, where they act as activators (GliA) of transcription. By reducing Gli expression then, the HD transcription factors reduce the total amount of Gli (GliT), so a higher proportion of GliT can be stabilized as GliA for the same concentration of Shh.

Immunology

There is some evidence that T cells exhibit hysteresis in that it takes a lower signal threshold to activate T cells that have been previously activated. Ras activation is required for downstream effector functions of activated T cells. Triggering of the T cell receptor induces high levels of Ras activation, which results in higher levels of GTP-bound (active) Ras at the cell surface. Since higher levels of active Ras have accumulated at the cell surface in T cells that have been previously stimulated by strong engagement of the T cell receptor, weaker subsequent T cell receptor signals received shortly afterwards will deliver the same level of activation due to the presence of higher levels of already activated Ras as compared to a naïve cell.

Neuroscience

The property by which some neurons do not return to their basal conditions from a stimulated condition immediately after removal of the stimulus is an example of hysteresis.

Respiratory Physiology

Lung hysteresis is evident when observing the compliance of a lung on inspiration versus expiration. The difference in compliance (volume/pressure) is due to the additional energy required during inspiration to recruit and inflate additional alveoli.

The transpulmonary pressure vs Volume curve of inhalation is different from the Pressure vs Volume curve of exhalation, the difference being described as hysteresis. Lung volume at any given pressure during inhalation is less than the lung volume at any given pressure during exhalation.

Voice and Speech Physiology

A hysteresis effect may be observed in voicing onset versus offset. The threshold value of the subglottal pressure required to start the vocal fold vibration is lower than the threshold value at which the vibration stops, when other parameters are kept constant. In utterances of vowel-voiceless consonant-vowel sequences during speech, the intraoral pressure is lower at the voice onset of the second vowel compared to the voice offset of the first vowel, the oral airflow is lower, the transglottal pressure is larger and the glottal width is smaller.

Ecology and Epidemiology

Hysteresis is a commonly encountered phenomenon in ecology and epidemiology, where the observed equilibrium of a system can not be predicted solely based on environmental variables, but also requires knowledge of the system's past history. Notable examples include the theory of spruce budworm outbreaks and behavioral-effects on disease transmission.

In Economics

Economic systems can exhibit hysteresis. For example, export performance is subject to strong hysteresis effects: because of the fixed transportation costs it may take a big push to start a country's exports, but once the transition is made, not much may be required to keep them going.

When some negative shock reduces employment in a company or industry, fewer employed workers then remain. As usually the employed workers have the power to set wages, their reduced number incentivizes them to bargain for even higher wages when the economy again gets better instead of letting the wage be at the equilibrium wage level, where the supply and demand of workers would match. This causes hysteresis: the unemployment becomes permanently higher after negative shocks.

Permanently Higher Unemployment

The idea of hysteresis is used extensively in the area of labor economics, specifically with reference to the unemployment rate. According to theories based on hysteresis, severe economic downturns (recession) and/or persistent stagnation (slow demand growth, usually after a recession) cause unemployed individuals to lose their job skills (commonly developed on the job) or to find them become obsolete or become demotivated/disillusioned/depressed or lose job-seeking skills. In addition, employers may use time spent in unemployment as a screening tool, i.e., to weed out less desired employees in hiring decisions. Then, in times of an economic upturn, recovery, or 'boom', the affected workers will not share in the prosperity, remaining unemployed for long periods (e.g., over 52 weeks). This makes unemployment "structural," i.e., extremely difficult to reduce simply by increasing the aggregate demand for products and labor without causing increased inflation. That is, it is possible that a ratchet effect in unemployment rates exists, so a short-term rise in unemployment rates tend to persist. For example, traditional anti-inflationary policy (the use of recession to fight inflation) leads to a permanently higher 'natural' rate of unemployment (more scientifically known as the NAIRU). This occurs first because inflationary expectations are 'sticky' downward due to wage and price rigidities (and so adapt slowly over time rather than being approximately correct as in theories of rational expectations) and second because labor markets do not clear instantly in response to unemployment.

The existence of hysteresis has been put forward as a possible explanation for the persistently high unemployment of many economies in the 1990s. Hysteresis has been invoked by Olivier Blanchard among others to explain the differences in long run unemployment rates between Europe and the United States. Labor market reform (usually meaning institutional change promoting more flexible wages, firing, and hiring) or strong demand-side economic growth may not therefore reduce this pool of long-term unemployed. Thus, specific targeted training programs are presented as a possible policy solution. However, the hysteresis hypothesis suggests such training programs are aided by persistently high demand for products (perhaps with incomes policies to avoid increased inflation), which reduces the transition costs out of unemployment and into paid employment easier.

Game Theory

Hysteresis occurs in applications of game theory to economics, in models with product quality, agent honesty or corruption of various institutions. Slightly different initial conditions can lead to opposite outcomes and resulting stable good and bad equilibria.

In Law

When hysteresis was first applied to the law, it was understood as the sluggishness in the production of international rules. More recently, legal scholars have affirmed that the law exhibits a hysteretic nature when it comes to the regulation of technology and the slowness can be worsened by a misconception of democracy as never-ending bargaining. One of the consequences of legal hysteresis is private ordering. In this context, by 'private ordering' neither one means the spontaneous enforcement of agreements (usually due to the so-called reputational mechanism), nor the abuse of contracts in order to elude the law. One means that, as a response to a legislator that intervenes always too late, trapped in old legal categories, there are many regulating gaps when it comes to technology and these gaps are filled by the contracts.

Additional Considerations

Models of Hysteresis

Each subject that involves hysteresis has models that are specific to the subject. In addition, there are models that capture general features of many systems with hysteresis. An example is the Preisach model of hysteresis, which represents a hysteresis nonlinearity as a linear superposition of square loops called non-ideal relays. Many complex models of hysteresis arise from the simple parallel connection, or superposition, of elementary carriers of hysteresis termed hysterons.

A simple parametric description of various hysteretic loops may be found in the Lapshin model of hysteresis. Along with the classical loop, substitution of trapezoidal or triangle pulses instead of the harmonic functions also allows piecewise-linear hysteresis loops frequently used in discrete automatics to be built in the model. There is an implementation of the hysteresis model in R programming language (package Hysteresis).

The Bouc–Wen model of hysteresis is often used to describe non-linear hysteretic systems. It was introduced by Bouc and extended by Wen, who demonstrated its versatility by producing a variety of hysteretic patterns. This model is able to capture in analytical form, a range of shapes of hysteretic cycles which match the behaviour of a wide class of hysteretical systems; therefore, given its versability and mathematical tractability, the Bouc–Wen model has quickly gained popularity and has been extended and applied to a wide variety of engineering problems, including multi-degree-of-freedom (MDOF) systems, buildings, frames, bidirectional and torsional response of hysteretic systems two- and three-dimensional continua, and soil liquefaction among others. The Bouc–Wen model and its variants/extensions have been used in applications of structural control, in particular in the modeling of the behaviour of magnetorheological dampers, base isolation devices for buildings and other kinds of damping devices; it has also been in the modelling and analysis of structures built of reinforced concrete, steel, masonry and timber.

Energy

When hysteresis occurs with extensive and intensive variables, the work done on the system is the area under the hysteresis graph.

References

- Harris, Ph.D., Edward D. (1 January 2014). Minerals in Food Nutrition, Metabolism, Bioactivity (1st ed.). Lancaster, PA: DEStech Publications, Inc. p. 378. ISBN 978-1-932078-97-8. Retrieved 30 January 2015.

- Astrid Sigel; Helmut Sigel; Roland K.O. Sigel, eds. (2008). Biomineralization: From Nature to Application. Metal Ions in Life Sciences. 4. Wiley. ISBN 978-0-470-03525-2.

- Jean-Pierre Cuif; Yannicke Dauphin; James E. Sorauf (2011). Biominerals and fossils through time. Cambridge. ISBN 978-0-521-87473-1.

- Schmalz, G.; Arenholdt-Bindslev, D. (2008). "Chapter 1: Basic Aspects". Biocompatibility of Dental Materials. Berlin: Springer-Verlag. pp. 1–12. ISBN 9783540777823. Retrieved 29 February 2016.

- Brown, Theodore L.; LeMay, H. Eugene; Bursten, Bruce E. (2000). Chemistry The Central Science. Prentice-Hall, Inc. pp. 451–452. ISBN 0-13-084090-4.

- Lancaster, Lynne (2005). Concrete Vaulted Construction in Imperial Rome. Innovations in Context. Cambridge University Press. ISBN 978-0-511-16068-4.

- Kosmatka, S.H.; Panarese, W.C. (1988). Design and Control of Concrete Mixtures. Skokie, IL, USA: Portland Cement Association. pp. 17, 42, 70, 184. ISBN 978-0-89312-087-0.

- Sotirios J. Vahaviolos (1999). Acoustic Emission: Standards and Technology Update. STP-1353. Philadelphia, PA: ASTM International (publishing). p. 81. ISBN 0-8031-2498-8.

- Blitz, Jack; G. Simpson (1991). Ultrasonic Methods of Non-Destructive Testing. Springer-Verlag New York, LLC. ISBN 978-0-412-60470-6.

- Stuart Hewerdine, ed. (1993). Plant Integrity Assessment by Acoustic Emission Testing (2 ed.). Rugby, UK: Institution of Chemical Engineers. ISBN 0-85295-316-X.

Atomic Arrangements in Solids

Aerogel is a material that is originated from a gel; the liquid constituent in the gel is substituted with a gas. Alternatively, the other emerging technologies in materials science are metamaterial, nanomaterial, programmable matter, quantum dot etc. The aspects elucidated in this text are of vital importance, and provide a better understanding of materials science.

Crystal

A crystal of amethyst quartz

A crystal or crystalline solid is a solid material whose constituents (such as atoms, molecules or ions) are arranged in a highly ordered microscopic structure, forming a crystal lattice that extends in all directions. In addition, macroscopic single crystals are usually identifiable by their geometrical shape, consisting of flat faces with specific, characteristic orientations. The scientific study of crystals and crystal formation is known as crystallography. The process of crystal formation via mechanisms of crystal growth is called crystallization or solidification.

Examples of large crystals include snowflakes, diamonds, and table salt. Most inorganic solids are not crystals but polycrystals, i.e. many microscopic crystals fused together into a single solid. Examples of polycrystals include most metals, rocks, ceramics, and ice. A third category of solids is amorphous solids, where the atoms have no periodic structure whatsoever. Examples of amorphous solids include glass, wax, and many plastics.

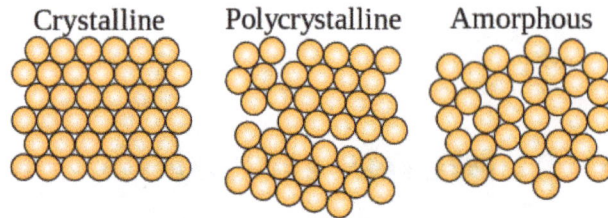

Crystalline   Polycrystalline   Amorphous

Microscopically, a single crystal has atoms in a near-perfect periodic arrangement; a polycrystal is composed of many microscopic crystals (called "crystallites" or "grains"); and an amorphous solid (such as glass) has no periodic arrangement even microscopically.

Crystal Structure (Microscopic)

Halite (Table Salt, NaCl): Microscopic and Macroscopic

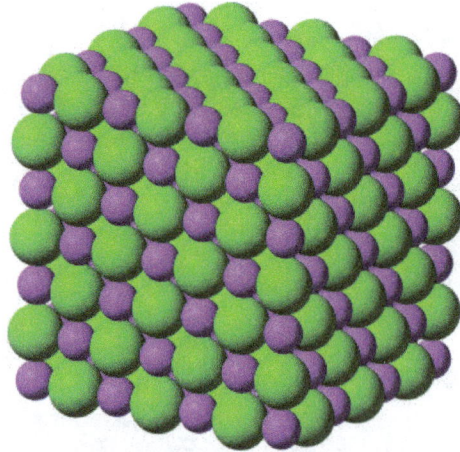

Microscopic structure of a halite crystal. (Purple is sodium ion, green is chlorine ion.) There is cubic symmetry in the atoms' arrangement.

Macroscopic (~16cm) halite crystal. The right-angles between crystal faces are due to the cubic symmetry of the atoms' arrangement.

The scientific definition of a "crystal" is based on the microscopic arrangement of atoms inside it, called the crystal structure. A crystal is a solid where the atoms form a periodic arrangement.

Not all solids are crystals. For example, when liquid water starts freezing, the phase change begins with small ice crystals that grow until they fuse, forming a *polycrystalline* structure. In the final block of ice, each of the small crystals (called "crystallites" or "grains") is a true crystal with a periodic arrangement of atoms, but the whole polycrystal does *not* have a periodic arrangement of atoms, because the periodic pattern is broken at the grain boundaries. Most macroscopic inorganic solids are polycrystalline, including almost all metals, ceramics, ice, rocks, etc. Solids that are neither crystalline nor polycrystalline, such as glass, are called *amorphous solids*, also called glassy, vitreous, or noncrystalline. These have no periodic order, even microscopically. There are distinct differences between crystalline solids and amorphous solids: most notably, the process of forming a glass does not release the latent heat of fusion, but forming a crystal does.

A crystal structure (an arrangement of atoms in a crystal) is characterized by its *unit cell*, a small imaginary box containing one or more atoms in a specific spatial arrangement. The unit cells are stacked in three-dimensional space to form the crystal.

The symmetry of a crystal is constrained by the requirement that the unit cells stack perfectly with no gaps. There are 219 possible crystal symmetries, called crystallographic space groups. These are grouped into 7 crystal systems, such as cubic crystal system (where the crystals may form cubes or rectangular boxes, such as halite shown at right) or hexagonal crystal system (where the crystals may form hexagons, such as ordinary water ice).

Crystal Faces and Shapes

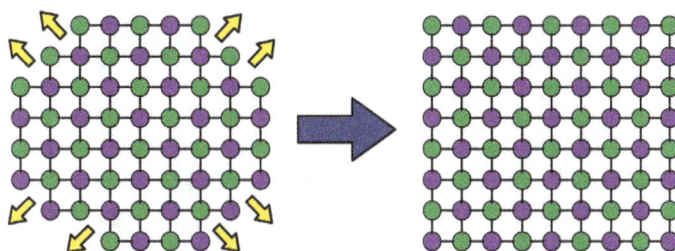

As a halite crystal is growing, new atoms can very easily attach to the parts of the surface with rough atomic-scale structure and many dangling bonds. Therefore, these parts of the crystal grow out very quickly (yellow arrows). Eventually, the whole surface consists of smooth, stable faces, where new atoms cannot as easily attach themselves.

Crystals are commonly recognized by their shape, consisting of flat faces with sharp angles. These shape characteristics are not *necessary* for a crystal—a crystal is scientifically defined by its microscopic atomic arrangement, not its macroscopic shape—but the characteristic macroscopic shape is often present and easy to see.

Euhedral crystals are those with obvious, well-formed flat faces. Anhedral crystals do not, usually because the crystal is one grain in a polycrystalline solid.

The flat faces (also called facets) of a euhedral crystal are oriented in a specific way relative to the underlying atomic arrangement of the crystal: they are planes of relatively low Miller index. This occurs because some surface orientations are more stable than others (lower surface energy). As a crystal grows, new atoms attach easily to the rougher and less stable parts of the surface, but less easily to the flat, stable surfaces. Therefore, the flat surfaces tend to grow larger and smoother, until the whole crystal surface consists of these plane surfaces.

One of the oldest techniques in the science of crystallography consists of measuring the three-dimensional orientations of the faces of a crystal, and using them to infer the underlying crystal symmetry.

A crystal's habit is its visible external shape. This is determined by the crystal structure (which restricts the possible facet orientations), the specific crystal chemistry and bonding (which may favor some facet types over others), and the conditions under which the crystal formed.

Occurrence in Nature

Ice crystals.

1.0 cm

Fossil shell with calcite crystals.

Rocks

By volume and weight, the largest concentrations of crystals in the Earth are part of its solid bedrock. Crystals found in rocks typically range in size from a fraction of a millimetre to several centimetres across, although exceptionally large crystals are occasionally found. As of 1999, the world's largest known naturally occurring crystal is a crystal of beryl from Malakialina, Madagascar, 18 m (59 ft) long and 3.5 m (11 ft) in diameter, and weighing 380,000 kg (840,000 lb).

Some crystals have formed by magmatic and metamorphic processes, giving origin to large masses of crystalline rock. The vast majority of igneous rocks are formed from molten magma and the degree of crystallization depends primarily on the conditions under which they solidified. Such rocks as granite, which have cooled very slowly and under great pressures, have completely crystallized; but many kinds of lava were poured out at the surface and cooled very rapidly, and in this latter group a small amount of amorphous or glassy matter is common. Other crystalline rocks, the metamorphic rocks such as marbles, mica-schists and quartzites, are recrystallized. This means that they were at first fragmental rocks like limestone, shale and sandstone and have never been in a molten condition nor entirely in solution, but the high temperature and pressure conditions of metamorphism have acted on them by erasing their original structures and inducing recrystallization in the solid state.

Other rock crystals have formed out of precipitation from fluids, commonly water, to form druses or quartz veins. The evaporites such as halite, gypsum and some limestones have been deposited from aqueous solution, mostly owing to evaporation in arid climates.

Ice

Water-based ice in the form of snow, sea ice and glaciers is a very common manifestation of crystalline or polycrystalline matter on Earth. A single snowflake is typically a single crystal, while an ice cube is a polycrystal.

Organigenic Crystals

Many living organisms are able to produce crystals, for example calcite and aragonite in the case of most molluscs or hydroxylapatite in the case of vertebrates.

Polymorphism and Allotropy

The same group of atoms can often solidify in many different ways. Polymorphism is the ability of a solid to exist in more than one crystal form. For example, water ice is ordinarily found in the hexagonal form Ice I_h, but can also exist as the cubic Ice I_c, the rhombohedral ice II, and many other forms. The different polymorphs are usually called different *phases*.

In addition, the same atoms may be able to form noncrystalline phases. For example, water can also form amorphous ice, while SiO_2 can form both fused silica (an amorphous glass) and quartz (a crystal). Likewise, if a substance can form crystals, it can also form polycrystals.

For pure chemical elements, polymorphism is known as allotropy. For example, diamond and graphite are two crystalline forms of carbon, while amorphous carbon is a noncrystalline form. Polymorphs, despite having the same atoms, may have wildly different properties. For example, diamond is among the hardest substances known, while graphite is so soft that it is used as a lubricant.

Polyamorphism is a similar phenomenon where the same atoms can exist in more than one amorphous solid form.

Crystallization

Crystallization is the process of forming a crystalline structure from a fluid or from materials dissolved in a fluid. (More rarely, crystals may be deposited directly from gas; see thin-film deposition and epitaxy.)

Crystallization is a complex and extensively-studied field, because depending on the conditions, a single fluid can solidify into many different possible forms. It can form a single crystal, perhaps with various possible phases, stoichiometries, impurities, defects, and habits. Or, it can form a polycrystal, with various possibilities for the size, arrangement, orientation, and phase of its grains. The final form of the solid is determined by the conditions under which the fluid is being solidified, such as the chemistry of the fluid, the ambient pressure, the temperature, and the speed with which all these parameters are changing.

Specific industrial techniques to produce large single crystals (called *boules*) include the Czochralski process and the Bridgman technique. Other less exotic methods of crystallization may be used, depending on the physical properties of the substance, including hydrothermal synthesis, sublimation, or simply solvent-based crystallization.

Large single crystals can be created by geological processes. For example, selenite crystals in excess of 10 meters are found in the Cave of the Crystals in Naica, Mexico.

Crystals can also be formed by biological processes. Conversely, some organisms have special techniques to *prevent* crystallization from occurring, such as antifreeze proteins.

Defects, Impurities, and Twinning

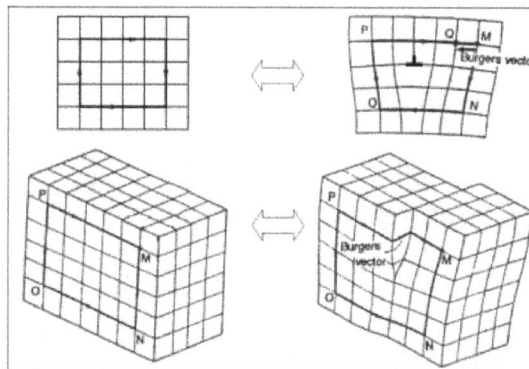

Two types of crystallographic defects. Top right: edge dislocation. Bottom right: screw dislocation.

An *ideal* crystal has every atom in a perfect, exactly repeating pattern. However, in reality, most crystalline materials have a variety of crystallographic defects, places where the crystal's pattern is interrupted. The types and structures of these defects may have a profound effect on the properties of the materials.

A few examples of crystallographic defects include vacancy defects (an empty space where an atom should fit), interstitial defects (an extra atom squeezed in where it does not fit), and dislocations. Dislocations are especially important in materials science, because they help determine the mechanical strength of materials.

Another common type of crystallographic defect is an impurity, meaning that the "wrong" type of atom is present in a crystal. For example, a perfect crystal of diamond would only contain carbon atoms, but a real crystal might perhaps contain a few boron atoms as well. These boron impurities change the diamond's color to slightly blue. Likewise, the only difference between ruby and sapphire is the type of impurities present in a corundum crystal.

Twinned pyrite crystal group.

In semiconductors, a special type of impurity, called a dopant, drastically changes the crystal's electrical properties. Semiconductor devices, such as transistors, are made possible largely by putting different semiconductor dopants into different places, in specific patterns.

Twinning is a phenomenon somewhere between a crystallographic defect and a grain boundary. Like a grain boundary, a twin boundary has different crystal orientations on its two sides. But unlike a grain boundary, the orientations are not random, but related in a specific, mirror-image way.

Mosaicity is a spread of crystal plane orientations. A mosaic crystal is supposed to consist of smaller crystalline units that are somewhat misaligned with respect to each other.

Chemical Bonds

In general, solids can be held together by various types of chemical bonds, such as metallic bonds, ionic bonds, covalent bonds, van der Waals bonds, and others. None of these are necessarily crystalline or non-crystalline. However, there are some general trends as follows.

Metals are almost always polycrystalline, though there are exceptions like amorphous metal and single-crystal metals. The latter are grown synthetically. (A microscopically-small piece of metal may naturally form into a single crystal, but larger pieces generally do not.) Ionically bonded solids are usually crystalline or polycrystalline. In practice, large salt crystals can be created by solidification of a molten fluid, or by crystallization out of a solution. Covalently bonded crystals are also very common, notable examples being diamond, quartz, and graphite. Polymer materials generally will form crystalline regions, but the lengths of the molecules usually prevent complete crystallization—and sometimes polymers are completely amorphous. Weak van der Waals forces also help hold together certain crystals, including graphite.

Quasicrystals

A quasicrystal consists of arrays of atoms that are ordered but not strictly periodic. They have many attributes in common with ordinary crystals, such as displaying a discrete pattern in x-ray diffraction, and the ability to form shapes with smooth, flat faces.

The material holmium–magnesium–zinc (Ho–Mg–Zn) forms quasicrystals, which can take on the macroscopic shape of a dodecahedron. (Only a quasicrystal, not a normal crystal, can take this shape.) The edges are 2 mm long.

Quasicrystals are most famous for their ability to show five-fold symmetry, which is impossible for an ordinary periodic crystal.

The International Union of Crystallography has redefined the term "crystal" to include both ordinary periodic crystals and quasicrystals ("any solid having an essentially discrete diffraction diagram").

Quasicrystals, first discovered in 1982, are quite rare in practice. Only about 100 solids are known to form quasicrystals, compared to about 400,000 periodic crystals known in 2004. The 2011 Nobel Prize in Chemistry was awarded to Dan Shechtman for the discovery of quasicrystals.

Special Properties from Anisotropy

Crystals can have certain special electrical, optical, and mechanical properties that glass and polycrystals normally cannot. These properties are related to the anisotropy of the crystal, i.e. the lack of rotational symmetry in its atomic arrangement. One such property is the piezoelectric effect, where a voltage across the crystal can shrink or stretch it. Another is birefringence, where a double image appears when looking through a crystal. Moreover, various properties of a crystal, including electrical conductivity, electrical permittivity, and Young's modulus, may be different in different directions in a crystal. For example, graphite crystals consist of a stack of sheets, and although each individual sheet is mechanically very strong, the sheets are rather loosely bound to each other. Therefore, the mechanical strength of the material is quite different depending on the direction of stress.

Not all crystals have all of these properties. Conversely, these properties are not quite exclusive to crystals. They can appear in glasses or polycrystals that have been made anisotropic by working or stress—for example, stress-induced birefringence.

Crystallography

Crystallography is the experimental science of determining the arrangement of atoms in the crystalline solids. The word "crystallography" derives from the Greek words *crystallon* "cold drop, frozen drop", with its meaning extending to all solids with some degree of transparency, and *grapho* "I write". In July 2012, the United Nations recognised the importance of the science of crystallography by proclaiming that 2014 would be the International Year of Crystallography. X-ray crystallography is used to determine the structure of large biomolecules such as proteins. Before the development of X-ray diffraction crystallography, the study of crystals was based on physical measurements of their geometry. This involved measuring the angles of crystal faces relative to each other and to theoretical reference axes (crystallographic axes), and establishing the symmetry of the crystal in question. This physical measurement is carried out using a goniometer. The position in 3D space of each crystal face is plotted on a stereographic net such as a Wulff net or Lambert net. The pole to each face is plotted on the net. Each point is labelled with its Miller index. The final plot allows the symmetry of the crystal to be established.

A crystalline solid: atomic resolution image of strontium titanate. Brighter atoms are strontium and darker ones are titanium.

Crystallographic methods now depend on analysis of the diffraction patterns of a sample targeted by a beam of some type. X-rays are most commonly used; other beams used include electrons or neutrons. This is facilitated by the wave properties of the particles. Crystallographers often explicitly state the type of beam used, as in the terms *X-ray crystallography, neutron diffraction* and *electron diffraction*. These three types of radiation interact with the specimen in different ways.

- X-rays interact with the spatial distribution of electrons in the sample.

- Electrons are charged particles and therefore interact with the total charge distribution of both the atomic nuclei and the electrons of the sample.

- Neutrons are scattered by the atomic nuclei through the strong nuclear forces, but in addition, the magnetic moment of neutrons is non-zero. They are therefore also scattered by magnetic fields. When neutrons are scattered from hydrogen-containing materials, they produce diffraction patterns with high noise levels. However, the material can sometimes be treated to substitute deuterium for hydrogen.

Because of these different forms of interaction, the three types of radiation are suitable for different crystallographic studies.

Theory

An image of a small object is made using a lens to focus the beam, similar to a lens in a microscope. However, the wavelength of visible light (about 4000 to 7000 ångström) is three orders of magnitude longer than the length of typical atomic bonds and atoms themselves (about 1 to 2 Å). Therefore, obtaining information about the spatial arrangement of atoms requires the use of radiation with shorter wavelengths, such as X-ray or neutron beams. Employing shorter wavelengths implied abandoning microscopy and true imaging, however, because there exists no material from which a lens capable of focusing this type of radiation can be created. (Nevertheless, scientists have had some success focusing X-rays with microscopic Fresnel zone plates made from gold, and by critical-angle reflection inside long tapered capillaries.) Diffracted X-ray or neutron beams cannot be focused to produce images, so the sample structure must be reconstructed from the diffraction pattern. Sharp features in the diffraction pattern arise from periodic, repeating structure in the sample, which are often very strong due to coherent reflection of many photons from many regularly spaced instances of similar structure, while non-periodic components of the structure result in diffuse (and usually weak) diffraction features - areas with a higher density and repetition of atom order tend to reflect more light toward one point in space when compared to those areas with fewer atoms and less repetition.

Because of their highly ordered and repetitive structure, crystals give diffraction patterns of sharp Bragg reflection spots, and are ideal for analyzing the structure of solids.

Notation

- Coordinates in *square brackets* such as denote a direction vector (in real space).

- Coordinates in *angle brackets* or *chevrons* such as <100> denote a *family* of directions which are related by symmetry operations. In the cubic crystal system for example, <100> would mean , , or the negative of any of those directions.

- Miller indices in *parentheses* such as (100) denote a plane of the crystal structure, and regular repetitions of that plane with a particular spacing. In the cubic system, the normal to the (hkl) plane is the direction [hkl], but in lower-symmetry cases, the normal to (hkl) is not parallel to [hkl].

- Indices in *curly brackets* or *braces* such as {100} denote a family of planes and their normals which are equivalent in cubic materials due to symmetry operations, much the way angle brackets denote a family of directions. In non-cubic materials, <hkl> is not necessarily perpendicular to {hkl}.

Techniques

Some materials that have been analyzed crystallographically, such as proteins, do not occur naturally as crystals. Typically, such molecules are placed in solution and allowed to slowly crystallize through vapor diffusion. A drop of solution containing the molecule, buffer, and precipitants is sealed in a container with a reservoir containing a hygroscopic solution. Water in the drop diffuses to the reservoir, slowly increasing the concentration and allowing a crystal to form. If the concentration were to rise more quickly, the molecule would simply precipitate out of solution, resulting in disorderly granules rather than an orderly and hence usable crystal.

Once a crystal is obtained, data can be collected using a beam of radiation. Although many universities that engage in crystallographic research have their own X-ray producing equipment, synchrotrons are often used as X-ray sources, because of the purer and more complete patterns such sources can generate. Synchrotron sources also have a much higher intensity of X-ray beams, so data collection takes a fraction of the time normally necessary at weaker sources. Complementary neutron crystallography techniques are used to identify the positions of hydrogen atoms, since X-rays only interact very weakly with light elements such as hydrogen.

Producing an image from a diffraction pattern requires sophisticated mathematics and often an iterative process of modelling and refinement. In this process, the mathematically predicted diffraction patterns of an hypothesized or "model" structure are compared to the actual pattern generated by the crystalline sample. Ideally, researchers make several initial guesses, which through refinement all converge on the same answer. Models are refined until their predicted patterns match to as great a degree as can be achieved without radical revision of the model. This is a painstaking process, made much easier today by computers.

The mathematical methods for the analysis of diffraction data only apply to *patterns,* which in turn result only when waves diffract from orderly arrays. Hence crystallography applies for the most part only to crystals, or to molecules which can be coaxed to crystallize for the sake of measurement. In spite of this, a certain amount of molecular information can be deduced from patterns that are generated by fibers and powders, which while not as perfect as a solid crystal, may exhibit a degree of order. This level of order can be sufficient to deduce the structure of simple molecules, or to determine the coarse features of more complicated molecules. For example, the double-helical structure of DNA was deduced from an X-ray diffraction pattern that had been generated by a fibrous sample.

In Materials Science

Crystallography is used by materials scientists to characterize different materials. In single crystals, the effects of the crystalline arrangement of atoms is often easy to see macroscopically, because the natural shapes of crystals reflect the atomic structure. In addition, physical properties are often controlled by crystalline defects. The understanding of crystal structures is an important prerequisite for understanding crystallographic defects. Mostly, materials do not occur as a single crystal, but in poly-crystalline form (i.e., as an aggregate of small crystals with different orientations). Because of this, the powder diffraction method, which takes diffraction patterns of polycrystalline samples with a large number of crystals, plays an important role in structural determination.

Other physical properties are also linked to crystallography. For example, the minerals in clay form small, flat, platelike structures. Clay can be easily deformed because the platelike particles can slip along each other in the plane of the plates, yet remain strongly connected in the direction perpendicular to the plates. Such mechanisms can be studied by crystallographic texture measurements.

In another example, iron transforms from a body-centered cubic (bcc) structure to a face-centered cubic (fcc) structure called austenite when it is heated. The fcc structure is a close-packed structure unlike the bcc structure; thus the volume of the iron decreases when this transformation occurs.

Crystallography is useful in phase identification. When manufacturing or using a material, it is generally desirable to know what compounds and what phases are present in the material, as their composition, structure and proportions will influence the material's properties. Each phase has a characteristic arrangement of atoms. X-ray or neutron diffraction can be used to identify which patterns are present in the material, and thus which compounds are present. Crystallography covers the enumeration of the symmetry patterns which can be formed by atoms in a crystal and for this reason is related to group theory and geometry.

Biology

X-ray crystallography is the primary method for determining the molecular conformations of biological macromolecules, particularly protein and nucleic acids such as DNA and RNA. In fact, the double-helical structure of DNA was deduced from crystallographic data. The first crystal structure of a macromolecule was solved in 1958, a three-dimensional model of the myoglobin molecule obtained by X-ray analysis. The Protein Data Bank (PDB) is a freely accessible repository for the structures of proteins and other biological macromolecules. Computer programs such as RasMol or Pymol can be used to visualize biological molecular structures. Neutron crystallography is often used to help refine structures obtained by X-ray methods or to solve a specific bond; the methods are often viewed as complementary, as X-rays are sensitive to electron positions and scatter most strongly off heavy atoms, while neutrons are sensitive to nucleus positions and scatter strongly even off many light isotopes, including hydrogen and deuterium. Electron crystallography has been used to determine some protein structures, most notably membrane proteins and viral capsids.

Reference Literature in Crystallography

The International Tables of Crystallography is a eight book series that outlines the standard notations for formatting, describing and testing crystals. The series contains books that covers analysis methods and the mathematical procedures for determining organic structure though x-ray crystallography, electron diffraction, and neutron diffraction. The International tables are focused on procedures, techniques and descriptions and does not list the physical properties of individual crystals themselves. Each book is about 1000 pages and the titles of the books are:

Vol A - Space Group Symmetry

Vol A1 - Symmetry Relations Between Space Groups

Vol B - Reciprocal Space

Vol C - Mathmatical, Physical, and Chemical Tables

Vol D - Physical Properties of Crystals

Vol E - Subperiodic Groups

Vol F - Crystallography of Biological Macromolecues

Vol G - Definition and Exchange of Crystallographic Data

Amorphous Solid

In condensed matter physics and materials science, an amorphous (from the Greek *a*, without, *morphé*, shape, form) or non-crystalline solid is a solid that lacks the long-range order characteristic of a crystal. In some older books, the term has been used synonymously with glass. Nowadays, "Glassy solid" or "Amorphous solid" is considered to be the overarching concept, and glass the more special case: A glass is an amorphous solid that exhibits a glass transition. Polymers are often amorphous. Other types of amorphous solids include gels, thin films, and nanostructured materials such as glass.

Amorphous metals have low toughness, but high strength

Amorphous materials have an internal structure made of interconnected structural blocks. Whether a material is liquid or solid depends primarily on the connectivity between its elementary building blocks so that solids are characterized by a high degree of connectivity whereas structural blocks in fluids have lower connectivity.

States of crystalline and amorphous materials as a function of connectivity

Nano-structured Materials

Even amorphous materials have some shortrange order at the atomic length scale due to the nature of chemical bonding. Furthermore, in very small crystals a large fraction of the atoms are the crystal; relaxation of the surface and interfacial effects distort the atomic positions, decreasing the structural order. Even the most advanced structural characterization techniques, such as x-ray diffraction and transmission electron microscopy, have difficulty in distinguishing between amorphous and crystalline structures on these length scales.

Amorphous Thin Films

Amorphous phases are important constituents of thin films, which are solid layers of a few nm to some tens of μm thickness deposited upon a substrate. So-called structure zone models were developed to describe the micro structure and ceramics of thin films as a function of the homologous temperature T_h that is the ratio of deposition temperature over melting temperature. According to these models, a necessary (but not sufficient) condition for the occurrence of amorphous phases is that T_h has to be smaller than 0.3, that is the deposition temperature must be below 30% of the melting temperature. For higher values, the surface diffusion of deposited atomic species would allow for the formation of crystallites with long range atomic order.

Regarding their applications, amorphous metallic layers played an important role in the discussion of a suspected superconductivity in amorphous metals. Today, optical coatings made from TiO_2, SiO_2, Ta_2O_5 etc. and combinations of them in most cases consist of amorphous phases of these compounds. Much research is carried out into thin amorphous films as a gas separating membrane layer. The technologically most important thin amorphous film is probably represented by few nm thin SiO_2 layers serving as isolator above the conducting channel of a metal-oxide semiconductor field-effect transistor (MOSFET). Also, hydrogenated amorphous silicon, a-Si:H in short, is of technical significance for thin film solar cells. In case of a-Si:H the missing long-range order between silicon atoms is partly induced by the presence by hydrogen in the percent range.

The occurrence of amorphous phases turned out as a phenomenon of particular interest for studying thin film growth. Remarkably, the growth of polycrystalline films is often used and preceded by an initial amorphous layer, the thickness of which may amount to only a few nm. The most investigated example is represented by thin multicrystalline silicon films, where such as the unoriented molecule. An initial amorphous layer was observed in many studies. Wedge-shaped polycrystals were identified by transmission electron microscopy to grow out of the amorphous phase only after the latter has exceeded a certain thickness, the precise value of which depends on deposition temperature, background pressure and various other process parameters. The phenomenon has been interpreted in the framework of Ostwald's rule of stages that predicts the formation of phases to proceed with increasing condensation time towards increasing stability. Experimental studies of the phenomenon require a clearly defined state of the substrate surface and its contaminant density etc., upon which the thin film is deposited.

Crystal Structure

In crystallography, crystal structure is a description of the ordered arrangement of atoms, ions or molecules in a crystalline material. Ordered structures occur from the intrinsic nature of the constituent particles to form symmetric patterns that repeat along the principal directions of three-dimensional space in matter.

The smallest group of particles in the material that constitutes the repeating pattern is the unit cell of the structure. The unit cell completely defines the symmetry and structure of the entire crystal lattice, which is built up by repetitive translation of the unit cell along its principal axes. The repeating patterns are said to be located at the points of the Bravais lattice.

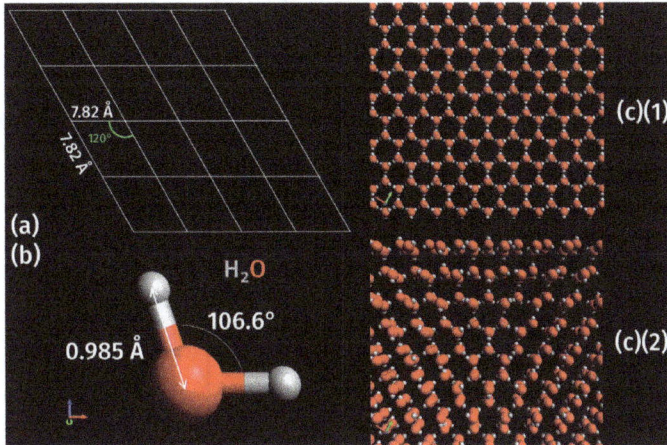

The (3-D) crystal structure of H2O ice Ih (c) consists of bases of H2O ice molecules (b) located on lattice points within the (2-D) hexagonal space lattice (a). The values for the H–O–H angle and O–H distance have come from Physics of Ice with uncertainties of ±1.5° and ±0.005 Å, respectively. The white box in (c) is the unit cell defined by Bernal and Fowler

The lengths of the principal axes, or edges, of the unit cell and the angles between them are the lattice constants, also called *lattice parameters*. The symmetry properties of the crystal are described by the concept of space groups. All possible symmetric arrangements of particles in space may be described by the set of seven space groups.

The crystal structure and symmetry play a role in determining many physical properties, such as cleavage, electronic band structure, and optical transparency.

Unit Cell

The crystal structure of a material (the arrangement of atoms within a given type of crystal) can be described in terms of its unit cell. The unit cell is a small box containing one or more atoms arranged in 3 dimensions. The unit cells stacked in three-dimensional space describe the bulk arrangement of atoms of the crystal. The unit cell is represented in terms of its lattice parameters, which are the lengths of the cell edges (a, b and c) and the angles between them (alpha, beta and gamma), while the positions of the atoms inside the unit cell are described by the set of atomic positions (x_i, y_i, z_i) measured from a lattice point. Commonly, atomic positions are represented in terms of fractional coordinates, relative to the unit cell lengths.

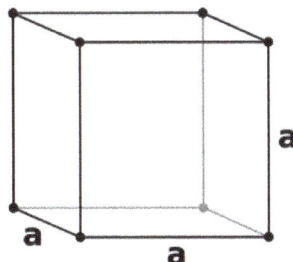

Simple cubic (P)

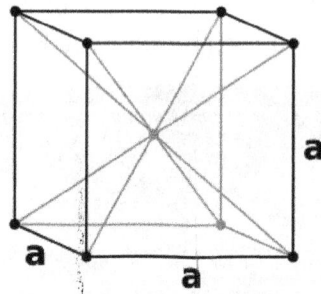

Body-centered cubic (I)

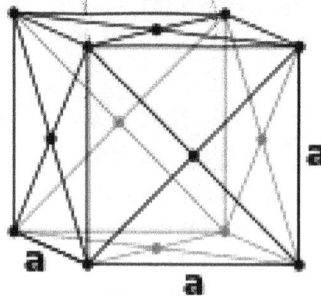

Face-centered cubic (F)

The atom positions within the unit cell can be calculated through application of symmetry operations to the asymmetric unit. The asymmetric unit refers to the smallest possible occupation of space within the unit cell. This does not, however imply that the entirety of the asymmetric unit must lie within the boundaries of the unit cell. Symmetric transformations of atom positions are calculated from the space group of the crystal structure, and this is usually a black box operation performed by computer programs. However, manual calculation of the atomic positions within the unit cell can be performed from the asymmetric unit, through the application of the symmetry operators described within the *International Tables for Crystallography: Volume A*.

Miller Indices

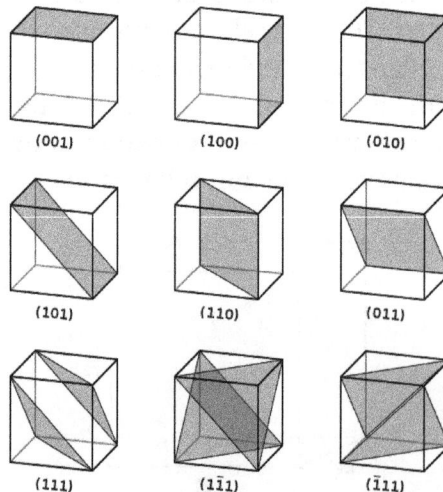

Planes with different Miller indices in cubic crystals

Vectors and planes in a crystal lattice are described by the three-value Miller index notation. It uses the indices ℓ, m, and n as directional parameters, which are separated by 90°, and are thus orthogonal.

By definition, the syntax (ℓmn) denotes a plane that intercepts the three points a_1/ℓ, a_2/m, and a_3/n, or some multiple thereof. That is, the Miller indices are proportional to the inverses of the intercepts of the plane with the unit cell (in the basis of the lattice vectors). If one or more of the indices is zero, it means that the planes do not intersect that axis (i.e., the intercept is "at infinity"). A plane containing a coordinate axis is translated so that it no longer contains that axis before its Miller indices are determined. The Miller indices for a plane are integers with no common factors. Negative indices are indicated with horizontal bars, as in (123). In an orthogonal coordinate system for a cubic cell, the Miller indices of a plane are the Cartesian components of a vector normal to the plane.

Considering only (ℓmn) planes intersecting one or more lattice points (the *lattice planes*), the distance d between adjacent lattice planes is related to the (shortest) reciprocal lattice vector orthogonal to the planes by the formula

$$d = \frac{2\pi}{|\mathbf{g}_{\ell mn}|}$$

Planes and Directions

The crystallographic directions are geometric lines linking nodes (atoms, ions or molecules) of a crystal. Likewise, the crystallographic planes are geometric *planes* linking nodes. Some directions and planes have a higher density of nodes. These high density planes have an influence on the behavior of the crystal as follows:

- Optical properties: Refractive index is directly related to density (or periodic density fluctuations).

- Adsorption and reactivity: Physical adsorption and chemical reactions occur at or near surface atoms or molecules. These phenomena are thus sensitive to the density of nodes.

- Surface tension: The condensation of a material means that the atoms, ions or molecules are more stable if they are surrounded by other similar species. The surface tension of an interface thus varies according to the density on the surface.

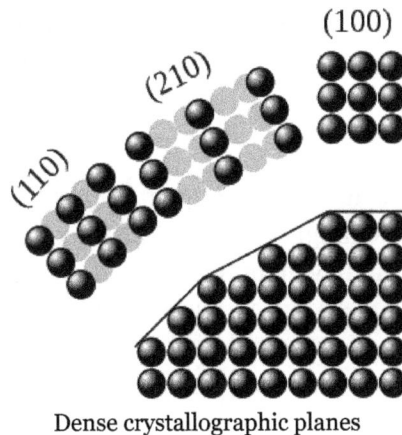

Dense crystallographic planes

- Microstructural defects: Pores and crystallites tend to have straight grain boundaries following higher density planes.

- Cleavage: This typically occurs preferentially parallel to higher density planes.

- Plastic deformation: Dislocation glide occurs preferentially parallel to higher density planes. The perturbation carried by the dislocation (Burgers vector) is along a dense direction. The shift of one node in a more dense direction requires a lesser distortion of the crystal lattice.

Some directions and planes are defined by symmetry of the crystal system. In monoclinic, rhombohedral, tetragonal, and trigonal/hexagonal systems there is one unique axis (sometimes called the principal axis) which has higher rotational symmetry than the other two axes. The basal plane is the plane perpendicular to the principal axis in these crystal systems. For triclinic, orthorhombic, and cubic crystal systems the axis designation is arbitrary and there is no principal axis.

Cubic Structures

For the special case of simple cubic crystals, the lattice vectors are orthogonal and of equal length (usually denoted a); similarly for the reciprocal lattice. So, in this common case, the Miller indices (ℓmn) and $[\ell mn]$ both simply denote normals/directions in Cartesian coordinates. For cubic crystals with lattice constant a, the spacing d between adjacent (ℓmn) lattice planes is (from above):

$$d_{\ell mn} = \frac{a}{\sqrt{\ell^2 + m^2 + n^2}}$$

Because of the symmetry of cubic crystals, it is possible to change the place and sign of the integers and have equivalent directions and planes:

- Coordinates in *angle brackets* such as $\langle 100 \rangle$ denote a *family* of directions that are equivalent due to symmetry operations, such as , , or the negative of any of those directions.

- Coordinates in *curly brackets* or *braces* such as $\{100\}$ denote a family of plane normals that are equivalent due to symmetry operations, much the way angle brackets denote a family of directions.

For face-centered cubic (fcc) and body-centered cubic (bcc) lattices, the primitive lattice vectors are not orthogonal. However, in these cases the Miller indices are conventionally defined relative to the lattice vectors of the cubic supercell and hence are again simply the Cartesian directions.

Classification

The defining property of a crystal is its inherent symmetry, by which we mean that under certain 'operations' the crystal remains unchanged. All crystals have translational symmetry in three directions, but some have other symmetry elements as well. For example, rotating the crystal 180° about a certain axis may result in an atomic configuration that is identical to the original configuration. The crystal is then said to have a twofold rotational symmetry about this axis. In addition to rotational symmetries like this, a crystal may have symmetries in the form of mirror planes and translational symmetries, and also the so-called "compound symmetries," which are a combination of translation and rotation/mirror symmetries. A full classification of a crystal is achieved when all of these inherent symmetries of the crystal are identified.

Lattice Systems

These lattice systems are a grouping of crystal structures according to the axial system used to describe their lattice. Each lattice system consists of a set of three axes in a particular geometric arrangement. There are seven lattice systems. They are similar to but not quite the same as the seven crystal systems and the six crystal families.

The simplest and most symmetric, the cubic (or isometric) system, has the symmetry of a cube, that is, it exhibits four threefold rotational axes oriented at 109.5° (the tetrahedral angle) with respect to each other. These threefold axes lie along the body diagonals of the cube. The other six lattice systems, are hexagonal, tetragonal, rhombohedral (often confused with the trigonal crystal system), orthorhombic, monoclinic and triclinic.

Atomic Coordination

By considering the arrangement of atoms relative to each other, their coordination numbers (or number of nearest neighbors), interatomic distances, types of bonding, etc., it is possible to form a general view of the structures and alternative ways of visualizing them.

Close Packing

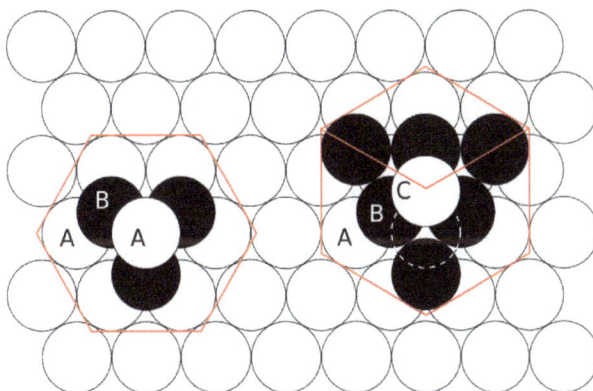

The hcp lattice (left) and the fcc lattice (right)

The principles involved can be understood by considering the most efficient way of packing together equal-sized spheres and stacking close-packed atomic planes in three dimensions. For example, if plane A lies beneath plane B, there are two possible ways of placing an additional atom on top of layer B. If an additional layer was placed directly over plane A, this would give rise to the following series:

...ABABABAB...

This arrangement of atoms in a crystal structure is known as hexagonal close packing (hcp).

If, however, all three planes are staggered relative to each other and it is not until the fourth layer is positioned directly over plane A that the sequence is repeated, then the following sequence arises:

...ABCABCABC...

This type of structural arrangement is known as cubic close packing (ccp).

The unit cell of a ccp arrangement of atoms is the face-centered cubic (fcc) unit cell. This is not immediately obvious as the closely packed layers are parallel to the {111} planes of the fcc unit cell. There are four different orientations of the close-packed layers.

The packing efficiency can be worked out by calculating the total volume of the spheres and dividing by the volume of the cell as follows:

$$\frac{4 \times \frac{4}{3}\pi r^3}{16\sqrt{2}r^3} = \frac{\pi}{3\sqrt{2}} = 0.7405...$$

The 74% packing efficiency is the maximum density possible in unit cells constructed of spheres of only one size. Most crystalline forms of metallic elements are hcp, fcc, or bcc (body-centered cubic). The coordination number of atoms in hcp and fcc structures is 12 and its atomic packing factor (APF) is the number mentioned above, 0.74. This can be compared to the APF of a bcc structure, which is 0.68.

Bravais Lattices

Bravais lattices, also referred to as *space lattices*, describe the geometric arrangement of the lattice points, and therefore the translational symmetry of the crystal. The three dimensions of space afford 14 distinct Bravais lattices describing the translational symmetry. All crystalline materials recognized today, not including quasicrystals, fit in one of these arrangements. The fourteen three-dimensional lattices, classified by crystal system.

The crystal structure consists of the same group of atoms, the *basis*, positioned around each and every lattice point. This group of atoms therefore repeats indefinitely in three dimensions according to the arrangement of one of the Bravais lattices. The characteristic rotation and mirror symmetries of the unit cell is described by its crystallographic point group.

Point Groups

The crystallographic point group or *crystal class* is the mathematical group comprising the symmetry operations that leave at least one point unmoved and that leave the appearance of the crystal structure unchanged. These symmetry operations include

- *Reflection*, which reflects the structure across a *reflection plane*

- *Rotation*, which rotates the structure a specified portion of a circle about a *rotation axis*

- *Inversion*, which changes the sign of the coordinate of each point with respect to a *center of symmetry* or *inversion point*

- *Improper rotation*, which consists of a rotation about an axis followed by an inversion.

Rotation axes (proper and improper), reflection planes, and centers of symmetry are collectively called *symmetry elements*. There are 32 possible crystal classes. Each one can be classified into one of the seven crystal systems.

Space Groups

In addition to the operations of the point group, the space group of the crystal structure contains translational symmetry operations. These include:

- Pure *translations*, which move a point along a vector

- *Screw axes*, which rotate a point around an axis while translating parallel to the axis.

- *Glide planes*, which reflect a point through a plane while translating it parallel to the plane.

There are 230 distinct space groups.

Grain Boundaries

Grain boundaries are interfaces where crystals of different orientations meet. A grain boundary is a single-phase interface, with crystals on each side of the boundary being identical except in orientation. The term "crystallite boundary" is sometimes, though rarely, used. Grain boundary areas contain those atoms that have been perturbed from their original lattice sites, dislocations, and impurities that have migrated to the lower energy grain boundary.

Treating a grain boundary geometrically as an interface of a single crystal cut into two parts, one of which is rotated, we see that there are five variables required to define a grain boundary. The first two numbers come from the unit vector that specifies a rotation axis. The third number designates the angle of rotation of the grain. The final two numbers specify the plane of the grain boundary (or a unit vector that is normal to this plane).

Grain boundaries disrupt the motion of dislocations through a material, so reducing crystallite size is a common way to improve strength, as described by the Hall–Petch relationship. Since grain boundaries are defects in the crystal structure they tend to decrease the electrical and thermal conductivity of the material. The high interfacial energy and relatively weak bonding in most grain boundaries often makes them preferred sites for the onset of corrosion and for the precipitation of new phases from the solid. They are also important to many of the mechanisms of creep.

Grain boundaries are in general only a few nanometers wide. In common materials, crystallites are large enough that grain boundaries account for a small fraction of the material. However, very small grain sizes are achievable. In nanocrystalline solids, grain boundaries become a significant volume fraction of the material, with profound effects on such properties as diffusion and plasticity. In the limit of small crystallites, as the volume fraction of grain boundaries approaches 100%, the material ceases to have any crystalline character, and thus becomes an amorphous solid.

Defects and Impurities

Real crystals feature defects or irregularities in the ideal arrangements described above and it is these defects that critically determine many of the electrical and mechanical properties of real materials. When one atom substitutes for one of the principal atomic components within the crystal structure, alteration in the electrical and thermal properties of the material may ensue. Impurities may also manifest as spin impurities in certain materials. Research on magnetic impurities demonstrates that substantial alteration of certain properties such as specific heat may be affected by small concentrations of an impurity, as for example impurities in semiconducting ferromagnetic alloys may lead to different properties as first predicted in the late 1960s. Dislocations in the crystal lattice allow shear at lower stress than that needed for a perfect crystal structure.

Prediction of Structure

The difficulty of predicting stable crystal structures based on the knowledge of only the chemical composition has long been a stumbling block on the way to fully computational materials design. Now, with more powerful algorithms and high-performance computing, structures of medium complexity can be predicted using such approaches as evolutionary algorithms, random sampling, or metadynamics.

The crystal structures of simple ionic solids (e.g., NaCl or table salt) have long been rationalized in terms of Pauling's rules, first set out in 1929 by Linus Pauling, referred to by many since as the "father of the chemical bond". Pauling also considered the nature of the interatomic forces in metals, and concluded that about half of the five d-orbitals in the transition metals are involved in bonding, with the remaining nonbonding d-orbitals being responsible for the magnetic properties. He, therefore, was able to correlate the number of d-orbitals in bond formation with the bond length as well as many of the physical properties of the substance. He subsequently introduced the metallic orbital, an extra orbital necessary to permit uninhibited resonance of valence bonds among various electronic structures.

In the resonating valence bond theory, the factors that determine the choice of one from among alternative crystal structures of a metal or intermetallic compound revolve around the energy of resonance of bonds among interatomic positions. It is clear that some modes of resonance would make larger contributions (be more mechanically stable than others), and that in particular a simple ratio of number of bonds to number of positions would be exceptional. The resulting principle is that a special stability is associated with the simplest ratios or "bond numbers": $\frac{1}{2}$, $\frac{1}{3}$, $\frac{2}{3}$, $\frac{1}{4}$, $\frac{3}{4}$, etc. The choice of structure and the value of the axial ratio (which determines the relative bond lengths) are thus a result of the effort of an atom to use its valency in the formation of stable bonds with simple fractional bond numbers.

After postulating a direct correlation between electron concentration and crystal structure in beta-phase alloys, Hume-Rothery analyzed the trends in melting points, compressibilities and bond lengths as a function of group number in the periodic table in order to establish a system of valencies of the transition elements in the metallic state. This treatment thus emphasized the increasing bond strength as a function of group number. The operation of directional forces were emphasized in one article on the relation between bond hybrids and the metallic structures. The resulting correlation between electronic and crystalline structures is summarized by a single parameter, the weight of the d-electrons per hybridized metallic orbital. The "d-weight" calculates out to 0.5, 0.7 and 0.9 for the fcc, hcp and bcc structures respectively. The relationship between d-electrons and crystal structure thus becomes apparent.

In crystal structure predictions/simulations, the periodicity is usually applied, since the system is imagined as unlimited big in all directions. Starting from a triclinic structure with no further symmetry property assumed, the system may be driven to show some additional symmetry properties by applying Newton's Second Law on particles in the unit cell and a recently developed dynamical equation for the system period vectors (lattice parameters including angles), even if the system is subject to external stress.

Polymorphism

Polymorphism is the occurrence of multiple crystalline forms of a material. It is found in many crystalline materials including polymers, minerals, and metals. According to Gibbs' rules of phase equilibria, these unique crystalline phases are dependent on intensive variables such as pressure and temperature. Polymorphism is related to allotropy, which refers to elemental solids. The complete morphology of a material is described by polymorphism and other variables such as crystal habit, amorphous fraction or crystallographic defects. Polymorphs have different stabilities and may spontaneously convert from a metastable form (or thermodynamically unstable form) to the stable form at a particular temperature. They also exhibit different melting points, solubilities, and X-ray diffraction patterns.

One good example of this is the quartz form of silicon dioxide, or SiO_2. In the vast majority of silicates, the Si atom shows tetrahedral coordination by 4 oxygens. All but one of the crystalline forms involve tetrahedral $\{SiO_4\}$ units linked together by shared vertices in different arrangements. In different minerals the tetrahedra show different degrees of networking and polymerization. For example, they occur singly, joined together in pairs, in larger finite clusters including rings, in chains, double chains, sheets, and three-dimensional frameworks. The minerals are classified into groups based on these structures. In each of its 7 thermodynamically stable crystalline forms or polymorphs of crystalline quartz, only 2 out of 4 of each the edges of the $\{SiO_4\}$ tetrahedra are shared with others, yielding the net chemical formula for silica: SiO_2.

Another example is elemental tin (Sn), which is malleable near ambient temperatures but is brittle when cooled. This change in mechanical properties due to existence of its two major allotropes, α- and β-tin. The two allotropes that are encountered at normal pressure and temperature, α-tin and β-tin, are more commonly known as *gray tin* and *white tin* respectively. Two more allotropes, γ and σ, exist at temperatures above 161 °C and pressures above several GPa. White tin is metallic, and is the stable crystalline form at or above room temperature. Below 13.2 °C, tin exists in the gray form, which has a diamond cubic crystal structure, similar to diamond, silicon or germanium. Gray tin has no metallic properties at all, is a dull gray powdery material, and has few uses, other than a few specialized semiconductor applications. Although the α−β transformation temperature of tin is nominally 13.2 °C, impurities (e.g. Al, Zn, etc.) lower the transition temperature well below 0 °C, and upon addition of Sb or Bi the transformation may not occur at all.

Physical Properties

Twenty of the 32 crystal classes are piezoelectric, and crystals belonging to one of these classes (point groups) display piezoelectricity. All piezoelectric classes lack a center of symmetry. Any material develops a dielectric polarization when an electric field is applied, but a substance that has such a natural charge separation even in the absence of a field is called a polar material. Whether or not a material is polar is determined solely by its crystal structure. Only 10 of the 32 point groups are polar. All polar crystals are pyroelectric, so the 10 polar crystal classes are sometimes referred to as the pyroelectric classes.

There are a few crystal structures, notably the perovskite structure, which exhibit ferroelectric behavior. This is analogous to ferromagnetism, in that, in the absence of an electric field during production, the ferroelectric crystal does not exhibit a polarization. Upon the application of an electric field of sufficient magnitude, the crystal becomes permanently polarized. This polarization can be reversed by a sufficiently large counter-charge, in the same way that a ferromagnet can be reversed. However, although they are called ferroelectrics, the effect is due to the crystal structure (not the presence of a ferrous metal).

Miller Index

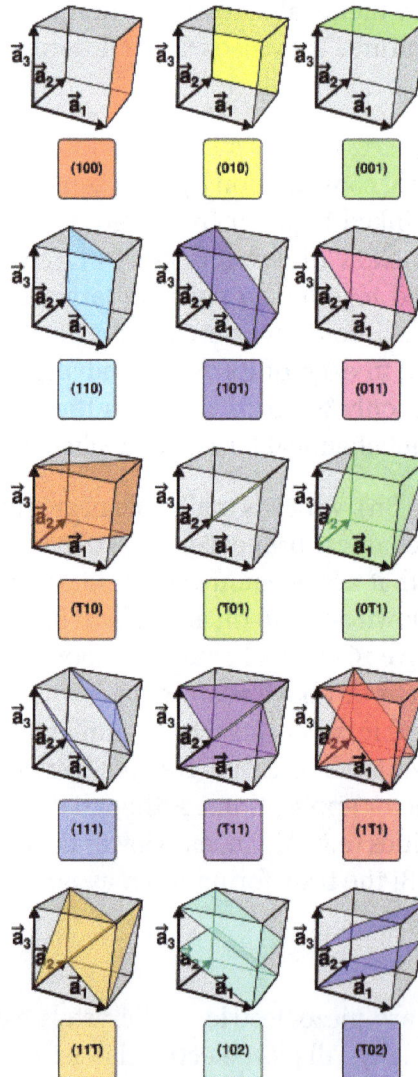

Planes with different Miller indices in cubic crystals

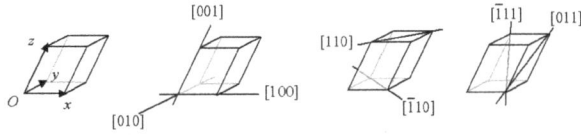

Examples of directions

Miller indices form a notation system in crystallography for planes in crystal (Bravais) lattices.

In particular, a family of lattice planes is determined by three integers h, k, and ℓ, the *Miller indices*. They are written (hkℓ), and denote the family of planes orthogonal to $h\mathbf{b}_1 + k\mathbf{b}_2 + \ell\mathbf{b}_3$, where $\mathbf{b}_i$ are the basis of the reciprocal lattice vectors. (Note that the plane is not always orthogonal to the linear combination of direct lattice vectors $h\mathbf{a}_1 + k\mathbf{a}_2 + \ell\mathbf{a}_3$ because the reciprocal lattice vectors need not be mutually orthogonal.) By convention, negative integers are written with a bar, as in 3 for −3. The integers are usually written in lowest terms, i.e. their greatest common divisor should be 1.

There are also several related notations:

- the notation {hkℓ} denotes the set of all planes that are equivalent to (hkℓ) by the symmetry of the lattice.

In the context of crystal directions (not planes), the corresponding notations are:

- [hkℓ], with square instead of round brackets, denotes a direction in the basis of the *direct* lattice vectors instead of the reciprocal lattice; and

- similarly, the notation <hkℓ> denotes the set of all directions that are equivalent to [hkℓ] by symmetry.

Miller indices were introduced in 1839 by the British mineralogist William Hallowes Miller. The method was also historically known as the Millerian system, and the indices as Millerian, although this is now rare.

The Miller indices are defined with respect to any choice of unit cell and not only with respect to primitive basis vectors, as is sometimes stated.

Definition

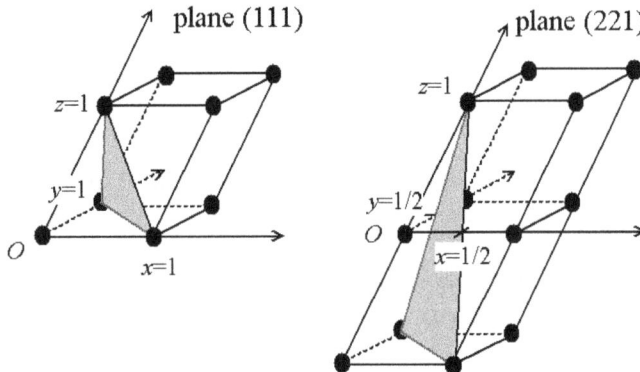

Examples of determining indices for a plane using intercepts with axes; left (111), right (221)

There are two equivalent ways to define the meaning of the Miller indices: via a point in the reciprocal lattice, or as the inverse intercepts along the lattice vectors. Both definitions are given below. In either case, one needs to choose the three lattice vectors a_1, a_2, and a_3 that define the unit cell (note that the conventional unit cell may be larger than the primitive cell of the Bravais lattice, as the examples below illustrate). Given these, the three primitive reciprocal lattice vectors are also determined (denoted b_1, b_2, and b_3).

Then, given the three Miller indices h, k, ℓ, (hkℓ) denotes planes orthogonal to the reciprocal lattice vector:

$$\mathbf{g}_{hk\ell} = h\mathbf{b}_1 + k\mathbf{b}_2 + \ell\mathbf{b}_3.$$

That is, (hkℓ) simply indicates a normal to the planes in the basis of the primitive reciprocal lattice vectors. Because the coordinates are integers, this normal is itself always a reciprocal lattice vector. The requirement of lowest terms means that it is the *shortest* reciprocal lattice vector in the given direction.

Equivalently, (hkℓ) denotes a plane that intercepts the three points a_1/h, a_2/k, and a_3/ℓ, or some multiple thereof. That is, the Miller indices are proportional to the *inverses* of the intercepts of the plane, in the basis of the lattice vectors. If one of the indices is zero, it means that the planes do not intersect that axis (the intercept is "at infinity").

Considering only (hkℓ) planes intersecting one or more lattice points (the *lattice planes*), the

That is, it uses the direct lattice basis instead of the reciprocal lattice. Note that [hkℓ] is *not* generally normal to the (hkℓ) planes, except in a cubic lattice as described below.

Case of Cubic Structures

For the special case of simple cubic crystals, the lattice vectors are orthogonal and of equal length (usually denoted a), as are those of the reciprocal lattice. Thus, in this common case, the Miller indices (hkℓ) and [hkℓ] both simply denote normals/directions in Cartesian coordinates.

For cubic crystals with lattice constant a, the spacing d between adjacent (hkℓ) lattice planes is (from above)

$$d_{hk\ell} = \frac{a}{\sqrt{h^2 + k^2 + \ell^2}}.$$

Because of the symmetry of cubic crystals, it is possible to change the place and sign of the integers and have equivalent directions and planes:

- Indices in *angle brackets* such as ⟨100⟩ denote a *family* of directions which are equivalent due to symmetry operations, such as , , or the negative of any of those directions.

- Indices in *curly brackets* or *braces* such as {100} denote a family of plane normals which are equivalent due to symmetry operations, much the way angle brackets denote a family of directions.

For face-centered cubic and body-centered cubic lattices, the primitive lattice vectors are not orthogonal. However, in these cases the Miller indices are conventionally defined relative to the lattice vectors of the cubic supercell and hence are again simply the Cartesian directions.

Crystallographic Planes and Directions

The crystallographic directions are fictitious lines linking nodes (atoms, ions or molecules) of a crystal. Similarly, the crystallographic planes are fictitious *planes* linking nodes. Some directions and planes have a higher density of nodes; these dense planes have an influence on the behaviour of the crystal:

- optical properties: in condensed matter, the light "jumps" from one atom to the other with the Rayleigh scattering; the velocity of light thus varies according to the directions, whether the atoms are close or far; this gives the birefringence

- adsorption and reactivity: the adsorption and the chemical reactions occur on atoms or molecules, these phenomena are thus sensitive to the density of nodes;

- surface tension: the condensation of a material means that the atoms, ions or molecules are more stable if they are surrounded by other similar species; the surface tension of an interface thus varies according to the density on the surface

  o the pores and crystallites tend to have straight grain boundaries following dense planes

  o cleavage

- dislocations (plastic deformation)

  o the dislocation core tends to spread on dense planes (the elastic perturbation is "diluted"); this reduces the friction (Peierls–Nabarro force), the sliding occurs more frequently on dense planes;

  o the perturbation carried by the dislocation (Burgers vector) is along a dense direction: the shift of one node in a dense direction is a lesser distortion;

  o the dislocation line tends to follow a dense direction, the dislocation line is often a straight line, a dislocation loop is often a polygon.

For all these reasons, it is important to determine the planes and thus to have a notation system.

Integer Vs. Irrational Miller Indices: Lattice Planes and Quasicrystals

Ordinarily, Miller indices are always integers by definition, and this constraint is physically significant. To understand this, suppose that we allow a plane (abc) where the Miller "indices" a, b and c (defined as above) are not necessarily integers.

If a, b and c have rational ratios, then the same family of planes can be written in terms of integer indices (hkℓ) by scaling a, b and c appropriately: divide by the largest of the three numbers, and then multiply by the least common denominator. Thus, integer Miller indices implicitly include indices with all rational ratios. The reason why planes where the components (in the reciprocal-lattice basis) have rational ratios are of special interest is that these are the lattice planes: they are the only planes whose intersections with the crystal are 2d-periodic.

For a plane (abc) where a, b and c have irrational ratios, on the other hand, the intersection of the plane with the crystal is *not* periodic. It forms an aperiodic pattern known as a quasicrystal. This construction corresponds precisely to the standard "cut-and-project" method of defining a quasic-rystal, using a plane with irrational-ratio Miller indices. (Although many quasicrystals, such as the Penrose tiling, are formed by "cuts" of periodic lattices in more than three dimensions, involving the intersection of more than one such hyperplane.)

Crystallographic Defect

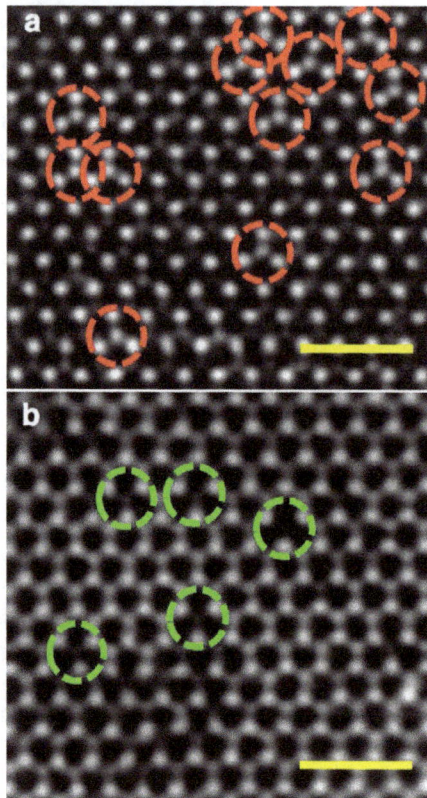

Electron microscopy of antisites (a, Mo substitutes for S) and vacancies (b, missing S atoms) in a monolayer of molybdenum disulfide. Scale bar: 1 nm.

Crystalline solids exhibit a periodic crystal structure. The positions of atoms or molecules occur on repeating fixed distances, determined by the unit cell parameters. However, the arrangement of atoms or molecules in most crystalline materials is not perfect. The regular patterns are interrupt-ed by crystallographic defects.

Point Defects

Point defects are defects that occur only at or around a single lattice point. They are not extended in space in any dimension. Strict limits for how small a point defect is are generally not defined explicitly, typically, however, these defects involve at most a few extra or missing atoms. Larger defects in an ordered structure are usually considered dislocation loops. For historical reasons, many point defects, especially in ionic crystals, are called *centers*: for example a vacancy in many ionic solids is called a luminescence center, a color center, or F-center. These dislocations permit ionic transport through crystals leading to electrochemical reactions. These are frequently specified using Kröger–Vink Notation.

- Vacancy defects are lattice sites which would be occupied in a perfect crystal, but are vacant. If a neighboring atom moves to occupy the vacant site, the vacancy moves in the opposite direction to the site which used to be occupied by the moving atom. The stability of the surrounding crystal structure guarantees that the neighboring atoms will not simply collapse around the vacancy. In some materials, neighboring atoms actually move away from a vacancy, because they experience attraction from atoms in the surroundings. A vacancy (or pair of vacancies in an ionic solid) is sometimes called a Schottky defect.

- Interstitial defects are atoms that occupy a site in the crystal structure at which there is usually not an atom. They are generally high energy configurations. Small atoms in some crystals can occupy interstices without high energy, such as hydrogen in palladium.

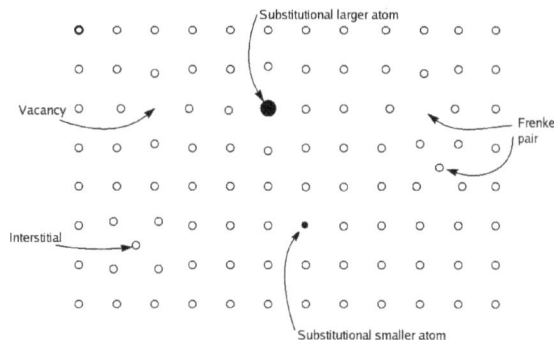

Schematic illustration of some simple point defect types in a monatomic solid

- A nearby pair of a vacancy and an interstitial is often called a Frenkel defect or Frenkel pair. This is caused when an ion moves into an interstitial site and creates a vacancy.

- Due to fundamental limitations of material purification methods, materials are never 100% pure, which by definition induces defects in crystal structure. In the case of an impurity, the atom is often incorporated at a regular atomic site in the crystal structure. This is neither a vacant site nor is the atom on an interstitial site and it is called a *substitutional* defect. The atom is not supposed to be anywhere in the crystal, and is thus an impurity. In some cases where the radius of the substitutional atom (ion) is substantially smaller than that of the atom (ion) it is replacing, its equilibrium position can be shifted away from the lattice site. These types of substitutional defects are often referred to as off-center ions. There are two different types of substitutional defects: Isovalent substitution and aliovalent substitution. Isovalent substitution is where the ion that is substituting the original ion is of the same oxidation state as the ion it is replacing. Aliovalent substitution is where the ion that is substituting the original ion is of a different oxidation state than the ion it is replacing. Aliovalent substitutions change the overall charge within the ionic compound, but the ionic compound must be neutral. Therefore, a charge compensation mechanism is required. Hence either one of the metals is partially or fully oxidised or reduced, or ion vacancies are created.

- Antisite defects occur in an ordered alloy or compound when atoms of different type exchange positions. For example, some alloys have a regular structure in which every other atom is a different species; for illustration assume that type A atoms sit on the corners of a cubic lattice, and type B atoms sit in the center of the cubes. If one cube has an A atom at its center, the atom is on a site usually occupied by a B atom, and is thus an antisite defect. This is neither a vacancy nor an interstitial, nor an impurity.

- Topological defects are regions in a crystal where the normal chemical bonding environment is topologically different from the surroundings. For instance, in a perfect sheet of graphite (graphene) all atoms are in rings containing six atoms. If the sheet contains regions where the number of atoms in a ring is different from six, while the total number of atoms remains the same, a topological defect has formed. An example is the Stone Wales defect in nanotubes, which consists of two adjacent 5-membered and two 7-membered atom rings.

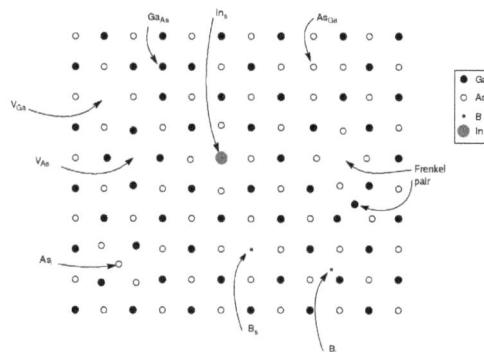

Schematic illustration of defects in a compound solid, using GaAs as an example.

- Also amorphous solids may contain defects. These are naturally somewhat hard to define, but sometimes their nature can be quite easily understood. For instance, in ideally bonded amorphous silica all Si atoms have 4 bonds to O atoms and all O atoms have 2 bonds to Si atom. Thus e.g. an O atom with only one Si bond (a dangling bond) can be considered a defect in silica. Moreover, defects can also be defined in amorphous solids based on empty or densely packed local atomic neighbourhoods, and the properties of such 'defects' can be shown to be similar to normal vacancies and interstitials in crystals,.

- Complexes can form between different kinds of point defects. For example, if a vacancy encounters an impurity, the two may bind together if the impurity is too large for the lattice. Interstitials can form 'split interstitial' or 'dumbbell' structures where two atoms effectively share an atomic site, resulting in neither atom actually occupying the site.

Line Defects

Line defects can be described by gauge theories.

Dislocations are linear defects around which some of the atoms of the crystal lattice are misaligned. There are two basic types of dislocations, the *edge* dislocation and the *screw* dislocation. "Mixed" dislocations, combining aspects of both types, are also common.

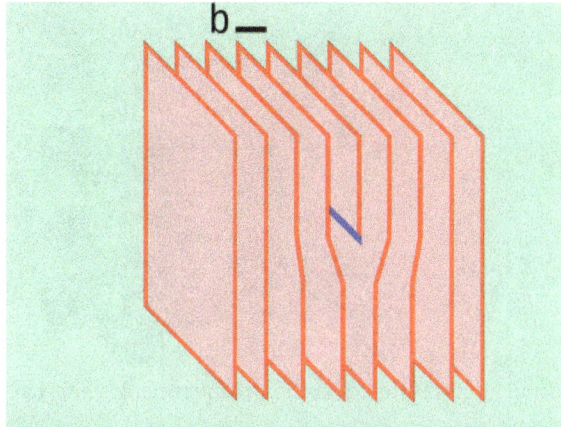

An edge dislocation is shown. The dislocation line is presented in blue, the Burgers vector b in black.

Edge dislocations are caused by the termination of a plane of atoms in the middle of a crystal. In such a case, the adjacent planes are not straight, but instead bend around the edge of the terminating plane so that the crystal structure is perfectly ordered on either side. The analogy with a stack of paper is apt: if a half a piece of paper is inserted in a stack of paper, the defect in the stack is only noticeable at the edge of the half sheet.

The screw dislocation is more difficult to visualise, but basically comprises a structure in which a helical path is traced around the linear defect (dislocation line) by the atomic planes of atoms in the crystal lattice.

The presence of dislocation results in lattice strain (distortion). The direction and magnitude of such distortion is expressed in terms of a Burgers vector (b). For an edge type, b is perpendicular to the dislocation line, whereas in the cases of the screw type it is parallel. In metallic materials, b is aligned with close-packed crystallographic directions and its magnitude is equivalent to one interatomic spacing.

Dislocations can move if the atoms from one of the surrounding planes break their bonds and re-bond with the atoms at the terminating edge.

It is the presence of dislocations and their ability to readily move (and interact) under the influence of stresses induced by external loads that leads to the characteristic malleability of metallic materials.

Dislocations can be observed using transmission electron microscopy, field ion microscopy and atom probe techniques. Deep level transient spectroscopy has been used for studying the electrical activity of dislocations in semiconductors, mainly silicon.

Disclinations are line defects corresponding to "adding" or "subtracting" an angle around a line. Basically, this means that if you track the crystal orientation around the line defect, you get a rotation. Usually, they were thought to play a role only in liquid crystals, but recent developments suggest that they might have a role also in solid materials, e.g. leading to the self-healing of cracks.

Planar Defects

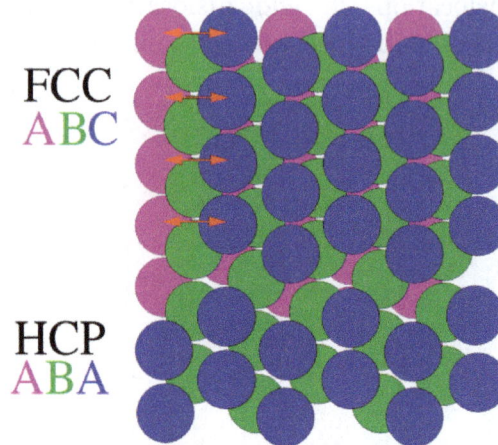

Origin of stacking faults: Different stacking sequences of close-packed crystals

- Grain boundaries occur where the crystallographic direction of the lattice abruptly changes. This usually occurs when two crystals begin growing separately and then meet.

- Antiphase boundaries occur in ordered alloys: in this case, the crystallographic direction remains the same, but each side of the boundary has an opposite phase: For example, if the ordering is usually ABABABAB (hexagonal close-packed crystal), an antiphase boundary takes the form of ABABBABA.

- Stacking faults occur in a number of crystal structures, but the common example is in close-packed structures. They are formed by a local deviation of the stacking sequence of layers in a crystal. An example would be the ABABCABAB stacking sequence.

- A twin boundary is a defect that introduces a plane of mirror symmetry in the ordering of a crystal. For example, in cubic close-packed crystals, the stacking sequence of a twin boundary would be ABCABCBACBA.

- On surfaces of single crystals, steps between atomically flat terraces can also be regarded as planar defects. It has been shown that such defects and their geometry have significant influence on the adsorption of organic molecules

Bulk Defects

- three-dimensional macroscopic or bulk defects, such as pores, cracks, or inclusions

- Voids — small regions where there are no atoms, and which can be thought of as clusters of vacancies

- Impurities can cluster together to form small regions of a different phase. These are often called precipitates.

Mathematical Classification Methods

A successful mathematical classification method for physical lattice defects, which works not only with the theory of dislocations and other defects in crystals but also, e.g., for disclinations in liquid crystals and for excitations in superfluid ^{3}He, is the topological homotopy theory.

Computer Simulation Methods

Density functional theory, classical molecular dynamics and kinetic Monte Carlo simulations are widely used to study the properties of defects in solids with computer simulations. Simulating jamming of hard spheres of different sizes and/or in containers with non-commeasurable sizes using the Lubachevsky–Stillinger algorithm can be an effective technique for demonstrating some types of crystallographic defects.

Crystal Structure Prediction

Crystal structure prediction (CSP) is the calculation of the crystal structures of solids from first principles. Reliable methods of predicting the crystal structure of a compound, based only on its molecular structure, has been a goal of the physical sciences since the 1950s. Computational methods employed include simulated annealing, evolutionary algorithms, distributed multipole analysis, random sampling, basin hopping, data mining, density functional theory and molecular mechanics.

History

The crystal structures of simple ionic solids have long been rationalised in terms of Pauling's rules, first set out in 1929 by Linus Pauling. For metals and semiconductors one has different rules involving valence electron concentration. However, prediction and rationalization are rather different things. Most commonly, by crystal structure prediction one understands search for minimum-energy arrangement of the atoms (or, for molecular crystals, of the molecules) in space. The problem has two facets - combinatorial (the "search" problem, in practice most acute for inorganic crystals), and energetic ("ranking", most acute for molecular organic crystals). For complex non-molecular crystals ("search problem"), major recent advances have been the Martonak version of metadynamics, the Oganov-Glass evolutionary algorithm USPEX,. The latter is capable of solving the global optimization problem with up to a few hundred degrees of freedom, while the approach of metadynamics is to reduce all structural variables to a handful of "slow" collective variables (which often works).

Molecular Crystals

Predicting organic crystal structures is important to academic and industrial science, particularly for pharmaceuticals and pigments, where understanding polymorphism is beneficial. The crystal structures of molecular substances, particularly organic compounds, are very hard to predict and rank in stability. Intermolecular interactions are relatively weak and non-directional and long range. This results in typical lattice and free energy differences between polymorphs that are often only a few kJ/mol, very rarely exceeding 10 kJ/mol. Crystal structure prediction methods often locate many possible structures within this small energy range. These small energy differences are challenging to predict reliably and with a reasonable computational effort.

Since 2007, significant progress has been made in the CSP of small organic molecules, with several different methods proving effective. The most widely-discussed method first ranks the energies of all possible crystal structures using a customised MM force field, and finishes by using a dispersion-corrected DFT step to estimate the lattice energy and stability of each short-listed candidate structure.

Crystal Structure Prediction Software

The following codes can predict stable and metastable structures given chemical composition and external conditions (pressure, temperature):

- CALYPSO - The Crystal structure AnaLYsis by Particle Swarm Optimization, implementing the particle swarm optimization (PSO) algorithm to identify/determine the crystal structure. As with other codes, the knowledge of structure can be used to design the multi-functional materials (e.g.,superconductive, thermoelectric, superhard, and energetic materials,etc.). Free for academic researchers. Regularly updated.

- GASP - predicts the structure and composition of stable and metastable phases of crystals, molecules, atomic clusters and defects from first-principles. Can be interfaced to other energy codes including: VASP, LAMMPS, MOPAC, Gulp, JDFTx etc. Free to use and regularly updated.

- GRACE - for predicting molecular crystal structures, especially for the pharmaceutical industry. Based on dispersion-corrected density functional theory. Commercial software under active development.

- GULP - Monte Carlo and genetic algorithms for atomic crystals. GULP is based on classical force fields and works with many types of force fields. Free for academic researchers. Regularly updated.

- USPEX - multi-method software that includes evolutionary algorithm and other methods (random sampling, evolutionary metadynamics, improved PSO, variable-cell NEB method for phase transition mechanisms). Can be used for atomic or molecular crystals; bulk crystals, nanoparticles, polymers, surface reconstructions; can optimize the energy or other physical properties. In addition to finding the structure at given composition, can identify all stable compositions in a multicomponent variable-composition system. Free for academic researchers. Used by >2000 researchers. Regularly updated.

- XtalOpt - open source code implementing an evolutionary algorithm.

References

- Prince, E. (2006). International Tables for Crystallography Vol. C: Mathmatical, Physical and Chemical Tables. Wiley. ISBN 978-1-4020-4969-9.

- Encyclopaedia of Physics (2nd Edition), R.G. Lerner, G.L. Trigg, VHC publishers, 1991, ISBN (Verlagsgesellschaft) 3-527-26954-1, ISBN (VHC Inc.) 0-89573-752-3

- Hook, J.R.; Hall, H.E. (2010). Solid State Physics. Manchester Physics Series (2nd ed.). John Wiley & Sons. ISBN 9780471928041.

- Donald E. Sands (1994). "§4-2 Screw axes and §4-3 Glide planes". Introduction to Crystallography (Reprint of WA Benjamin corrected 1975 ed.). Courier-Dover. pp. 70–71. ISBN 0486678393.

- Holleman, Arnold F.; Wiberg, Egon; Wiberg, Nils (1985). "Tin". Lehrbuch der Anorganischen Chemie (in German) (91–100 ed.). Walter de Gruyter. pp. 793–800. ISBN 3-11-007511-3.

- Schwartz, Mel (2002). "Tin and Alloys, Properties". Encyclopedia of Materials, Parts and Finishes (2nd ed.). CRC Press. ISBN 1-56676-661-3.

- J. W. Edington (1976) Practical electron microscopy in materials science (N. V. Philips' Gloeilampenfabrieken, Eindhoven) ISBN 1-878907-35-2, Appendix 2

- Liu, Gang (2015). "Dynamical equations for the period vectors in a periodic system under constant external stress". Can. J. Phys. 93 (9): 974. Bibcode:2015CaJPh..93..974L. doi:10.1139/cjp-2014-0518.

- Magnuson et al. Electronic Structure and Chemical Bonding of Amorphous Chromium Carbide Thin Films; J. Phys. - Cond. Mat. 24 , 225004 (2012).

- M.I. Ojovan, W.E. Lee. Connectivity and glass transition in disordered oxide systems. J. Non-Cryst. Solids, 356, 2534-2540 (2010). doi:10.1016/j.jnoncrysol.2010.05.012

Emerging Technologies in Materials Science

A crystal is a material that is in a proper structure. They can be recognized by their geometrical shape. The other atomic arrangements discussed in this chapter are crystal structure, Miller index, crystal structure prediction, crystallographic defect etc. This section serves as a source to understand the basic aspects of atomic arrangements present in solids.

Aerogel

A block of aerogel in a person's hand

Aerogel is a synthetic porous ultralight material derived from a gel, in which the liquid component of the gel has been replaced with a gas. The result is a solid with extremely low density and low thermal conductivity. Nicknames include *frozen smoke, solid smoke, solid air,* or *blue smoke* owing to its translucent nature and the way light scatters in the material. It feels like fragile expanded polystyrene to the touch. Aerogels can be made from a variety of chemical compounds.

Aerogel was first created by Samuel Stephens Kistler in 1931, as a result of a bet with Charles Learned over who could replace the liquid in "jellies" with gas without causing shrinkage.

Aerogels are produced by extracting the liquid component of a gel through supercritical drying. This allows the liquid to be slowly dried off without causing the solid matrix in the gel to collapse from capillary action, as would happen with conventional evaporation. The first aerogels were produced from silica gels. Kistler's later work involved aerogels based on alumina, chromia and tin dioxide. Carbon aerogels were first developed in the late 1980s.

Aerogel does not have a designated material with set chemical formula but the term is used to group all the material with a certain geometric structure.

Properties

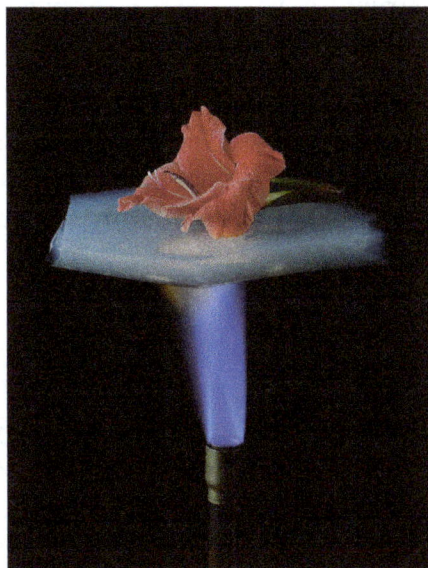

A flower is on a piece of aerogel which is suspended over a flame from a Bunsen burner. Aerogel has excellent insulating properties, and the flower is protected from the flame.

Despite their name, aerogels are solid, rigid, and dry materials that do not resemble a gel in their physical properties: The name comes from the fact that they are made from gels. Pressing softly on an aerogel typically does not leave even a minor mark; pressing more firmly will leave a permanent depression. Pressing extremely firmly will cause a catastrophic breakdown in the sparse structure, causing it to shatter like glass – a property known as *friability*; although more modern variations do not suffer from this. Despite the fact that it is prone to shattering, it is very strong structurally. Its impressive load bearing abilities are due to the dendritic microstructure, in which spherical particles of average size (2–5 nm) are fused together into clusters. These clusters form a three-dimensional highly porous structure of almost fractal chains, with pores just under 100 nm. The average size and density of the pores can be controlled during the manufacturing process.

Aerogel is a material that is 98.2% air. Aerogel has a porous solid network that contains air pockets, with the air pockets taking up majority of space within the material. The lack of solid material allows aerogel to be almost weightless.

Aerogels are good thermal insulators because they almost nullify two of the three methods of heat transfer (convection, conduction, and radiation). They are good conductive insulators because they are composed almost entirely of gas, which are very poor heat conductors. (Silica aerogel is an especially good insulator because silica is also a poor conductor of heat; a metallic or carbon aerogel, on the other hand, would be less effective.) They are good convective inhibitors because air cannot circulate through the lattice. Aerogels are poor radiative insulators because infrared radiation (which transfers heat) passes through them.

Owing to its hygroscopic nature, aerogel feels dry and acts as a strong desiccant. People handling aerogel for extended periods should wear gloves to prevent the appearance of dry brittle spots on their skin.

The slight color it does have is due to Rayleigh scattering of the shorter wavelengths of visible light by the nano-sized dendritic structure. This causes it to appear smoky blue against dark backgrounds and yellowish against bright backgrounds.

Aerogels by themselves are hydrophilic, but chemical treatment can make them hydrophobic. If they absorb moisture they usually suffer a structural change, such as contraction, and deteriorate, but degradation can be prevented by making them hydrophobic. Aerogels with hydrophobic interiors are less susceptible to degradation than aerogels with only an outer hydrophobic layer, even if a crack penetrates the surface. Hydrophobic treatment facilitates processing because it allows the use of a water jet cutter.

Knudsen Effect

Aerogels may have a thermal conductivity smaller than that of the gas they contain. This is caused by the Knudsen effect, a reduction of thermal conductivity in gases when the size of the cavity encompassing the gas becomes comparable to the mean free path. Effectively, the cavity restricts the movement of the gas particles, decreasing the thermal conductivity in addition to eliminating convection. For example, thermal conductivity of air is about 25 mW/m·K at STP and in a large container, but decreases to about 5 mW/m·K in a pore 30 nanometers in diameter.

Structure

Aerogel structure is the result from a sol-gel polymerization, which is when monomers (simple molecules) reacts with other monomers to form a sol or a substance that consists of bonded, cross-linked macromolecules with deposits of liquid solution between them. When the material is critically heated the liquid is evaporated out and the bonded, cross-linked macromolecule frame is left behind. The result of the polymerization and critical heating is the creation of a material that has a porous strong structure classified as aerogel. Variations in synthesis can alter the surface area and pore size of the aerogel. The smaller the pore size the more suceptible the aerogel is to fracture.

Waterproofing

Aerogel contains particles that are 2–5 nm in diameter. After the process of creating aerogel, it will contain a large amount of hydroxyl groups on the surface. The hydroxyl groups can cause a strong reaction when placing it in water. The aerogel will catastrophically dissolve in the water. One way to waterproof the hydrophilic aerogel is by soaking the aerogel with some chemical base that will replace hydroxyl groups with non-polar groups on the surface, the non-polar groups (-OR) is most effective when R is an aliphatic group.

Porosity of Aerogel

There are several ways to determine the porosity of aerogel; the three main methods are gas adsorption, Mercury Porosimetry, and Scattering Method. In gas adsorption, nitrogen at its boiling point is adsorbed into the aerogel sample. The gas being adsorbed is dependent on the size of the pores within the sample and on the partial pressure of the gas relative to its saturation pressure. The volume of the gas adsorbed is measured by using the Brunauer, Emmit and, Teller formula (BET) gives the specific surface area of the sample. At high partial pressure in the adsorption/desorption the Kelvin equation gives the pore size distribution of the sample. In Mercury Porosimetry, the mercury is forced into the aerogel porous system to determine the pores size, but this method is highly inefficient since the solid frame of aerogel will collapse from the high compressive force. The Scattering Methods involves the angle dependent deflection of radiation within the aerogel sample. The sample can be solid particles or pores. The radiation goes into the material and determines the fractal geometry of the aerogel pore network. The best radiation wavelengths to use are X-rays and neutrons. Aerogel is also an open porous network, the difference between an open porous network and a closed porous network is that in the open network, gasses can enter and leave the substance without any limitation. While a closed porous network traps the gases within the material forcing it to stay within the pores. The high porosity and surface area of silica aerogels allow for it to be used in a variety of environmental filtration applications.

Materials

A 2.5 kg brick is supported by a piece of aerogel with a mass of 2 g.

Silica

Silica aerogel is the most common type of aerogel, and the most extensively studied and used. It is silica-based, derived from silica gel. The lowest-density silica nanofoam weighs 1,000 g/m³, which is the evacuated version of the record-aerogel of 1,900 g/m³. The density of air is 1,200 g/m³ (at 20 °C and 1 atm). As of 2013, aerographene had a lower density at 160 g/m³, or 13% the density of air at room temperature.

The silica solidifies into three-dimensional, intertwined clusters that comprise only 3% of the volume. Conduction through the solid is therefore very low. The remaining 97% of the volume is composed of air in extremely small nanopores. The air has little room to move, inhibiting both convection and gas-phase conduction.

Silica aerogels also have a high optical transmission of ~99% and a low refractive index of ~1.05.

It has remarkable thermal insulative properties, having an extremely low thermal conductivity: from 0.03 W/m·K in atmospheric pressure down to 0.004 W/m·K in modest vacuum, which correspond to R-values of 14 to 105 (US customary) or 3.0 to 22.2 (metric) for 3.5 in (89 mm) thickness. For comparison, typical wall insulation is 13 (US customary) or 2.7 (metric) for the same thickness. Its melting point is 1,473 K (1,200 °C; 2,192 °F).

Until 2011, silica aerogel held 15 entries in *Guinness World Records* for material properties, including best insulator and lowest-density solid, though it was ousted from the latter title by the even lighter materials aerographite in 2012 and then aerographene in 2013.

Carbon

Carbon aerogels are composed of particles with sizes in the nanometer range, covalently bonded together. They have very high porosity (over 50%, with pore diameter under 100 nm) and surface areas ranging between 400–1,000 m²/g. They are often manufactured as composite paper: non-woven paper made of carbon fibers, impregnated with resorcinol–formaldehyde aerogel, and pyrolyzed. Depending on the density, carbon aerogels may be electrically conductive, making composite aerogel paper useful for electrodes in capacitors or deionization electrodes. Due to their extremely high surface area, carbon aerogels are used to create supercapacitors, with values ranging up to thousands of farads based on a capacitance density of 104 F/g and 77 F/cm³. Carbon aerogels are also extremely "black" in the infrared spectrum, reflecting only 0.3% of radiation between 250 nm and 14.3 μm, making them efficient for solar energy collectors.

The term "aerogel" to describe airy masses of carbon nanotubes produced through certain chemical vapor deposition techniques is incorrect. Such materials can be spun into fibers with strength greater than Kevlar, and unique electrical properties. These materials are not aerogels, however, since they do not have a monolithic internal structure and do not have the regular pore structure characteristic of aerogels.

Metal Oxide

Metal oxide aerogels are used as catalysts in various chemical reactions/transformations or as precursors for other materials.

Aerogels made with aluminium oxide are known as alumina aerogels. These aerogels are used as catalysts, especially when "doped" with a metal other than aluminium. Nickel–alumina aerogel is the most common combination. Alumina aerogels are also being considered by NASA for capturing hypervelocity particles; a formulation doped with gadolinium and terbium could fluoresce at the particle impact site, with the amount of fluorescence dependent on impact energy.

One of the most notable difference between silica aerogels and metal oxide aerogel is that metal oxide aerogels are often variedly colored.

Aerogel	Color
Silica, Alumina, Titania, Zirconia	Clear with Rayleigh scattering blue or white
Iron Oxide	Rust red or yellow, opaque
Chromia	Deep green or deep blue, opaque
Vanadia	Olive green, opaque
Neodymium Oxide	Purple, transparent
Samarium Oxide	Yellow, transparent
Holmium Oxide, Erbium Oxide	Pink, transparent

Other

Organic polymers can be used to create aerogels. SEAgel is made of agar. Cellulose from plants can be used to create a flexible aerogel.

Chalcogel is an aerogel made of chalcogens (the column of elements on the periodic table beginning with oxygen) such as sulfur, selenium and other elements. Metals less expensive than platinum have been used in its creation.

Aerogels made of cadmium selenide quantum dots in a porous 3-D network have been developed for use in the semiconductor industry.

Aerogel performance may be augmented for a specific application by the addition of dopants, reinforcing structures and hybridizing compounds. Aspen Aerogels makes products such as Spaceloft which are composites of aerogel with some kind of fibrous batting.

Applications

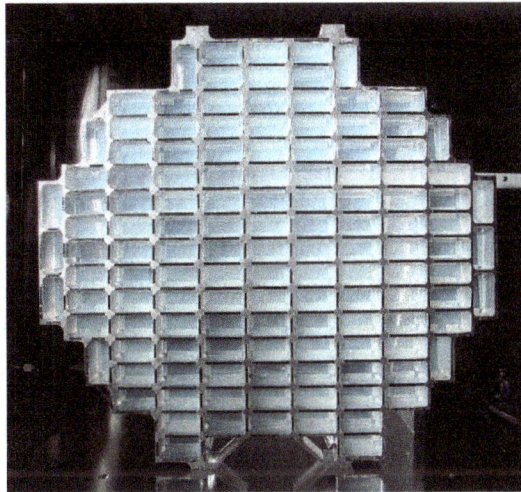

The Stardust dust collector with aerogel blocks. (NASA)

Aerogels are used for a variety of applications:

- In 2004 about US$25 million of aerogel insulation product were sold, which had risen to about US$500 million by 2013. This represents the most substantial economic impact of these materials today. The potential to replace conventional insulation by aerogel solutions in the building and construction sector as well as in industrial insulation is quite significant.

- In granular form to add insulation to skylights. Georgia Institute of Technology's 2007 Solar Decathlon House project used an aerogel as an insulator in the semi-transparent roof.

- A chemical adsorber for cleaning up spills.

- A catalyst or a catalyst carrier.

- Silica aerogels can be used in imaging devices, optics, and light guides.

- A material for filtration due to its high surface area and porosity, to be used for the removal of heavy metals

- Thickening agents in some paints and cosmetics.

- As components in energy absorbers.

- Laser targets for the National Ignition Facility.

- A material used in impedance matchers for transducers, speakers and range finders.

- Commercial manufacture of aerogel 'blankets' began around the year 2000, combining silica aerogel and fibrous reinforcement that turns the brittle aerogel into a durable, flexible material. The mechanical and thermal properties of the product may be varied based upon the choice of reinforcing fibers, the aerogel matrix and opacification additives included in the composite.

- NASA used an aerogel to trap space dust particles aboard the Stardust spacecraft. The particles vaporize on impact with solids and pass through gases, but can be trapped in aerogels. NASA also used aerogel for thermal insulation of the Mars Rover and space suits.

- The US Navy is evaluating aerogel undergarments as passive thermal protection for divers.

- In particle physics as radiators in Cherenkov effect detectors, such as the ACC system of the Belle detector, used in the Belle Experiment at KEKB. The suitability of aerogels is determined by their low index of refraction, filling the gap between gases and liquids, and their transparency and solid state, making them easier to use than cryogenic liquids or compressed gases. Their low mass is also advantageous for space missions.

- Resorcinol–formaldehyde aerogels (polymers chemically similar to phenol formaldehyde resins) are used as precursors for manufacture of carbon aerogels, or when an organic insulator with large surface is desired. They come as high-density material, with surface area about 600 m^2/g.

- Metal–aerogel nanocomposites prepared by impregnating the hydrogel with solution containing ions of a transition metal and irradiating the result with gamma rays, precipitates nanoparticles of the metal. Such composites can be used as catalysts, sensors, electromagnetic shielding, and in waste disposal. A prospective use of platinum-on-carbon catalysts is in fuel cells.

- As a drug delivery system owing to its biocompatibility. Due to its high surface area and porous structure, drugs can be adsorbed from supercritical CO_2. The release rate of the drugs can be tailored by varying the properties of the aerogel.

- Carbon aerogels are used in the construction of small electrochemical double layer supercapacitors. Due to the high surface area of the aerogel, these capacitors can be 1/2000th to 1/5000th the size of similarly rated electrolytic capacitors. Aerogel supercapacitors can have a very low impedance compared to normal supercapacitors and can absorb or produce very high peak currents. At present, such capacitors are polarity-sensitive and need to be wired in series to achieve a working voltage of greater than about 2.75 V.

- Dunlop Sport uses aerogel in some of its racquets for tennis, squash and badminton.

- In water purification, chalcogels have shown promise in absorbing the heavy metal pollutants mercury, lead, and cadmium from water.

- Aerogel can introduce disorder into superfluid helium-3.

- In aircraft de-icing, a new proposal uses a carbon nanotube aerogel. A thin filament is spun on a winder to create a 10 micron-thick film, equivalent to an A4 sheet of paper. The amount of material needed to cover the wings of a jumbo jet weighs 80 grams (2.8 oz). Aerogel heaters could be left on continuously at low power, to prevent ice from forming.

- Thermal insulation transmission tunnel of the Chevrolet Corvette (C7).

- CamelBak uses aerogel as insulation in a thermal sport bottle.

- 45 North uses aerogel as palm insulation in its Sturmfist 5 cycling gloves.

Production

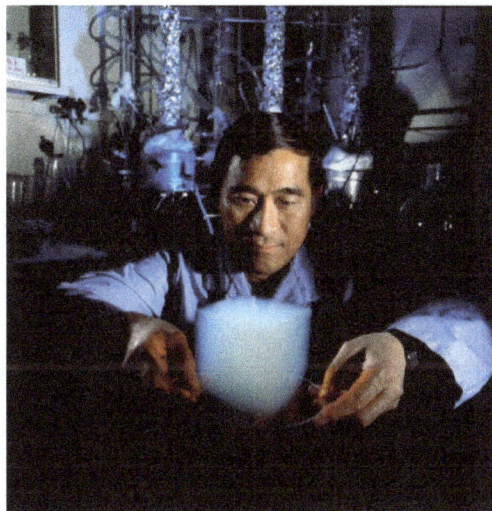

Peter Tsou with a sample of aerogel at Jet Propulsion Laboratory, California Institute of Technology

Silica aerogels are typically synthesized by using a sol-gel process. The first step is the creation of a colloidal suspension of solid particles known as a "sol". The precursors are a liquid alcohol such as ethanol which is mixed with a silicon alkoxide, such as tetramethoxysilane (TMOS), tetraethoxysilane (TEOS), and polyethoxydisiloxane (PEDS) (earlier work used sodium silicates). The solution of silica is mixed with a catalyst and allowed to gel during a hydrolysis reaction which forms particles of silicon dioxide. The oxide suspension begins to undergo condensation reactions which result in the creation of metal oxide bridges (either M–O–M, "oxo" bridges or M–OH–M, "ol" bridges) linking the dispersed colloidal particles. These reactions generally have moderately slow reaction rates, and as a result either acidic or basic catalysts are used to improve the processing speed. Basic catalysts tend to produce more transparent aerogels and minimize the shrinkage during the drying process and also strengthen it to prevent pore collapse during drying.

Finally, during the drying process of the aerogel, the liquid surrounding the silica network is carefully removed and replaced with air, while keeping the aerogel intact. Gels where the liquid is allowed to evaporate at a natural rate are known as xerogels. As the liquid evaporates, forces caused by surface tensions of the liquid-solid interfaces are enough to destroy the fragile gel network. As a result, xerogels cannot achieve the high porosities and instead peak at lower porosities and exhibit large amounts of shrinkage after drying.

In 1931, to develop the first aerogels, Kistler used a process known as supercritical drying which avoids a direct phase change. By increasing the temperature and pressure he forced the liquid into a supercritical fluid state where by dropping the pressure he could instantly gasify and remove the liquid inside the aerogel, avoiding damage to the delicate three-dimensional network. While this can be done with ethanol, the high temperatures and pressures lead to dangerous processing conditions. A safer, lower temperature and pressure method involves a solvent exchange. This is typically done by exchanging the initial aqueous pore liquid for a CO_2 miscible liquid such as ethanol or acetone, then onto liquid carbon dioxide and then bringing the carbon dioxide above its critical point. A variant on this process involves the direct injection of supercritical carbon dioxide into the pressure vessel containing the aerogel. The end result of either process exchanges the initial liquid from the gel with carbon dioxide, without allowing the gel structure to collapse or lose volume.

Resorcinol–formaldehyde aerogel (RF aerogel) is made in a way similar to production of silica aerogel. A carbon aerogel can then be made from this resorcinol–formaldehyde aerogel by pyrolysis in an inert gas atmosphere, leaving a matrix of carbon. It is commercially available as solid shapes, powders, or composite paper. Additives have been successful in enhancing certain properties of the aerogel for the use of specific applications. Aerogel composites have been made using a variety of continuous and discontinuous reinforcements. The high aspect ratio of fibers such as fiberglass have been used to reinforce aerogel composites with significantly improved mechanical properties.

Safety

Silica-based aerogels are not known to be carcinogenic or toxic. However, they are a mechanical irritant to the eyes, skin, respiratory tract, and digestive system. Small silica particles can potentially cause silicosis when inhaled. They can also induce dryness of the skin, eyes, and mucous membranes. Therefore, it is recommended that protective gear including respiratory protection, gloves and eye goggles be worn whenever handling aerogels.

High-temperature Superconductivity

High-temperature superconductors (abbreviated high-T_c or HTS) are materials that behave as superconductors at unusually high temperatures. The first high-T_c superconductor was discovered in 1986 by IBM researchers Georg Bednorz and K. Alex Müller, who were awarded the 1987 Nobel Prize in Physics "for their important break-through in the discovery of superconductivity in ceramic materials".

A small sample of the high-temperature superconductor BSCCO-2223.

Whereas "ordinary" or metallic superconductors usually have transition temperatures (temperatures below which they are superconductive) below 30 K (−243.2 °C), and must be cooled using liquid helium in order to achieve superconductivity, HTS have been observed with transition temperatures as high as 138 K (−135 °C), and can be cooled to superconductivity using liquid nitrogen. Until 2008, only certain compounds of copper and oxygen (so-called "cuprates") were believed to have HTS properties, and the term high-temperature superconductor was used interchangeably with cuprate superconductor for compounds such as bismuth strontium calcium copper oxide (BSCCO) and yttrium barium copper oxide (YBCO). Several iron-based compounds (the iron pnictides) are now known to be superconducting at high temperatures.

In 2015, hydrogen sulfide (H_2S) under extremely high pressure (around 150 gigapascals) was found to undergo superconducting transition near 203 K (-70 °C), the highest temperature superconductor known to date.

History

The phenomenon of superconductivity was discovered by Kamerlingh Onnes in 1911, in metallic mercury below 4 K (−269.15 °C). For seventy-five years after that, researchers attempted to observe superconductivity at higher and higher temperatures. In the late 1970s, superconductivity was observed in certain metal oxides at temperatures as high as 13 K (−260.1 °C), which were much higher than those for elemental metals. In 1986, J. Georg Bednorz and K. Alex Müller, working at the IBM research lab near Zurich, Switzerland were exploring a new class of ceramics for superconductivity. Bednorz encountered a barium-doped compound of lanthanum and copper oxide whose resistance dropped down to zero at a temperature around 35 K (−238.2 °C). Their results were soon confirmed by many groups, notably Paul Chu at the University of Houston and Shoji Tanaka at the University of Tokyo.

Shortly after, P. W. Anderson, at Princeton University came up with the first theoretical description of these materials, using the resonating valence bond theory, but a full understanding of these materials is still developing today. These superconductors are now known to possess a d-wave pair symmetry. The first proposal that high-temperature cuprate superconductivity involves d-wave pairing was made in 1987 by Bickers, Scalapino and Scalettar, followed by three subsequent theories in 1988 by Inui, Doniach, Hirschfeld and Ruckenstein, using spin-fluctuation theory, and by Gros, Poilblanc, Rice and Zhang, and by Kotliar and Liu identifying d-wave pairing as a natural consequence of the RVB theory. The confirmation of the d-wave nature of the cuprate superconductors was made by a variety of experiments, including the direct observation of the d-wave nodes in the excitation spectrum through Angle Resolved Photoemission Spectroscopy, the observation of a half-integer flux in tunneling experiments, and indirectly from the temperature dependence of the penetration depth, specific heat and thermal conductivity.

The superconductor with the highest transition temperature that has been confirmed by multiple independent research groups (a prerequisite to be called a discovery, verified by peer review) is mercury barium calcium copper oxide ($HgBa_2Ca_2Cu_3O_8$) at around 133 K.

After more than twenty years of intensive research, the origin of high-temperature superconductivity is still not clear, but it seems that instead of *electron-phonon* attraction mechanisms, as in conventional superconductivity, one is dealing with genuine *electronic* mechanisms (e.g. by antiferromagnetic correlations), and instead of conventional, purely s-wave pairing, more exotic pairing symmetries are thought to be involved (d-wave in the case of the cuprates; primarily extended s-wave, but occasionally d-wave, in the case of the iron-based superconductors). One goal of all this research is room-temperature superconductivity. In 2014, evidence showing that fractional particles can happen in quasi two-dimensional magnetic materials, was found by EPFL scientists lending support for Anderson's theory of high-temperature superconductivity.

Crystal Structures of High-temperature Ceramic Superconductors

The structure of high-T_c copper oxide or cuprate superconductors are often closely related to perovskite structure, and the structure of these compounds has been described as a distorted, oxygen deficient multi-layered perovskite structure. One of the properties of the crystal structure of oxide superconductors is an alternating multi-layer of CuO_2 planes with superconductivity taking place between these layers. The more layers of CuO_2, the higher T_c. This structure causes a large anisotropy in normal conducting and superconducting properties, since electrical currents are carried by holes induced in the oxygen sites of the CuO_2 sheets. The electrical conduction is highly anisotropic, with a much higher conductivity parallel to the CuO_2 plane than in the perpendicular direction. Generally, critical temperatures depend on the chemical compositions, cations substitutions and oxygen content. They can be classified as superstripes; i.e., particular realizations of superlattices at atomic limit made of superconducting atomic layers, wires, dots separated by spacer layers, that gives multiband and multigap superconductivity.

YBaCuO Superconductors

The first superconductor found with $T_c > 77$ K (liquid nitrogen boiling point) is yttrium barium copper oxide ($YBa_2Cu_3O_{7-x}$); the proportions of the three different metals in the $YBa_2Cu_3O_7$ superconductor are in the mole ratio of 1 to 2 to 3 for yttrium to barium to copper, respectively. Thus, this particular superconductor is often referred to as the 123 superconductor.

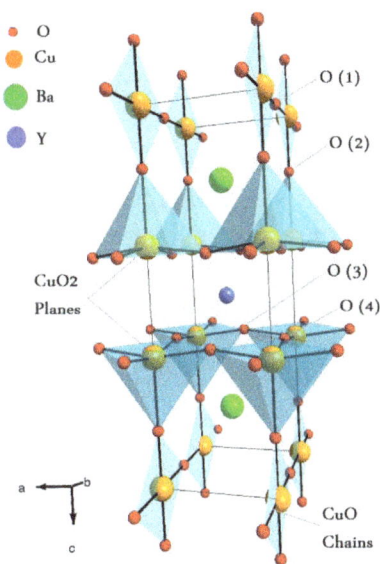

YBCO unit cell

The unit cell of $YBa_2Cu_3O_7$ consists of three pseudocubic elementary perovskite unit cells. Each perovskite unit cell contains a Y or Ba atom at the center: Ba in the bottom unit cell, Y in the middle one, and Ba in the top unit cell. Thus, Y and Ba are stacked in the sequence [Ba–Y–Ba] along the c-axis. All corner sites of the unit cell are occupied by Cu, which has two different coordinations, Cu(1) and Cu(2), with respect to oxygen. There are four possible crystallographic sites for oxygen: O(1), O(2), O(3) and O(4). The coordination polyhedra of Y and Ba with respect to oxygen are different. The tripling of the perovskite unit cell leads to nine oxygen atoms, whereas $YBa_2Cu_3O_7$ has seven oxygen atoms and, therefore, is referred to as an oxygen-deficient perovskite structure. The structure has a stacking of different layers: (CuO)(BaO)(CuO$_2$)(Y)(CuO$_2$)(BaO)(CuO). One of the key feature of the unit cell of $YBa_2Cu_3O_{7-x}$ (YBCO) is the presence of two layers of CuO_2. The role of the Y plane is to serve as a spacer between two CuO_2 planes. In YBCO, the Cu–O chains are known to play an important role for superconductivity. T_c is maximal near 92 K when $x \approx 0.15$ and the structure is orthorhombic. Superconductivity disappears at $x \approx 0.6$, where the structural transformation of YBCO occurs from orthorhombic to tetragonal.

Bi-, Tl- and Hg-based High-T_c Superconductors

The crystal structure of Bi-, Tl- and Hg-based high-T_c superconductors are very similar. Like YBCO, the perovskite-type feature and the presence of CuO_2 layers also exist in these superconductors. However, unlike YBCO, Cu–O chains are not present in these superconductors. The YBCO superconductor has an orthorhombic structure, whereas the other high-T_c superconductors have a tetragonal structure.

The Bi–Sr–Ca–Cu–O system has three superconducting phases forming a homologous series as $Bi_2Sr_2Ca_{n-1}Cu_nO_{4+2n+x}$ (n = 1, 2 and 3). These three phases are Bi-2201, Bi-2212 and Bi-2223, having transition temperatures of 20, 85 and 110 K, respectively, where the numbering system represent number of atoms for Bi, Sr, Ca and Cu respectively. The two phases have a tetragonal structure which consists of two sheared crystallographic unit cells. The unit cell of these phases has double Bi–O planes which are stacked in a way that the Bi atom of one plane sits below the oxygen atom of the next consecutive plane. The Ca atom forms a layer within the interior of the CuO_2 layers in both Bi-2212 and Bi-2223; there is no Ca layer in the Bi-2201 phase. The three phases differ with each other in the number of CuO_2 planes; Bi-2201, Bi-2212 and Bi-2223 phases have one, two and three CuO_2 planes, respectively. The c axis lattice constants of these phases increases with the number of CuO_2 planes. The coordination of the Cu atom is different in the three phases. The Cu atom forms an octahedral coordination with respect to oxygen atoms in the 2201 phase, whereas in 2212, the Cu atom is surrounded by five oxygen atoms in a pyramidal arrangement. In the 2223 structure, Cu has two coordinations with respect to oxygen: one Cu atom is bonded with four oxygen atoms in square planar configuration and another Cu atom is coordinated with five oxygen atoms in a pyramidal arrangement.

Tl–Ba–Ca–Cu–O superconductor: The first series of the Tl-based superconductor containing one Tl–O layer has the general formula $TlBa_2Ca_{n-1}Cu_nO_{2n+3}$, whereas the second series containing two Tl–O layers has a formula of $Tl_2Ba_2Ca_{n-1}Cu_nO_{2n+4}$ with n = 1, 2 and 3. In the structure of $Tl_2Ba_2CuO_6$ (Tl-2201), there is one CuO_2 layer with the stacking sequence (Tl–O) (Tl–O) (Ba–O) (Cu–O) (Ba–O) (Tl–O) (Tl–O). In $Tl_2Ba_2CaCu_2O_8$ (Tl-2212), there are two Cu–O layers with a Ca layer in between. Similar to the $Tl_2Ba_2CuO_6$ structure, Tl–O layers are present outside the Ba–O layers. In $Tl_2Ba_2Ca_2Cu_3O_{10}$ (Tl-2223), there are three CuO2 layers enclosing Ca layers between each of these. In Tl-based superconductors, T_c is found to increase with the increase in CuO_2 layers. However, the value of T_c decreases after four CuO_2 layers in $TlBa_2Ca_{n-1}Cu_nO_{2n+3}$, and in the $Tl_2Ba_2Ca_{n-1}Cu_nO_{2n+4}$ compound, it decreases after three CuO_2 layers.

Hg–Ba–Ca–Cu–O superconductor: The crystal structure of $HgBa_2CuO_4$ (Hg-1201), $HgBa_2CaCu_2O_6$ (Hg-1212) and $HgBa_2Ca_2Cu_3O_8$ (Hg-1223) is similar to that of Tl-1201, Tl-1212 and Tl-1223, with Hg in place of Tl. It is noteworthy that the T_c of the Hg compound (Hg-1201) containing one CuO_2 layer is much larger as compared to the one-CuO_2-layer compound of thallium (Tl-1201). In the Hg-based superconductor, T_c is also found to increase as the CuO_2 layer increases. For Hg-1201, Hg-1212 and Hg-1223, the values of T_c are 94, 128 and the record value at ambient pressure 134 K, respectively, as shown in table below. The observation that the T_c of Hg-1223 increases to 153 K under high pressure indicates that the T_c of this compound is very sensitive to the structure of the compound.

Critical temperature (T_c), crystal structure and lattice constants of some high-T_c superconductors				
Formula	Notation	T_c (K)	No. of Cu-O planes in unit cell	Crystal structure
$YBa_2Cu_3O_7$	123	92	2	Orthorhombic
$Bi_2Sr_2CuO_6$	Bi-2201	20	1	Tetragonal
$Bi_2Sr_2CaCu_2O_8$	Bi-2212	85	2	Tetragonal
$Bi_2Sr_2Ca_2Cu_3O_{10}$	Bi-2223	110	3	Tetragonal
$Tl_2Ba_2CuO_6$	Tl-2201	80	1	Tetragonal
$Tl_2Ba_2CaCu_2O_8$	Tl-2212	108	2	Tetragonal
$Tl_2Ba_2Ca_2Cu_3O_{10}$	Tl-2223	125	3	Tetragonal
$TlBa_2Ca_3Cu_4O_{11}$	Tl-1234	122	4	Tetragonal
$HgBa_2CuO_4$	Hg-1201	94	1	Tetragonal
$HgBa_2CaCu_2O_6$	Hg-1212	128	2	Tetragonal
$HgBa_2Ca_2Cu_3O_8$	Hg-1223	134	3	Tetragonal

Preparation of High-T_c Superconductors

The simplest method for preparing high-T_c superconductors is a solid-state thermochemical reaction involving mixing, calcination and sintering. The appropriate amounts of precursor powders, usually oxides and carbonates, are mixed thoroughly using a Ball mill. Solution chemistry processes such as coprecipitation, freeze-drying and sol-gel methods are alternative ways for preparing a homogeneous mixture. These powders are calcined in the temperature range from 800 °C to 950 °C for several hours. The powders are cooled, reground and calcined again. This process is repeated several times to get homogeneous material. The powders are subsequently compacted to pellets and sintered. The sintering environment such as temperature, annealing time, atmosphere and cooling rate play a very important role in getting good high-T_c superconducting materials. The $YBa_2Cu_3O_{7-x}$ compound is prepared by calcination and sintering of a homogeneous mixture of Y_2O_3, $BaCO_3$ and CuO in the appropriate atomic ratio. Calcination is done at 900–950 °C, whereas sintering is done at 950 °C in an oxygen atmosphere. The oxygen stoichiometry in this material is very crucial for obtaining a superconducting $YBa_2Cu_3O_{7-x}$ compound. At the time of sintering, the semiconducting tetragonal $YBa_2Cu_3O_6$ compound is formed, which, on slow cooling in oxygen atmosphere, turns into superconducting $YBa_2Cu_3O_{7-x}$. The uptake and loss of oxygen are reversible in $YBa_2Cu_3O_{7-x}$. A fully oxygenated orthorhombic $YBa_2Cu_3O_{7-x}$ sample can be transformed into tetragonal $YBa_2Cu_3O_6$ by heating in a vacuum at temperature above 700 °C.

The preparation of Bi-, Tl- and Hg-based high-T_c superconductors is difficult compared to YBCO. Problems in these superconductors arise because of the existence of three or more phases having a similar layered structure. Thus, syntactic intergrowth and defects such as stacking faults occur during synthesis and it becomes difficult to isolate a single superconducting phase. For Bi–Sr–Ca–Cu–O, it is relatively simple to prepare the Bi-2212 ($T_c \approx 85$ K) phase, whereas it is very difficult to prepare a single phase of Bi-2223 ($T_c \approx 110$ K). The Bi-2212 phase appears only after few hours of sintering at 860–870 °C, but the larger fraction of the Bi-2223 phase is formed after a long reaction time of more than a week at 870 °C. Although the substitution of Pb in the Bi–Sr–Ca–Cu–O compound has been found to promote the growth of the high-T_c phase, a long sintering time is still required.

Properties

"High-temperature" has two common definitions in the context of superconductivity:

1. Above the temperature of 30 K that had historically been taken as the upper limit allowed by BCS theory(1957). This is also above the 1973 record of 23 K that had lasted until copper-oxide materials were discovered in 1986.

2. Having a transition temperature that is a larger fraction of the Fermi temperature than for conventional superconductors such as elemental mercury or lead. This definition encompasses a wider variety of unconventional superconductors and is used in the context of theoretical models.

The label high-T_c may be reserved by some authors for materials with critical temperature greater than the boiling point of liquid nitrogen (77 K or −196 °C). However, a number of materials – including the original discovery and recently discovered pnictide superconductors – had critical temperatures below 77 K but are commonly referred to in publication as being in the high-T_c class.

Technological applications could benefit from both the higher critical temperature being above the boiling point of liquid nitrogen and also the higher critical magnetic field (and critical current density) at which superconductivity is destroyed. In magnet applications, the high critical magnetic field may prove more valuable than the high T_c itself. Some cuprates have an upper critical field of about 100 tesla. However, cuprate materials are brittle ceramics which are expensive to manufacture and not easily turned into wires or other useful shapes. Also, high-temperature superconductors do not form large, continuous superconducting domains, but only clusters of microdomains within which superconducting occurs. They are therefore unsuitable for applications requiring actual superconducted currents, such as magnets for magnetic resonance spectrometers.

After two decades of intense experimental and theoretical research, with over 100,000 published papers on the subject, several common features in the properties of high-temperature superconductors have been identified. As of 2011, no widely accepted theory explains their properties. Relative to conventional superconductors, such as elemental mercury or lead that are adequately explained by the BCS theory, cuprate superconductors (and other unconventional superconductors) remain distinctive. There also has been much debate as to high-temperature superconductivity coexisting with magnetic ordering in YBCO, iron-based superconductors, several ruthenocuprates and other exotic superconductors, and the search continues for other families of materials. HTS are Type-II superconductors, which allow magnetic fields to penetrate their interior in quantized units of flux, meaning that much higher magnetic fields are required to suppress superconductivity. The layered structure also gives a directional dependence to the magnetic field response.

Cuprates

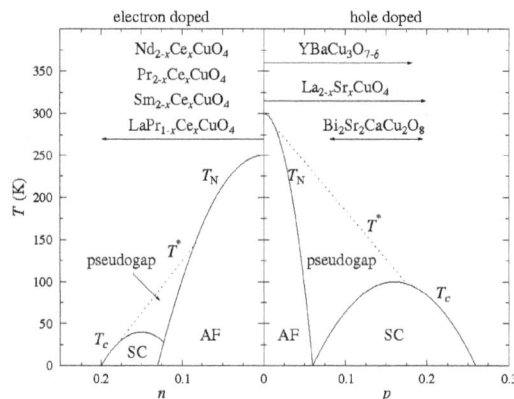

Simplified doping dependent phase diagram of cuprate superconductors for both electron (n) and hole (p) doping. The phases shown are the antiferromagnetic (AF) phase close to zero doping, the superconducting phase around optimal doping, and the pseudogap phase. Doping ranges possible for some common compounds are also shown. After.

Cuprate superconductors are generally considered to be quasi-two-dimensional materials with their superconducting properties determined by electrons moving within weakly coupled copper-oxide (CuO_2) layers. Neighbouring layers containing ions such as lanthanum, barium, strontium, or other atoms act to stabilize the structure and dope electrons or holes onto the copper-oxide layers. The undoped "parent" or "mother" compounds are Mott insulators with long-range antiferromagnetic order at low enough temperature. Single band models are generally considered to be sufficient to describe the electronic properties.

The cuprate superconductors adopt a perovskite structure. The copper-oxide planes are checkerboard lattices with squares of O^{2-} ions with a Cu^{2+} ion at the centre of each square. The unit cell is rotated by 45° from these squares. Chemical formulae of superconducting materials generally contain fractional numbers to describe the doping required for superconductivity. There are several families of cuprate superconductors and they can be categorized by the elements they contain and the number of adjacent copper-oxide layers in each superconducting block. For example, YBCO and BSCCO can alternatively be referred to as Y123 and Bi2201/Bi2212/Bi2223 depending on the number of layers in each superconducting block (n). The superconducting transition temperature has been found to peak at an optimal doping value (p = 0.16) and an optimal number of layers in each superconducting block, typically n = 3.

Possible mechanisms for superconductivity in the cuprates are still the subject of considerable debate and further research. Certain aspects common to all materials have been identified. Similarities between the antiferromagnetic low-temperature state of the undoped materials and the superconducting state that emerges upon doping, primarily the $d_{x^2-y^2}$ orbital state of the Cu^{2+} ions, suggest that electron-electron interactions are more significant than electron-phonon interactions in cuprates – making the superconductivity unconventional. Recent work on the Fermi surface has shown that nesting occurs at four points in the antiferromagnetic Brillouin zone where spin waves exist and that the superconducting energy gap is larger at these points. The weak isotope effects observed for most cuprates contrast with conventional superconductors that are well described by BCS theory.

Similarities and differences in the properties of hole-doped and electron doped cuprates:

- Presence of a pseudogap phase up to at least optimal doping.

- Different trends in the Uemura plot relating transition temperature to the superfluid density. The inverse square of the London penetration depth appears to be proportional to the critical temperature for a large number of underdoped cuprate superconductors, but the constant of proportionality is different for hole- and electron-doped cuprates. The linear trend implies that the physics of these materials is strongly two-dimensional.

- Universal hourglass-shaped feature in the spin excitations of cuprates measured using inelastic neutron diffraction.

- Nernst effect evident in both the superconducting and pseudogap phases.

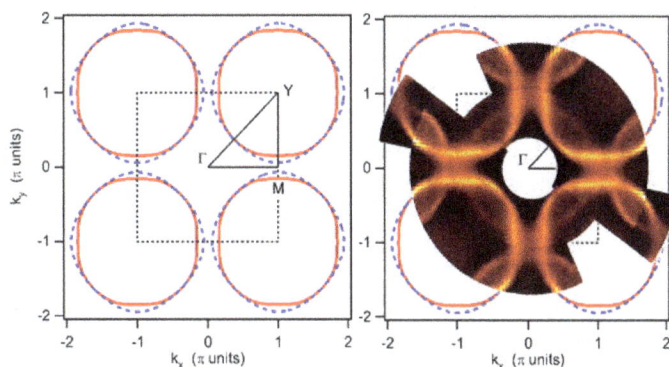

Fig. 1. The Fermi surface of bi-layer BSCCO, calculated (left) and measured by ARPES (right).
The dashed rectangle represents the first Brillouin zone.

The electronic structure of superconducting cuprates is highly anisotropic. Therefore, the Fermi surface of HTSC is very close to the Fermi surface of the doped CuO_2 plane (or multi-planes, in case of multi-layer cuprates) and can be presented on the 2D reciprocal space (or momentum space) of the CuO_2 lattice. The typical Fermi surface within the first CuO_2 Brillouin zone is sketched in Fig. It can be derived from the band structure calculations or measured by angle resolved photoemission spectroscopy (ARPES). Fig. shows the Fermi surface of BSCCO measured by ARPES. In a wide range of charge carrier concentration (doping level), in which the hole-doped HTSC are superconducting, the Fermi surface is hole-like (*i.e.* open, as shown in Fig.). This results in an inherent in-plane anisotropy of the electronic properties of HTSC.

Iron-based Superconductors

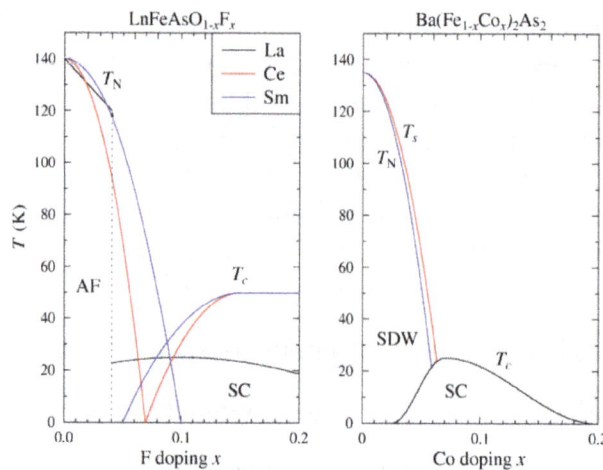

Simplified doping dependent phase diagrams of iron-based superconductors for both Ln-1111 and Ba-122 materials. The phases shown are the antiferromagnetic/spin density wave (AF/SDW) phase close to zero doping and the superconducting phase around optimal doping. The Ln-1111 phase diagrams for La and Sm were determined using muon spin spectroscopy, the phase diagram for Ce was determined using neutron diffraction. The Ba-122 phase diagram is based on.

Iron-based superconductors contain layers of iron and a pnictogen—such as arsenic or phosphorus—or a chalcogen. This is currently the family with the second highest critical temperature, behind the cuprates. Interest in their superconducting properties began in 2006 with the discovery of superconductivity in LaFePO at 4 K and gained much greater attention in 2008 after the analogous material LaFeAs(O,F) was found to superconduct at up to 43 K under pressure. The highest critical temperatures in the iron-based superconductor family exist in thin films of FeSe, where a critical temperature in excess of 100 K has recently been reported.

Since the original discoveries several families of iron-based superconductors have emerged:

- LnFeAs(O,F) or $LnFeAsO_{1-x}$ (Ln = lanthanide) with T_c up to 56 K, referred to as 1111 materials. A fluoride variant of these materials was subsequently found with similar T_c values.

- $(Ba,K)Fe_2As_2$ and related materials with pairs of iron-arsenide layers, referred to as 122 compounds. T_c values range up to 38 K. These materials also superconduct when iron is replaced with cobalt

- LiFeAs and NaFeAs with T_c up to around 20 K. These materials superconduct close to stoichiometric composition and are referred to as 111 compounds.

- FeSe with small off-stoichiometry or tellurium doping.

Most undoped iron-based superconductors show a tetragonal-orthorhombic structural phase transition followed at lower temperature by magnetic ordering, similar to the cuprate superconductors. However, they are poor metals rather than Mott insulators and have five bands at the Fermi surface rather than one. The phase diagram emerging as the iron-arsenide layers are doped is remarkably similar, with the superconducting phase close to or overlapping the magnetic phase. Strong evidence that the T_c value varies with the As-Fe-As bond angles has already emerged and shows that the optimal T_c value is obtained with undistorted $FeAs_4$ tetrahedra. The symmetry of the pairing wavefunction is still widely debated, but an extended s-wave scenario is currently favoured.

Hydrogen Sulfide

At pressures above 90 GPa (Gigapascal), hydrogen sulfide becomes a metallic conductor of electricity. When cooled below a critical temperature this high-pressure phase exhibits superconductivity. The critical temperature increases with pressure, ranging from 23 K at 100 GPa to 150 K at 200 GPa. If hydrogen sulfide is pressurized at higher temperatures, then cooled, the critical temperature reaches 203 K (–70 °C), the highest accepted superconducting critical temperature as of 2015. It has been predicted that by substituting a small part of sulfur with phosphorus and using even higher pressures it may be possible to raise the critical temperature to above 0 °C (273 K) and achieve room-temperature superconductivity.

Other Materials Sometimes Referred to as High-temperature Superconductors

Magnesium diboride is occasionally referred to as a high-temperature superconductor because its T_c value of 39 K is above that historically expected for BCS superconductors. However, it is more generally regarded as the highest-T_c conventional superconductor, the increased T_c resulting from two separate bands being present at the Fermi level.

Fulleride superconductors where alkali-metal atoms are intercalated into C_{60} molecules show superconductivity at temperatures of up to 38 K for Cs_3C_{60}.

Some organic superconductors and heavy fermion compounds are considered to be high-temperature superconductors because of their high T_c values relative to their Fermi energy, despite the T_c values being lower than for many conventional superconductors. This description may relate better to common aspects of the superconducting mechanism than the superconducting properties.

Metallic Hydrogen

Theoretical work by Neil Ashcroft in 1968 predicted that solid metallic hydrogen at extremely high pressure should become superconducting at approximately room-temperature because of its extremely high speed of sound and expected strong coupling between the conduction electrons and the lattice vibrations. This prediction is yet to be experimentally verified.

Magnetic Properties

All known high-T_c superconductors are Type-II superconductors. In contrast to Type-I superconductors, which expel all magnetic fields due to the Meissner effect, Type-II superconductors allow magnetic fields to penetrate their interior in quantized units of flux, creating "holes" or "tubes" of normal metallic regions in the superconducting bulk called vortices. Consequently, high-T_c superconductors can sustain much higher magnetic fields.

Ongoing Research

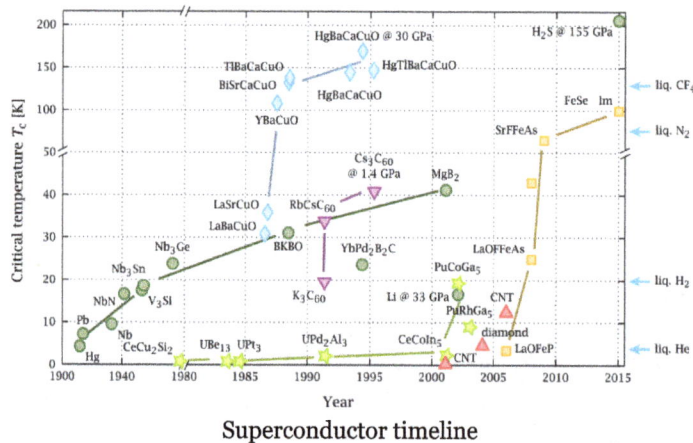

Superconductor timeline

The question of how superconductivity arises in high-temperature superconductors is one of the major unsolved problems of theoretical condensed matter physics. The mechanism that causes the electrons in these crystals to form pairs is not known. Despite intensive research and many promising leads, an explanation has so far eluded scientists. One reason for this is that the materials in question are generally very complex, multi-layered crystals (for example, BSCCO), making theoretical modelling difficult.

Improving the quality and variety of samples also gives rise to considerable research, both with the aim of improved characterisation of the physical properties of existing compounds, and synthesizing new materials, often with the hope of increasing T_c. Technological research focuses on making HTS materials in sufficient quantities to make their use economically viable and optimizing their properties in relation to applications.

Possible Mechanism

There have been two representative theories for high-temperature or unconventional superconductivity. Firstly, weak coupling theory suggests superconductivity emerges from antiferromagnetic spin fluctuations in a doped system. According to this theory, the pairing wave function of the cuprate HTS should have a $d_{x^2-y^2}$ symmetry. Thus, determining whether the pairing wave function has d-wave symmetry is essential to test the spin fluctuation mechanism. That is, if the HTS order parameter (pairing wave function) does not have d-wave symmetry, then a pairing mechanism related to spin fluctuations can be ruled out. (Similar arguments can be made for iron-based superconductors but the different material properties allow a different pairing symmetry.) Secondly, there was the interlayer coupling model, according to which a layered structure consisting of BCS-type (s-wave symmetry) superconductors can enhance the superconductivity by itself. By introducing an additional tunnelling interaction between each layer, this model successfully explained the anisotropic symmetry of the order parameter as well as the emergence of the HTS. Thus, in order to solve this unsettled problem, there have been numerous experiments such as photoemission spectroscopy, NMR, specific heat measurements, etc. Up to date the results were ambiguous, some reports supported the d symmetry for the HTS whereas others supported the s symmetry. This muddy situation possibly originated from the indirect nature of the experimental evidence, as well as experimental issues such as sample quality, impurity scattering, twinning, etc.

Junction Experiment Supporting the d Symmetry

The Meissner effect or a magnet levitating above a superconductor (cooled by liquid nitrogen)

An experiment based on flux quantization of a three-grain ring of $YBa_2Cu_3O_7$ (YBCO) was proposed to test the symmetry of the order parameter in the HTS. The symmetry of the order parameter could best be probed at the junction interface as the Cooper pairs tunnel across a Josephson junction or weak link. It was expected that a half-integer flux, that is, a spontaneous magnetization could only occur for a junction of d symmetry superconductors. But, even if the junction experiment is the strongest method to determine the symmetry of the HTS order parameter, the results have been ambiguous. J. R. Kirtley and C. C. Tsuei thought that the ambiguous results came from the defects inside the HTS, so that they designed an experiment where both clean limit (no defects) and dirty limit (maximal defects) were considered simultaneously. In the experiment, the spontaneous magnetization was clearly observed in YBCO, which supported the d symmetry of the order parameter in YBCO. But, since YBCO is orthorhombic, it might inherently have an admixture of s symmetry. So, by tuning their technique further, they found that there was an admixture of s symmetry in YBCO within about 3%. Also, they found that there was a pure $d_{x^2-y^2}$ order parameter symmetry in the tetragonal $Tl_2Ba_2CuO_6$.

Qualitative Explanation of the Spin-fluctuation Mechanism

Despite all these years, the mechanism of high-T_c superconductivity is still highly controversial, mostly due to the lack of exact theoretical computations on such strongly interacting electron systems. However, most rigorous theoretical calculations, including phenomenological and diagrammatic approaches, converge on magnetic fluctuations as the pairing mechanism for these systems. The qualitative explanation is as follows:

In a superconductor, the flow of electrons cannot be resolved into individual electrons, but instead consists of many pairs of bound electrons, called Cooper pairs. In conventional superconductors, these pairs are formed when an electron moving through the material distorts the surrounding crystal lattice, which in turn attracts another electron and forms a bound pair. This is sometimes called the "water bed" effect. Each Cooper pair requires a certain minimum energy to be displaced, and if the thermal fluctuations in the crystal lattice are smaller than this energy the pair can flow without dissipating energy. This ability of the electrons to flow without resistance leads to superconductivity.

In a high-T_c superconductor, the mechanism is extremely similar to a conventional superconductor, except, in this case, phonons virtually play no role and their role is replaced by spin-density waves. Just as all known conventional superconductors are strong phonon systems, all known high-T_c superconductors are strong spin-density wave systems, within close vicinity of a magnetic transition to, for example, an antiferromagnet. When an electron moves in a high-T_c superconductor, its spin creates a spin-density wave around it. This spin-density wave in turn causes a nearby electron to fall into the spin depression created by the first electron (water-bed effect again). Hence, again, a Cooper pair is formed. When the system temperature is lowered, more spin density waves and Cooper pairs are created, eventually leading to superconductivity. Note that in high-T_c systems, as these systems are magnetic systems due to the Coulomb interaction, there is a strong Coulomb repulsion between electrons. This Coulomb repulsion prevents pairing of the Cooper pairs on the same lattice site. The pairing of the electrons occur at near-neighbor lattice sites as a result. This is the so-called d-wave pairing, where the pairing state has a node (zero) at the origin.

Examples

Examples of high-T_c cuprate superconductors include $La_{1.85}Ba_{0.15}CuO_4$, and YBCO (Yttrium-Barium-Copper-Oxide), which is famous as the first material to achieve superconductivity above the boiling point of liquid nitrogen.

Transition temperatures of well-known superconductors (Boiling point of liquid nitrogen for comparison)			
Transition temperature (in kelvin)	Transition temperature (in Celsius)	Material	Class
203	−70	H_2S (at 150 GPa pressure)	Hydrogen-based superconductor
195	−78	Sublimation point of dry ice	
184	−89.2	Lowest temperature recorded on Earth	
133	−140	$HgBa_2Ca_2Cu_3O_x$ (HBCCO)	Copper-oxide superconductors
110	−163	$Bi_2Sr_2Ca_2Cu_3O_{10}$ (BSCCO)	
93	−180	$YBa_2Cu_3O_7$ (YBCO)	
90	−183	Boiling point of liquid oxygen	
77	−196	Boiling point of liquid nitrogen	

55	−218	SmFeAs(O,F)	Iron-based superconductors
41	−232	CeFeAs(O,F)	
26	−247	LaFeAs(O,F)	
20	−253	Boiling point of liquid hydrogen	
18	−255	Nb_3Sn	Metallic low-temperature superconductors
10	−263	NbTi	
9.2	−263.8	Nb	
4.2	−268.8	Boiling point of liquid helium	
4.2	−268.8	Hg (mercury)	Metallic low-temperature superconductors

Metamaterial

A metamaterial is a material engineered to have a property that is not found in nature. They are made from assemblies of multiple elements fashioned from composite materials such as metals or plastics. The materials are usually arranged in repeating patterns, at scales that are smaller than the wavelengths of the phenomena they influence. Metamaterials derive their properties not from the properties of the base materials, but from their newly designed structures. Their precise shape, geometry, size, orientation and arrangement gives them their smart properties capable of manipulating electromagnetic waves: by blocking, absorbing, enhancing, or bending waves, to achieve benefits that go beyond what is possible with conventional materials.

Appropriately designed metamaterials can affect waves of electromagnetic radiation or sound in a manner not observed in bulk materials. Those that exhibit a negative index of refraction for particular wavelengths have attracted significant research. These materials are known as negative-index metamaterials.

Potential applications of metamaterials are diverse and include optical filters, medical devices, remote aerospace applications, sensor detection and infrastructure monitoring, smart solar power management, crowd control, radomes, high-frequency battlefield communication and lenses for high-gain antennas, improving ultrasonic sensors, and even shielding structures from earthquakes. Metamaterials offer the potential to create superlenses. Such a lens could allow imaging below the diffraction limit that is the minimum resolution that can be achieved by conventional glass lenses. A form of 'invisibility' was demonstrated using gradient-index materials. Acoustic and seismic metamaterials are also research areas.

Metamaterial research is interdisciplinary and involves such fields as electrical engineering, electromagnetics, classical optics, solid state physics, microwave and antennae engineering, optoelectronics, material sciences, nanoscience and semiconductor engineering.

History

Explorations of artificial materials for manipulating electromagnetic waves began at the end of the

19th century. Some of the earliest structures that may be considered metamaterials were studied by Jagadish Chandra Bose, who in 1898 researched substances with chiral properties. Karl Ferdinand Lindman studied wave interaction with metallic helices as artificial chiral media in the early twentieth century.

Winston E. Kock developed materials that had similar characteristics to metamaterials in the late 1940s. In the 1950s and 1960s, artificial dielectrics were studied for lightweight microwave antennas. Microwave radar absorbers were researched in the 1980s and 1990s as applications for artificial chiral media.

Negative-index materials were first described theoretically by Victor Veselago in 1967. He proved that such materials could transmit light. He showed that the phase velocity could be made anti-parallel to the direction of Poynting vector. This is contrary to wave propagation in naturally-occurring materials.

John Pendry was the first to identify a practical way to make a left-handed metamaterial, a material in which the right-hand rule is not followed. Such a material allows an electromagnetic wave to convey energy (have a group velocity) against its phase velocity. Pendry's idea was that metallic wires aligned along the direction of a wave could provide negative permittivity (dielectric function $\varepsilon < 0$). Natural materials (such as ferroelectrics) display negative permittivity; the challenge was achieving negative permeability ($\mu < 0$). In 1999 Pendry demonstrated that a split ring (C shape) with its axis placed along the direction of wave propagation could do so. In the same paper, he showed that a periodic array of wires and rings could give rise to a negative refractive index. Pendry also proposed a related negative-permeability design, the Swiss roll.

In 2000, Smith et al. reported the experimental demonstration of functioning electromagnetic metamaterials by horizontally stacking, periodically, split-ring resonators and thin wire structures. A method was provided in 2002 to realize negative-index metamaterials using artificial lumped-element loaded transmission lines in microstrip technology. In 2003, complex (both real and imaginary parts of) negative refractive index and imaging by flat lens using left handed metamaterials were demonstrated. By 2007, experiments that involved negative refractive index had been conducted by many groups. At microwave frequencies, the first, imperfect invisibility cloak was realized in 2006.

Electromagnetic Metamaterials

An electromagnetic metamaterial affects electromagnetic waves that impinge on or interact with its structural features, which are smaller than the wavelength. To behave as a homogeneous material accurately described by an effective refractive index, its features must be much smaller than the wavelength.

For microwave radiation, the features are on the order of millimeters. Microwave frequency metamaterials are usually constructed as arrays of electrically conductive elements (such as loops of wire) that have suitable inductive and capacitive characteristics. One microwave metamaterial uses the split-ring resonator.

Photonic metamaterials, nanometer scale, manipulate light at optical frequencies. To date, subwavelength structures have shown only a few, questionable, results at visible wavelengths.

Photonic crystals and frequency-selective surfaces such as diffraction gratings, dielectric mirrors and optical coatings exhibit similarities to subwavelength structured metamaterials. However, these are usually considered distinct from subwavelength structures, as their features are structured for the wavelength at which they function and thus cannot be approximated as a homogeneous material. However, material structures such as photonic crystals are effective in the visible light spectrum. The middle of the visible spectrum has a wavelength of approximately 560 nm (for sunlight). Photonic crystal structures are generally half this size or smaller, that is <280 nm.

Plasmonic metamaterials utilize surface plasmons, which are packets of electrical charge that collectively oscillate at the surfaces of metals at optical frequencies.

Frequency selective surfaces (FSS) can exhibit subwavelength characteristics and are known variously as artificial magnetic conductors (AMC) or High Impedance Surfaces (HIS). FSS display inductive and capacitive characteristics that are directly related to their subwavelength structure.

Negative Refractive Index

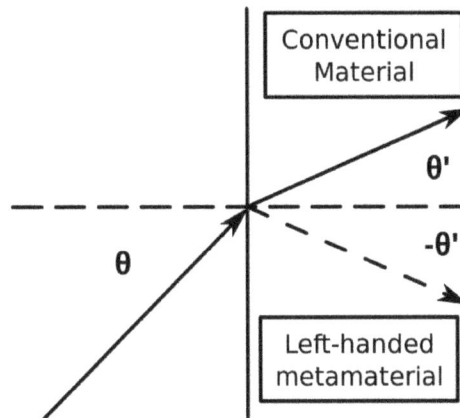

A comparison of refraction in a left-handed metamaterial to that in a normal material

Almost all materials encountered in optics, such as glass or water, have positive values for both permittivity ε and permeability μ. However, metals such as silver and gold have negative permittivity at shorter wavelengths. A material such as a surface plasmon that has either (but not both) ε or μ negative is often opaque to electromagnetic radiation. However, anisotropic materials with only negative permittivity can produce negative refraction due to chirality.

Although the optical properties of a transparent material are fully specified by the parameters εr and μr, refractive index n is often used in practice, which can be determined from All known non-metamaterial transparent materials possess positive εr and μr. By convention the positive square root is used for n.

However, some engineered metamaterials have $\varepsilon r < 0$ and $\mu r < 0$. Because the product $\varepsilon r \mu r$ is positive, n is real. Under such circumstances, it is necessary to take the negative square root for n.

Video representing negative refraction of light at uniform planar interface.

The foregoing considerations are simplistic for actual materials, which must have complex-valued εr and μr. The real parts of both εr and μr do not have to be negative for a passive material to display negative refraction. Metamaterials with negative n have numerous interesting properties:

- Snell's law ($n_1 \sin\theta_1 = n_2 \sin\theta_2$), but as n_2 is negative, the rays are refracted on the *same* side of the normal on entering the material.

- Cherenkov radiation points the other way.

- The time-averaged Poynting vector is antiparallel to phase velocity. However, for waves (energy) to propagate, a $-\mu$ must be paired with a $-\varepsilon$ in order to satisfy the wave number dependence on the material parameters

Negative index of refraction derives mathematically from the vector triplet E, H and k.

For plane waves propagating in electromagnetic metamaterials, the electric field, magnetic field and wave vector follow a left-hand rule, the reverse of the behavior of conventional optical materials.

Classification

Electromagnetic metamaterials are divided into different classes, as follows:

Negative Index

In negative-index metamaterials (NIM), both permittivity and permeability are negative, resulting in a negative index of refraction. These are also known as double negative metamaterials or double negative materials (DNG). Other terms for NIMs include "left-handed media", "media with a negative refractive index", and "backward-wave media".

In optical materials, if both permittivity ε and permeability μ are positive, wave propagation travels in the forward direction. If both ε and μ are negative, a backward wave is produced. If ε and μ have different polarities, waves do not propagate.

Mathematically, quadrant II and quadrant IV have coordinates (0,0) in a coordinate plane where ε is the horizontal axis, and μ is the vertical axis.

To date, only metamaterials exhibit a negative index of refraction.

Single negative

Single negative (SNG) metamaterials have either negative relative permittivity (εr) or negative relative permeability (μr), but not both. They act as metamaterials when combined with a different, complementary SNG, jointly acting as a DNG.

Epsilon negative media (ENG) display a negative εr while μr is positive. Many plasmas exhibit this characteristic. For example, noble metals such as gold or silver are ENG in the infrared and visible spectrums.

Mu-negative media (MNG) display a positive εr and negative μr. Gyrotropic or gyromagnetic materials exhibit this characteristic. A gyrotropic material is one that has been altered by the presence of a quasistatic magnetic field, enabling a magneto-optic effect. A magneto-optic effect is a phenomenon in which an electromagnetic wave propagates through such a medium. In such a material, left- and right-rotating elliptical polarizations can propagate at different speeds. When light is transmitted through a layer of magneto-optic material, the result is called the Faraday effect: the polarization plane can be rotated, forming a Faraday rotator. The results of such a reflection are known as the magneto-optic Kerr effect. Two gyrotropic materials with reversed rotation directions of the two principal polarizations are called optical isomers.

Joining a slab of ENG material and slab of MNG material resulted in properties such as resonances, anomalous tunneling, transparency and zero reflection. Like negative-index materials, SNGs are innately dispersive, so their εr, μr and refraction index n, are a function of frequency.

Bandgap

Electromagnetic bandgap metamaterials (EBM) control light propagation. This is accomplished either with photonic crystals (PC) or left-handed materials (LHM). PCs can prohibit light propagation altogether. Both classes can allow light to propagate in specific, designed directions and both can be designed with bandgaps at desired frequencies. The period size of EBGs is an appreciable fraction of the wavelength, creating constructive and destructive interference.

PC are distinguished from sub-wavelength structures, such as tunable metamaterials, because the PC derives its properties from its bandgap characteristics. PCs are sized to match the wavelength of light, versus other metamaterials that expose sub-wavelength structure. Furthermore, PCs function by diffracting light. In contrast, metamaterial does not use diffraction.

PCs have periodic inclusions that inhibit wave propagation due to the inclusions' destructive interference from scattering. The photonic bandgap property of PCs makes them the electromagnetic analog of electronic semi-conductor crystals.

EBGs have the goal of creating high quality, low loss, periodic, dielectric structures. An EBG affects photons in the same way semiconductor materials affect electrons. PCs are the perfect bandgap material, because they allow no light propagation. Each unit of the prescribed periodic structure acts like one atom, albeit of a much larger size.

EBGs are designed to prevent the propagation of an allocated bandwidth of frequencies, for certain arrival angles and polarizations. Various geometries and structures have been proposed to fabricate EBG's special properties. In practice it is impossible to build a flawless EBG device.

EBGs have been manufactured for frequencies ranging from a few gigahertz (GHz) to a few terahertz (THz), radio, microwave and mid-infrared frequency regions. EBG application developments include a transmission line, woodpiles made of square dielectric bars and several different types of low gain antennas.

Double Positive Medium

Double positive mediums (DPS) do occur in nature, such as naturally occurring dielectrics. Permittivity and magnetic permeability are both positive and wave propagation is in the forward direction. Artificial materials have been fabricated which combine DPS, ENG and MNG properties.

Bi-isotropic and Bianisotropic

Categorizing metamaterials into double or single negative, or double positive, normally assumes that the metamaterial has independent electric and magnetic responses described by ε and μ. However, in many cases, the electric field causes magnetic polarization, while the magnetic field induces electrical polarization, known as magnetoelectric coupling. Such media are denoted as bi-isotropic. Media that exhibit magnetoelectric coupling and that are anisotropic (which is the case for many metamaterial structures), are referred to as bi-anisotropic.

Four material parameters are intrinsic to magnetoelectric coupling of bi-isotropic media. They are the electric (E) and magnetic (H) field strengths, and electric (D) and magnetic (B) flux densities. These parameters are ε, μ, κ and χ or permittivity, permeability, strength of chirality, and the Tellegen parameter respectively. In this type of media, material parameters do not vary with changes along a rotated coordinate system of measurements. In this sense they are invariant or scalar.

The intrinsic magnetoelectric parameters, κ and χ, affect the phase of the wave. The effect of the chirality parameter is to split the refractive index. In isotropic media this results in wave propagation only if ε and μ have the same sign. In bi-isotropic media with χ assumed to be zero, and κ a non-zero value, different results appear. Either a backward wave or a forward wave can occur. Alternatively, two forward waves or two backward waves can occur, depending on the strength of the chirality parameter.

Chiral

Chiral metamaterials are constructed from chiral materials in which the effective parameter k is non-zero. This is a potential source of confusion as the metamaterial literature includes two conflicting uses of the terms *left-* and *right-handed*. The first refers to one of the two circularly polarized waves that are the propagating modes in chiral media. The second relates to the triplet of electric field, magnetic field and Poynting vector that arise in negative refractive index media, which in most cases are not chiral.

FSS Based

Frequency selective surface-based metamaterials block signals in one waveband and pass those at another waveband. They have become an alternative to fixed frequency metamaterials. They allow for optional changes of frequencies in a single medium, rather than the restrictive limitations of a fixed frequency response.

Other Types

Elastic

These metamaterials use different parameters to achieve a negative index of refraction in materials that are not electromagnetic. Furthermore, "a new design for elastic metamaterials that can behave either as liquids or solids over a limited frequency range may enable new applications based on the control of acoustic, elastic and seismic waves." They are also called mechanical metamaterials.

Acoustic

Acoustic metamaterials control, direct and manipulate sound in the form of sonic, infrasonic or ultrasonic waves in gases, liquids and solids. As with electromagnetic waves, sonic waves can exhibit negative refraction.

Control of sound waves is mostly accomplished through the bulk modulus β, mass density ρ and chirality. The bulk modulus and density are analogs of permittivity and permeability in electromagnetic metamaterials. Related to this is the mechanics of sound wave propagation in a lattice structure. Also materials have mass and intrinsic degrees of stiffness. Together, these form a resonant system and the mechanical (sonic) resonance may be excited by appropriate sonic frequencies (for example audible pulses).

Structural

Structural metamaterials provide properties such as crushability and light weight. Using projection micro-stereolithography, microlattices can be created using forms much like trusses and girders. Materials four orders of magnitude stiffer than conventional aerogel, but with the same density have been created. Such materials can withstand a load of at least 160,000 times their own weight by over-constraining the materials.

A ceramic nanotruss metamaterial can be flattened and revert to its original state.

Nonlinear

Metamaterials may be fabricated that include some form of nonlinear media, whose properties change with the power of the incident wave. Nonlinear media are essential for nonlinear optics. Most optical materials have a relatively weak response, meaning that their properties change by only a small amount for large changes in the intensity of the electromagnetic field. The local electromagnetic fields of the inclusions in nonlinear metamaterials can be much larger than the average value of the field. Besides, remarkable nonlinear effects have been predicted and observed if the metamaterial effective dielectric permittivity is very small (epsilon-near-zero media). In addition, exotic properties such as a negative refractive index, create opportunities to tailor the phase matching conditions that must be satisfied in any nonlinear optical structure.

Frequency Bands

Terahertz

Terahertz metamaterials interact at terahertz frequencies, usually defined as 0.1 to 10 THz. Terahertz radiation lies at the far end of the infrared band, just after the end of the microwave band. This corresponds to millimeter and submillimeter wavelengths between the 3 mm (EHF band) and 0.03 mm (long-wavelength edge of far-infrared light).

Photonic

Photonic metamaterial interact with optical frequencies (mid-infrared). The sub-wavelength period distinguishes them from photonic band gap structures.

Tunable

Tunable metamaterials allow arbitrary adjustments to frequency changes in the refractive index. A tunable metamaterial expands beyond the bandwidth limitations in left-handed materials by constructing various types of metamaterials.

Plasmonic

Plasmonic metamaterials exploit surface plasmons, which are produced from the interaction of light with metal-dielectrics. Under specific conditions, the incident light couples with the surface plasmons to create self-sustaining, propagating electromagnetic waves known as surface plasmon polaritons.

Applications

Metamaterials are under consideration for many applications. Metamaterial antennas are commercially available.

In 2007, one researcher stated that for metamaterial applications to be realized, energy loss must be reduced, materials must be extended into three-dimensional isotropic materials and production techniques must be industrialized.

Antennas

Metamaterial antennas are a class of antennas that use metamaterials to improve performance. Demonstrations showed that metamaterials could enhance an antenna's radiated power. Materials that can attain negative permeability allow for properties such as small antenna size, high directivity and tunable frequency.

Absorber

A metamaterial absorber manipulates the loss components of metamaterials' permittivity and magnetic permeability, to absorb large amounts of electromagnetic radiation. This is a useful feature for photodetection and solar photovoltaic applications. Loss components are also relevant in applications of negative refractive index (photonic metamaterials, antenna systems) or transformation optics (metamaterial cloaking, celestial mechanics), but often are not utilized in these applications.

Skin depth engineering can be used in metamaterial absorbers in photovoltaic applications as well as other optoelectronic devices, where optimizing the device performance demands minimizing resistive losses and power consumption, such as photodetectors, laser diodes, and light emitting diodes.

Superlens

A *superlens* is a two or three-dimensional device that uses metamaterials, usually with negative refraction properties, to achieve resolution beyond the diffraction limit (ideally, infinite resolution). Such a behaviour is enabled by the capability of double-negative materials to yield negative phase velocity. The diffraction limit is inherent in conventional optical devices or lenses.

Cloaking Devices

Metamaterials are a potential basis for a practical cloaking device. The proof of principle was demonstrated on October 19, 2006. No practical cloaks are publicly known to exist.

Seismic Protection

Seismic metamaterials counteract the adverse effects of seismic waves on man-made structures.

Sound Filtering

Metamaterials textured with nanoscale wrinkles could control sound or light signals, such as changing a material's color or improving ultrasound resolution. Uses include nondestructive material testing, medical diagnostics and sound suppression. The materials can be made through a high-precision, multi-layer deposition process. The thickness of each layer can be controlled within a fraction of a wavelength. The material is then compressed, creating precise wrinkles whose spacing can cause scattering of selected frequencies.

Theoretical Models

All materials are made of atoms, which are dipoles. These dipoles modify light velocity by a factor n (the refractive index). In a split ring resonator the ring and wire units act as atomic dipoles: the wire acts as a ferroelectric atom, while the ring acts as an inductor L, while the open section acts as a capacitor C. The ring as a whole acts as an LC circuit. When the electromagnetic field passes through the ring, an induced current is created. The generated field is perpendicular to the light's magnetic field. The magnetic resonance results in a negative permeability; the refraction index is negative as well. (The lens is not truly flat, since the structure's capacitance imposes a slope for the electric induction.)

Several (mathematical) material models frequency response in DNGs. One of these is the Lorentz model, which describes electron motion in terms of a driven-damped, harmonic oscillator. The Debye relaxation model applies when the acceleration component of the Lorentz mathematical model is small compared to the other components of the equation. The Drude model applies when the restoring force component is negligible and the coupling coefficient is generally the plasma frequency. Other component distinctions call for the use of one of these models, depending on its polarity or purpose.

Three-dimensional composites of metal/non-metallic inclusions periodically/randomly embedded in a low permittivity matrix are usually modeled by analytical methods, including mixing formulas and scattering-matrix based methods. The particle is modeled by either an electric dipole parallel to the electric field or a pair of crossed electric and magnetic dipoles parallel to the electric and magnetic fields, respectively, of the applied wave. These dipoles are the leading terms in the multipole series. They are the only existing ones for a homogeneous sphere, whose polarizability can be easily obtained from the Mie scattering coefficients. In general, this procedure is known as the "point-dipole approximation", which is a good approximation for metamaterials consisting of composites of electrically small spheres. Merits of these methods include low calculation cost and mathematical simplicity.

Other first principles techniques for analyzing regular 3D lattices of scatterers may be found in Computing photonic band structure

Institutional Networks

MURI

The Multidisciplinary University Research Initiative (MURI) encompasses dozens of Universities and a few government organizations. Participating universities include UC Berkeley, UC Los Angeles, UC San Diego, Massachusetts Institute of Technology, and Imperial College in London, UK. The sponsors are Office of Naval Research and the Defense Advanced Research Project Agency.

MURI supports research that intersects more than one traditional science and engineering discipline to accelerate both research and translation to applications. As of 2009, 69 academic institutions were expected to participate in 41 research efforts.

Metamorphose

The Virtual Institute for Artificial Electromagnetic Materials and Metamaterials "Metamorphose VI AISBL" is an international association to promote artificial electromagnetic materials and metamaterials. It organizes scientific conferences, supports specialized journals, creates and manages research programs, provides training programs (including PhD and training programs for industrial partners); and technology transfer to European Industry.

Nanomaterials

Nanomaterials describe, in principle, materials of which a single unit is sized (in at least one dimension) between 1 and 1000 nanometres (10^{-9} meter) but is usually 1—100 nm (the usual definition of nanoscale).

Nanomaterials research takes a materials science-based approach to nanotechnology, leveraging advances in materials metrology and synthesis which have been developed in support of microfabrication research. Materials with structure at the nanoscale often have unique optical, electronic, or mechanical properties.

Nanomaterials are slowly becoming commercialized and beginning to emerge as commodities.

Types

Natural Nanomaterials

Biological systems often feature natural, functional nanomaterials. The structure of foraminifera (mainly chalk) and viruses (protein, capsid), the wax crystals covering a lotus or nasturtium leaf, spider and spider-mite silk, the blue hue of tarantulas, the "spatulae" on the bottom of gecko feet, some butterfly wing scales, natural colloids (milk, blood), horny materials (skin, claws, beaks, feathers, horns, hair), paper, cotton, nacre, corals, and even our own bone matrix are all natural *organic* nanomaterials.

Natural *inorganic* nanomaterials occur through crystal growth in the diverse chemical conditions of the Earth's crust. For example, clays display complex nanostructures due to anisotropy of their underlying crystal structure, and volcanic activity can give rise to opals, which are an instance of a naturally occurring photonic crystals due to their nanoscale structure. Fires represent particularly complex reactions and can produce pigments, cement, fumed silica etc.

For the past decade, the chemical and physical properties of fullerenes have been a hot topic in the field of research and development, and are likely to continue to be for a long time. In April 2003, fullerenes were under study for potential medicinal use: binding specific antibiotics to the structure of resistant bacteria and even target certain types of cancer cells such as melanoma. The October 2005 issue of Chemistry and Biology contains an article describing the use of fullerenes as light-activated antimicrobial agents. In the field of nanotechnology, heat resistance and superconductivity are among the properties attracting intense research.

A common method used to produce fullerenes is to send a large current between two nearby graphite electrodes in an inert atmosphere. The resulting carbon plasma arc between the electrodes cools into sooty residue from which many fullerenes can be isolated.

There are many calculations that have been done using ab-initio Quantum Methods applied to fullerenes. By DFT and TDDFT methods one can obtain IR, Raman and UV spectra. Results of such calculations can be compared with experimental results.

Graphene Nanostructures

2D materials are crystalline materials consisting of a two-dimensional single layer of atoms. The most important representative graphene was discovered in 2004. Other 2D materials based on other elements have since been reported.

Box-shaped graphene (BSG) nanostructure is an example of 3D nanomaterial. BSG nanostructure has appeared after mechanical cleavage of pyrolytic graphite. This nanostructure is a multi-layer system of parallel hollow nanochannels located along the surface and having quadrangular cross-section. The thickness of the channel walls is approximately equal to 1 nm. The typical width of channel facets makes about 25 nm.

Nanoparticles

Inorganic nanomaterials, (e.g. quantum dots, nanowires and nanorods) because of their interesting optical and electrical properties, could be used in optoelectronics. Furthermore, the optical and electronic properties of nanomaterials which depend on their size and shape can be tuned via synthetic techniques. There are the possibilities to use those materials in organic material based optoelectronic devices such as Organic solar cells, OLEDs etc. The operating principles of such devices are governed by photoinduced processes like electron transfer and energy transfer. The performance of the devices depends on the efficiency of the photoinduced process responsible for their functioning. Therefore, better understanding of those photoinduced processes in organic/inorganic nanomaterial composite systems is necessary in order to use them in organic optoelectronic devices.

Nanoparticles or nanocrystals made of metals, semiconductors, or oxides are of particular interest for their mechanical, electrical, magnetic, optical, chemical and other properties. Nanoparticles have been used as quantum dots and as chemical catalysts such as nanomaterial-based catalysts. Recently, a range of nanoparticles are extensively investigated for biomedical applications including tissue engineering, drug delivery, biosensor.

Nanoparticles are of great scientific interest as they are effectively a bridge between bulk materials and atomic or molecular structures. A bulk material should have constant physical properties regardless of its size, but at the nano-scale this is often not the case. Size-dependent properties are observed such as quantum confinement in semiconductor particles, surface plasmon resonance in some metal particles and superparamagnetism in magnetic materials.

Nanoparticles exhibit a number of special properties relative to bulk material. For example, the bending of bulk copper (wire, ribbon, etc.) occurs with movement of copper atoms/clusters at about the 50 nm scale. Copper nanoparticles smaller than 50 nm are considered super hard materials that do not exhibit the same malleability and ductility as bulk copper. The change in properties is not always desirable. Ferroelectric materials smaller than 10 nm can switch their magnetisation direction using room temperature thermal energy, thus making them useless for memory storage. Suspensions of nanoparticles are possible because the interaction of the particle surface with the solvent is strong enough to overcome differences in density, which usually result in a material either sinking or floating in a liquid. Nanoparticles often have unexpected visual properties because they are small enough to confine their electrons and produce quantum effects. For example, gold nanoparticles appear deep red to black in solution.

The often very high surface area to volume ratio of nanoparticles provides a tremendous driving force for diffusion, especially at elevated temperatures. Sintering is possible at lower temperatures and over shorter durations than for larger particles. This theoretically does not affect the density of the final product, though flow difficulties and the tendency of nanoparticles to agglomerate do complicate matters. The surface effects of nanoparticles also reduces the incipient melting temperature.

Nanozymes

Nanozymes are nanomaterials with enzyme-like characteristics. They are an emerging type of artificial enzyme, which have been used for wide applications in such as biosensing, bioimaging, tumor diagnosis and therapy, antibiofouling, etc.

Synthesis

The goal of any synthetic method for nanomaterials is to yield a material that exhibits properties that are a result of their characteristic length scale being in the nanometer range (~1 – 100 nm). Accordingly, the synthetic method should exhibit control of size in this range so that one property or another can be attained. Often the methods are divided into two main types "Bottom Up" and "Top Down."

Bottom up Methods

Bottom up methods involve the assembly of atoms or molecules into nanostructured arrays. In these methods the raw material sources can be in the form of gases, liquids or solids. The latter requiring some sort of disassembly prior to their incorporation onto a nanostructure. Bottom methods generally fall into two categories: chaotic and controlled.

Chaotic Processes

Chaotic processes involve elevating the constituent atoms or molecules to a chaotic state and then suddenly changing the conditions so as to make that state unstable. Through the clever manipulation of any number of parameters, products form largely as a result of the insuring kinetics. The collapse from the chaotic state can be difficult or impossible to control and so ensemble statistics often govern the resulting size distribution and average size. Accordingly, nanoparticle formation is controlled through manipulation of the end state of the products.

Examples of Chaotic Processes are: Laser ablation, Exploding wire, Arc, Flame pyrolysis, Combustion, Precipitation synthesis techniques.

Controlled Processes

Controlled Processes involve the controlled delivery of the constituent atoms or molecules to the site(s) of nanoparticle formation such that the nanoparticle can grow to a prescribed sizes in a controlled manner. Generally the state of the constituent atoms or molecules are never far from that needed for nanoparticle formation. Accordingly, nanoparticle formation is controlled through the control of the state of the reactants.

Examples of controlled processes are self-limiting growth solution, self-limited chemical vapor deposition, shaped pulse femtosecond laser techniques, and molecular beam epitaxy.

Characterization

Novel effects can occur in materials when structures are formed with sizes comparable to any one of many possible length scales, such as the de Broglie wavelength of electrons, or the optical wavelengths of high energy photons. In these cases quantum mechanical effects can dominate material properties. One example is quantum confinement where the electronic properties of solids are altered with great reductions in particle size. The optical properties of nanoparticles, e.g. fluorescence, also become a function of the particle diameter. This effect does not come into play by going from macrosocopic to micrometer dimensions, but becomes pronounced when the nanometer scale is reached.

In addition to optical and electronic properties, the novel mechanical properties of many nanomaterials is the subject of nanomechanics research. When added to a bulk material, nanoparticles can strongly influence the mechanical properties of the material, such as the stiffness or elasticity. For example, traditional polymers can be reinforced by nanoparticles (such as carbon nanotubes) resulting in novel materials which can be used as lightweight replacements for metals. Such composite materials may enable a weight reduction accompanied by an increase in stability and improved functionality.

Finally, nanostructured materials with small particle size such as zeolites, and asbestos, are used as catalysts in a wide range of critical industrial chemical reactions. The further development of such catalysts can form the basis of more efficient, environmentally friendly chemical processes.

The first observations and size measurements of nano-particles were made during the first decade of the 20th century. Zsigmondy made detailed studies of gold sols and other nanomaterials with sizes down to 10 nm and less. He published a book in 1914. He used an ultramicroscope that employs a *dark field* method for seeing particles with sizes much less than light wavelength.

There are traditional techniques developed during 20th century in Interface and Colloid Science for characterizing nanomaterials. These are widely used for *first generation* passive nanomaterials specified in the next section.

These methods include several different techniques for characterizing particle size distribution. This characterization is imperative because many materials that are expected to be nano-sized are actually aggregated in solutions. Some of methods are based on light scattering. Others apply ultrasound, such as ultrasound attenuation spectroscopy for testing concentrated nano-dispersions and microemulsions.

There is also a group of traditional techniques for characterizing surface charge or zeta potential of nano-particles in solutions. This information is required for proper system stabilzation, preventing its aggregation or flocculation. These methods include microelectrophoresis, electrophoretic light scattering and electroacoustics. The last one, for instance colloid vibration current method is suitable for characterizing concentrated systems.

Uniformity

The chemical processing and synthesis of high performance technological components for the private, industrial and military sectors requires the use of high purity ceramics, polymers, glass-ceramics and material composites. In condensed bodies formed from fine powders, the irregular sizes and shapes of nanoparticles in a typical powder often lead to non-uniform packing morphologies that result in packing density variations in the powder compact.

Uncontrolled agglomeration of powders due to attractive van der Waals forces can also give rise to in microstructural inhomogeneities. Differential stresses that develop as a result of non-uniform drying shrinkage are directly related to the rate at which the solvent can be removed, and thus highly dependent upon the distribution of porosity. Such stresses have been associated with a plastic-to-brittle transition in consolidated bodies, and can yield to crack propagation in the unfired body if not relieved.

In addition, any fluctuations in packing density in the compact as it is prepared for the kiln are often amplified during the sintering process, yielding inhomogeneous densification. Some pores and other structural defects associated with density variations have been shown to play a detrimental role in the sintering process by growing and thus limiting end-point densities. Differential stresses arising from inhomogeneous densification have also been shown to result in the propagation of internal cracks, thus becoming the strength-controlling flaws.

It would therefore appear desirable to process a material in such a way that it is physically uniform with regard to the distribution of components and porosity, rather than using particle size distributions which will maximize the green density. The containment of a uniformly dispersed assembly of strongly interacting particles in suspension requires total control over particle-particle interactions. It should be noted here that a number of dispersants such as ammonium citrate (aqueous) and imidazoline or oleyl alcohol (nonaqueous) are promising solutions as possible additives for enhanced dispersion and deagglomeration. Monodisperse nanoparticles and colloids provide this potential.

Monodisperse powders of colloidal silica, for example, may therefore be stabilized sufficiently to ensure a high degree of order in the colloidal crystal or polycrystalline colloidal solid which results from aggregation. The degree of order appears to be limited by the time and space allowed for longer-range correlations to be established. Such defective polycrystalline colloidal structures would appear to be the basic elements of sub-micrometer colloidal materials science, and, therefore, provide the first step in developing a more rigorous understanding of the mechanisms involved in microstructural evolution in high performance materials and components.

Legal Definition

On 18 October 2011, the European Commission adopted the following definition of a nanomaterial: "A natural, incidental or manufactured material containing particles, in an unbound state or as an aggregate or as an agglomerate and where, for 50% or more of the particles in the number size distribution, one or more external dimensions is in the size range 1 nm – 100 nm. In specific cases and where warranted by concerns for the environment, health, safety or competitiveness the number size distribution threshold of 50% may be replaced by a threshold between 1 and 50%."

However, this differs from the definition adopted by the International Organization for Standardization (ISO), which is: "Material with any external dimension in the nanoscale or having internal structure in the nanoscale."

"Nanoscale" is, in turn, defined as: "Size range from approximately 1 nm to 100 nm."

It is not currently known which of these, if any, will prevail in courts of law.

Safety

Health Impact

While nanomaterials and nanotechnologies are expected to yield numerous health and health care advances, such as more targeted methods of delivering drugs, new cancer therapies, and methods of early detection of diseases, they also may have unwanted effects. Increased rate of absorption is the main concern associated with manufactured nanoparticles.

When materials are made into nanoparticles, their surface area to volume ratio increases. The greater specific surface area (surface area per unit weight) may lead to increased rate of absorption through the skin, lungs, or digestive tract and may cause unwanted effects to the lungs as well as other organs. However, the particles must be absorbed in sufficient quantities in order to pose health risks.

Nanoparticles created adventitiously (e.g., through the rubbing of prostheses) have long been known to be a health hazard, but as the use of nanomaterials increases worldwide, concerns for worker and user safety are mounting. To address such concerns, the Swedish Karolinska Institute conducted a study in which various nanoparticles were introduced to human lung epithelial cells.

The results, released in 2008, showed that iron oxide nanoparticles caused little DNA damage and were non-toxic. Zinc oxide nanoparticles were slightly worse. Titanium dioxide caused only DNA damage. Carbon nanotubes caused DNA damage at low levels. Copper oxide was found to be the worst offender, and was the only nanomaterial identified by the researchers as a clear health risk. Though nanomaterials are not confirmed as a health risk to workers who produce them, NIOSH recommends that exposure precautions and personal protective equipment be used to protect workers until the risks of nanomaterial manufacture are better understood.

Occupational Safety

Beyond normal chemical safety practices such as using personal protective equipment, The United States National Institute for Occupational Safety and Health recommends the use of HEPA filtration on local exhaust ventilation such as fume hoods. General laboratory ventilation is often not sufficient to effectively clear nanomaterials released into the general room air over a 30-minute period; therefore, researchers should leave the hood fan on even after synthesis is complete. Nanomaterials in dry powder form have the most risk of airborne dust generation and dermal contact, followed by liquid suspensions and then by nanomaterials embedded in a solid matrix.

Nanoparticles behave differently than other similarly sized particles. It is therefore necessary to develop specialized approaches to testing and monitoring their effects on human health and on the environment. The OECD Chemicals Committee has established the Working Party on Manufactured Nanomaterials to address this issue and to study the practices of OECD member countries in regards to nanomaterial safety.

Programmable Matter

Programmable matter is matter which has the ability to change its physical properties (shape, density, moduli, conductivity, optical properties, etc.) in a programmable fashion, based upon user input or autonomous sensing. Programmable matter is thus linked to the concept of a material which inherently has the ability to perform information processing.

History

Programmable matter is a term originally coined in 1991 by Toffoli and Margolus to refer to an ensemble of fine-grained computing elements arranged in space (Toffoli & Margolus 1991). Their paper describes a computing substrate that is composed of fine-grained compute nodes distributed throughout space which communicate using only nearest neighbor interactions. In this context, programmable matter refers to compute models similar to cellular automata and lattice gas automata (Rothman & Zaleski 1997). The CAM-8 architecture is an example hardware realization of this model. This function is also known as "digital referenced areas" (DRA) in some forms of self-replicating machine science.

In the early 1990s, there was a significant amount of work in reconfigurable modular robotics with a philosophy similar to programmable matter.

As semiconductor technology, nanotechnology, and self-replicating machine technology have advanced, the use of the term programmable matter has changed to reflect the fact that it is possible to build an ensemble of elements which can be "programmed" to change their physical properties in reality, not just in simulation. Thus, programmable matter has come to mean "any bulk substance which can be programmed to change its physical properties."

In the summer of 1998, in a discussion on artificial atoms and programmable matter, Wil McCarthy and G. Snyder coined the term "quantum wellstone" (or simply "wellstone") to describe this hypothetical but plausible form of programmable matter. McCarthy has used the term in his fiction.

In 2002, Seth Goldstein and Todd Mowry started the claytronics project at Carnegie Mellon University to investigate the underlying hardware and software mechanisms necessary to realize programmable matter.

In 2004, the DARPA Information Science and Technology group (ISAT) examined the potential of programmable matter. This resulted in the 2005–2006 study "Realizing Programmable Matter", which laid out a multi-year program for the research and development of programmable matter.

In 2007, programmable matter was the subject of a DARPA research solicitation and subsequent program.

Approaches

A 'simple' programmable matter where the programmable element is external to the material itself. Magnetized non-Newtonian fluid, forming support columns which resist impacts and sudden pressure.

In one school of thought the programming could be external to the material and might be achieved by the "application of light, voltage, electric or magnetic fields, etc." (McCarthy 2006). For example, a liquid crystal display is a form of programmable matter. A second school of thought is that the individual units of the ensemble can compute and the result of their computation is a change in the ensemble's physical properties. An example of this more ambitious form of programmable matter is claytronics.

There are many proposed implementations of programmable matter. Scale is one key differentiator between different forms of programmable matter. At one end of the spectrum reconfigurable modular robotics pursues a form of programmable matter where the individual units are in the centimeter size range. At the nanoscale end of the spectrum there are a tremendous number of different bases for programmable matter, ranging from shape changing molecules to quantum dots. Quantum dots are in fact often referred to as artificial atoms. In the micrometer to sub-millimeter range examples include MEMS-based units, cells created using synthetic biology, and the utility fog concept.

An important sub-group of programmable matter are robotic materials, which combine the structural aspects of a composite with the affordances offered by tight integration of sensors, actuators, computation and communication, while foregoing reconfiguration by particle motion.

Examples

There are many conceptions of programmable matter, and thus many discrete avenues of research using the name. Below are some specific examples of programmable matter.

"Simple"

These include materials that can change their properties based on some input, but do not have the ability to do complex computation by themselves.

Complex Fluids

The physical properties of several complex fluids can be modified by applying a current or voltage, as is the case with liquid crystals.

Metamaterials

Metamaterials are artificial composites that can be controlled to react in ways that do not occur in nature. One example developed by David Smith and then by John Pendry and David Schuri is of a material that can have its index of refraction tuned so that it can have a different index of refraction at different points in the material. If tuned properly this could result in an "invisibility cloak."

A further example of programmable -mechanical- metamaterial is presented by Bergamini et al. Here, a pass band within the phononic bandgap is introduced, by exploiting variable stiffness of piezoelectric elements linking aluminum stubs to the aluminum plate to create a phononic crystal as in the work of Wu et al. The piezoelectric elements are shunted to ground over synthetic inductors. Around the resonance frequency of the LC circuit formed by the piezoelectric and the inductors, the piezoelectric elements exhibit near zero stiffness, thus effectively disconnecting the stubs from the plate. This is considered an example of programmable mechanical metamaterial.

Shape-changing Molecules

An active area of research is in molecules that can change their shape, as well as other properties, in response to external stimuli. These molecules can be used individually or en masse to form new kinds of materials. For example, J Fraser Stoddart's group at UCLA has been developing molecules that can change their electrical properties.

Electropermanent Magnets

An electropermanent magnet is a type of magnet which consists of both an electromagnet and a dual material permanent magnet, in which the magnetic field produced by the electromagnet is used to change the magnetization of the permanent magnet. The permanent magnet consists of magnetically hard and soft materials, of which only the soft material can have its magnetization changed. When the magnetically soft and hard materials have opposite magnetizations the magnet has no net field, and when they are aligned the magnet displays magnetic behaviour.

They allow creating controllable permanent magnets where the magnetic effect can be maintained without requiring a continuous supply of electrical energy. For these reasons, electropermanent magnets are essential components of the research studies aiming to build programmable magnets that can give rise to self-building structures.

Robotics-based Approaches

Self-reconfiguring Modular Robotics

Self-reconfiguring modular robotics is a field of robotics in which a group of basic robot modules work together to dynamically form shapes and create behaviours suitable for many tasks. Like Programmable matter SRCMR aims to offer significant improvement to any kind of objects or system by introducing many new possibilities for example: 1. Most important is the incredible flexibility that comes from the ability to change the physical structure and behavior of a solution by changing the software that controls modules. 2. The ability to self-repair by automatically replacing a broken module will make SRCMR solution incredibly resilient. 3. Reducing the environmental foot print by reusing the same modules in many different solutions. Self-Reconfiguring Modular Robotics enjoys a vibrant and active research community.

Claytronics

Claytronics is an emerging field of engineering concerning reconfigurable nanoscale robots ('claytronic atoms', or *catoms*) designed to form much larger scale machines or mechanisms. The catoms will be sub-millimeter computers that will eventually have the ability to move around, communicate with other computers, change color, and electrostatically connect to other catoms to form different shapes.

Cellular Automata

Cellular automata are a useful concept to abstract some of the concepts of discrete units interacting to give a desired overall behavior.

Quantum Wells

Quantum wells can hold one or more electrons. Those electrons behave like artificial atoms which, like real atoms, can form covalent bonds, but these are extremely weak. Because of their larger sizes, other properties are also widely different.

Synthetic Biology

Synthetic biology is a field that aims to engineer cells with "novel biological functions." Such cells are usually used to create larger systems (e.g., biofilms) which can be "programmed" utilizing synthetic gene networks such as genetic toggle switches, to change their color, shape, etc. Such bioinspired approaches to materials production has been demonstrated, using self-assembling bacterial biofilm materials that can be programmed for specific functions, such as substrate adhesion, nanoparticle templating, and protein immobilization.

Quantum Dot

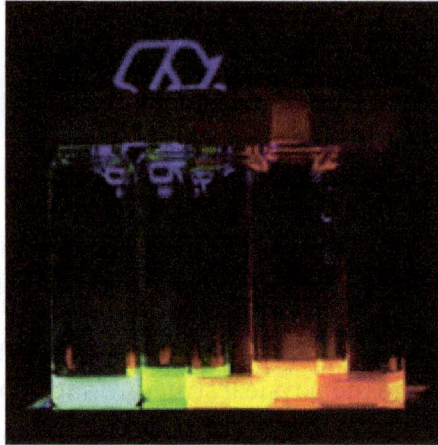

Colloidal quantum dots irradiated with a UV light. Different sized quantum dots emit different color light due to quantum confinement.

Quantum dots (QD) are very small semiconductor particles, only several nanometres in size, so small that their optical and electronic properties differ from those of larger particles. They are a central theme in nanotechnology. Many types of quantum dot will emit light of specific frequencies if electricity or light is applied to them, and these frequencies can be precisely tuned by changing the dots' size, shape and material, giving rise to many applications.

In the language of materials science, nanoscale semiconductor materials tightly confine either electrons or electron holes. Quantum dots are also sometimes referred to as artificial atoms, a term that emphasizes that a quantum dot is a single object with bound, discrete electronic states, as is the case with naturally occurring atoms or molecules.

Quantum dots exhibit properties that are intermediate between those of bulk semiconductors and those of discrete molecules. Their optoelectronic properties change as a function of both size and shape. Larger QDs (radius of 5–6 nm, for example) emit longer wavelengths resulting in emission colors such as orange or red. Smaller QDs (radius of 2–3 nm, for example) emit shorter wavelengths resulting in colors like blue and green, although the specific colors and sizes vary depending on the exact composition of the QD.

Because of their highly tunable properties, QDs are of wide interest. Potential applications include transistors, solar cells, LEDs, diode lasers and second-harmonic generation, quantum computing, and medical imaging. Additionally, their small size allows for QDs to be suspended in solution which leads to possible uses in inkjet printing and spin-coating. These processing techniques result in less-expensive and less time consuming methods of semiconductor fabrication.

Production

Quantum Dots with gradually stepping emission from violet to deep red are being produced in a kg scale at PlasmaChem GmbH

There are several ways to prepare quantum dots, the principal ones involving colloids.

Colloidal Synthesis

Colloidal semiconductor nanocrystals are synthesized from solutions, much like traditional chemical processes. The main difference is the product neither precipitates as a bulk solid nor remains dissolved. Heating the solution at high temperature, the precursors decompose forming monomers which then nucleate and generate nanocrystals. Temperature is a critical factor in determining optimal conditions for the nanocrystal growth. It must be high enough to allow for rearrangement and annealing of atoms during the synthesis process while being low enough to promote crystal growth. The concentration of monomers is another critical factor that has to be stringently controlled during nanocrystal growth. The growth process of nanocrystals can occur in two different regimes, "focusing" and "defocusing". At high monomer concentrations, the critical size (the size where nanocrystals neither grow nor shrink) is relatively small, resulting in growth of nearly all particles. In this regime, smaller particles grow faster than large ones (since larger crystals need more atoms to grow than small crystals) resulting in "focusing" of the size distribution to yield nearly monodisperse particles. The size focusing is optimal when the monomer concentration is kept such that the average nanocrystal size present is always slightly larger than the critical size. Over time, the monomer concentration diminishes, the critical size becomes larger than the average size present, and the distribution "defocuses".

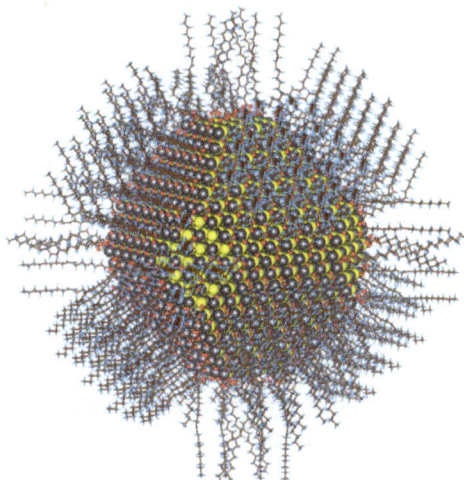

Ideallized image of colloidal nanoparticle of lead sulfide (selenide) with complete passivation by oleic acid,oleyl and hydroxyl (size ~5nm)

There are colloidal methods to produce many different semiconductors. Typical dots are made of binary compounds such as lead sulfide, lead selenide, cadmium selenide, cadmium sulfide, indium arsenide, and indium phosphide. Dots may also be made from ternary compounds such as cadmium selenide sulfide. These quantum dots can contain as few as 100 to 100,000 atoms within the quantum dot volume, with a diameter of ~ 10 to 50 atoms. This corresponds to about 2 to 10 nanometers, and at 10 nm in diameter, nearly 3 million quantum dots could be lined up end to end and fit within the width of a human thumb.

Large batches of quantum dots may be synthesized via colloidal synthesis. Due to this scalability and the convenience of benchtop conditions, colloidal synthetic methods are promising for commercial applications. It is acknowledged to be the least toxic of all the different forms of synthesis.

Plasma Synthesis

Plasma synthesis has evolved to be one of the most popular gas-phase approaches for the production of quantum dots, especially those with covalent bonds. For example, silicon (Si) and germanium (Ge) quantum dots have been synthesized by using nonthermal plasma. The size, shape, surface and composition of quantum dots can all be controlled in nonthermal plasma. Doping that seems quite challenging for quantum dots has also been realized in plasma synthesis. Quantum dots synthesized by plasma are usually in the form of powder, for which surface modification may be carried out. This can lead to excellent dispersion of quantum dots in either organic solvents or water (i. e., colloidal quantum dots).

Fabrication

- Self-assembled quantum dots are typically between 5 and 50 nm in size. Quantum dots defined by lithographically patterned gate electrodes, or by etching on two-dimensional electron gases in semiconductor heterostructures can have lateral dimensions between 20 and 100 nm.

- Some quantum dots are small regions of one material buried in another with a larger band gap. These can be so-called core–shell structures, e.g., with CdSe in the core and ZnS in the shell or from special forms of silica called ormosil.

- Quantum dots sometimes occur spontaneously in quantum well structures due to monolayer fluctuations in the well's thickness.

- Self-assembled quantum dots nucleate spontaneously under certain conditions during molecular beam epitaxy (MBE) and metallorganic vapor phase epitaxy (MOVPE), when a material is grown on a substrate to which it is not lattice matched. The resulting strain produces coherently strained islands on top of a two-dimensional wetting layer. This growth mode is known as Stranski–Krastanov growth. The islands can be subsequently buried to form the quantum dot. This fabrication method has potential for applications in quantum cryptography (i.e. single photon sources) and quantum computation. The main limitations of this method are the cost of fabrication and the lack of control over positioning of individual dots.

- Individual quantum dots can be created from two-dimensional electron or hole gases present in remotely doped quantum wells or semiconductor heterostructures called lateral quantum dots. The sample surface is coated with a thin layer of resist. A lateral pattern is then defined in the resist by electron beam lithography. This pattern can then be transferred to the electron or hole gas by etching, or by depositing metal electrodes (lift-off process) that allow the application of external voltages between the electron gas and the electrodes. Such quantum dots are mainly of interest for experiments and applications involving electron or hole transport, i.e., an electrical current.

- The energy spectrum of a quantum dot can be engineered by controlling the geometrical size, shape, and the strength of the confinement potential. Also, in contrast to atoms, it is relatively easy to connect quantum dots by tunnel barriers to conducting leads, which allows the application of the techniques of tunneling spectroscopy for their investigation.

The quantum dot absorption features correspond to transitions between discrete,three-dimensional particle in a box states of the electron and the hole, both confined to the same nanometer-size box.These discrete transitions are reminiscent of atomic spectra and have resulted in quantum dots also being called *artificial atoms*.

- Confinement in quantum dots can also arise from electrostatic potentials (generated by external electrodes, doping, strain, or impurities).

- CMOS technology can be employed to fabricate silicon quantum dots. Ultra small (L=20 nm, W=20 nm) CMOS transistors behave as single electron quantum dots when operated at cryogenic temperature over a range of $-269\ ^\circ$C (4 K) to about $-258\ ^\circ$C (15 K). The transistor displays Coulomb blockade due to progressive charging of electrons one by one. The number of electrons confined in the channel is driven by the gate voltage, starting from an occupation of zero electrons, and it can be set to 1 or many.

Viral Assembly

genetically engineered M13 bacteriophage viruses allow preparation of quantum dot biocomposite structures. It had previously been shown that genetically engineered viruses can recognize specific semiconductor surfaces through the method of selection by combinatorial phage display. Additionally, it is known that liquid crystalline structures of wild-type viruses (Fd, M13, and TMV) are adjustable by controlling the solution concentrations, solution ionic strength, and the external magnetic field applied to the solutions. Consequently, the specific recognition properties of the virus can be used to organize inorganic nanocrystals, forming ordered arrays over the length scale defined by liquid crystal formation. Using this information, Lee et al. (2000) were able to create self-assembled, highly oriented, self-supporting films from a phage and ZnS precursor solution. This system allowed them to vary both the length of bacteriophage and the type of inorganic material through genetic modification and selection.

Electrochemical Assembly

Highly ordered arrays of quantum dots may also be self-assembled by electrochemical techniques. A template is created by causing an ionic reaction at an electrolyte-metal interface which results in the spontaneous assembly of nanostructures, including quantum dots, onto the metal which is then used as a mask for mesa-etching these nanostructures on a chosen substrate.

Bulk-manufacture

Quantum dot manufacturing relies on a process called "high temperature dual injection" which has been scaled by multiple companies for commercial applications that require large quantities (hundreds of kilograms to tonnes) of quantum dots. This reproducible production method can be applied to a wide range of quantum dot sizes and compositions.

The bonding in certain cadmium-free quantum dots, such as III-V-based quantum dots, is more covalent than that in II-VI materials, therefore it is more difficult to separate nanoparticle nucleation and growth via a high temperature dual injection synthesis. An alternative method of quantum dot synthesis, the "molecular seeding" process, provides a reproducible route to the production of high quality quantum dots in large volumes. The process utilises identical molecules of a molecular cluster compound as the nucleation sites for nanoparticle growth, thus avoiding the need for a high temperature injection step. Particle growth is maintained by the periodic addition of precursors at moderate temperatures until the desired particle size is reached. The molecular seeding process is not limited to the production of cadmium-free quantum dots; for example, the process can be used to synthesise kilogram batches of high quality II-VI quantum dots in just a few hours.

Another approach for the mass production of colloidal quantum dots can be seen in the transfer of the well-known hot-injection methodology for the synthesis to a technical continuous flow system. The batch-to-batch variations arising from the needs during the mentioned methodology can be overcome by utilizing technical components for mixing and growth as well as transport and temperature adjustments. For the production of CdSe based semiconductor nanoparticles this method has been investigated and tuned to production amounts of kg per month. Since the use of technical components allows for easy interchange in regards of maximum through-put and size, it can be further enhanced to tens or even hundreds of kilograms.

In 2011 a consortium of U.S. and Dutch companies reported a "milestone" in high volume quantum dot manufacturing by applying the traditional high temperature dual injection method to a flow system.

On January 23, 2013 Dow entered into an exclusive licensing agreement with UK-based Nanoco for the use of their low-temperature molecular seeding method for bulk manufacture of cadmium-free quantum dots for electronic displays, and on September 24, 2014 Dow commenced work on the production facility in South Korea capable of producing sufficient quantum dots for "millions of cadmium-free televisions and other devices, such as tablets". Mass production is due to commence in mid-2015. On 24 March 2015 Dow announced a partnership deal with LG Electronics to develop the use of cadmium free quantum dots in displays.

Heavy Metal-free Quantum Dots

In many regions of the world there is now a restriction or ban on the use of heavy metals in many household goods, which means that most cadmium based quantum dots are unusable for consumer-goods applications.

For commercial viability, a range of restricted, heavy metal-free quantum dots has been developed showing bright emissions in the visible and near infra-red region of the spectrum and have similar optical properties to those of CdSe quantum dots. Among these systems are InP/ZnS and CuInS/ZnS, for example.

Peptides are being researched as potential quantum dot material. Since peptides occur naturally in all organisms, such dots would likely be nontoxic and easily biodegraded.

Safety

Toxicity

Some quantum dots pose risks to human health and the environment under certain conditions. Notably, the studies on quantum dots toxicity are focused on cadmium containing particles and has yet to be demonstrated in animal models after physiologically relevant dosing. In vitro studies, based on cell cultures, on quantum dots (QD) toxicity suggests that their toxicity may derive from multiple factors including its physicochemical characteristics (size, shape, composition, surface functional groups, and surface charges) and environment. Assessing their potential toxicity is complex as these factors include properties such as QD size, charge, concentration, chemical composition, capping ligands, and also on their oxidative, mechanical and photolytic stability.

Many studies have focused on the mechanism of QD cytotoxicity using model cell cultures. It has been demonstrated that after exposure to ultraviolet radiation or oxidized by air CdSe QDs release free cadmium ions causing cell death. Group II-VI QDs also have been reported to induce the formation of reactive oxygen species after exposure to light, which in turn can damage cellular components such as proteins, lipids and DNA. Some studies have also demonstrated that addition of a ZnS shell inhibit the process of reactive oxygen species in CdSe QDs. Another aspect of QD toxicity is the process of their size dependent intracellular pathways that concentrate these particles in cellular organelles that are inaccessible by metal ions, which may result in unique patterns of cytotoxicity compared to their constituent metal ions. The reports of QD localization in the cell nucleus present additional modes of toxicity because they may induce DNA mutation, which in turn will propagate through future generation of cells causing diseases.

Although concentration of QDs in certain organelles have been reported in in vivo studies using animal models, interestingly, no alterations in animal behavior, weight, hematological markers or organ damage has been found through either histological or biochemical analysis. These finding have led scientists to believe that intracellular dose is the most important deterring factor for QD toxicity. Therefore, factors determining the QD endocytosis that determine the effective intracellular concentration, such as QD size, shape and surface chemistry determine their toxicity. Excretion of QDs through urine in animal models also have demonstrated via injecting radio-labeled ZnS capped CdSe QDs where the ligand shell was labelled with ^{99m}Tc. Though multiple other studies have concluded retention of QDs in cellular levels, exocytosis of QDs is still poorly studied in the literature.

While significant research efforts have broadened the understanding of toxicity of QDs, there are large discrepancies in the literature and questions still remains to be answered. Diversity of this class material as compared to normal chemical substances makes the assessment of their toxicity very challenging. As their toxicity may also be dynamic depending on the environmental factors such as pH level, light exposure and cell type, traditional methods of assessing toxicity of chemicals such as LD_{50} are not applicable for QDs. Therefore, researchers are focusing on introducing novel approaches and adapting existing methods to include this unique class of materials. Furthermore, novel strategies to engineer safer QDs are still under exploration by the scientific community. A recent novelty in the field is the discovery of carbon quantum dots, a new generation of optically-active nanoparticles potentially capable of replacing semiconductor QDs, but with the advantage of much lower toxicity.

Optical Properties

In semiconductors, light absorption generally leads to an electron being excited from the valence to the conduction band, leaving behind a hole. The electron and the hole can bind to each other to form an exciton. When this exciton recombines (i.e. the electron resumes its ground state), the exciton's energy can be emitted as light. This is called Fluorescence. In a simplified model, the energy of the emitted photon can be understood as the sum of the band gap energy between the highest occupied level and the lowest unoccupied energy level, the confinement energies of the hole and the excited electron, and the bound energy of the exciton (the electron-hole pair):

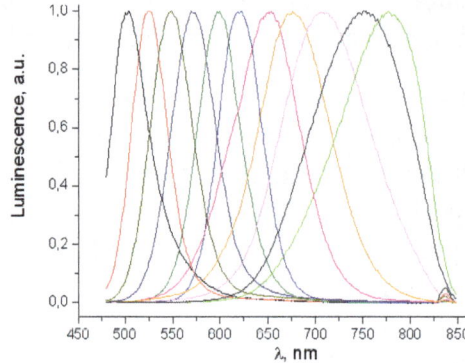

Fluorescence spectra of CdTe quantum dots of various sizes. Different sized quantum dots emit different color light due to quantum confinement.

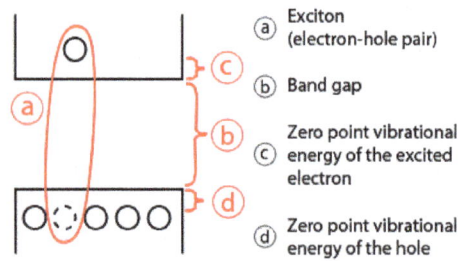

As the confinement energy depends on the quantum dot's size, both absorption onset and fluorescence emission can be tuned by changing the size of the quantum dot during its synthesis. The larger the dot, the redder (lower energy) its absorption onset and fluorescence spectrum. Conversely, smaller dots absorb and emit bluer (higher energy) light. Recent articles in *Nanotechnology* and in other journals have begun to suggest that the shape of the quantum dot may be a factor in the coloration as well, but as yet not enough information is available. Furthermore, it was shown that the lifetime of fluorescence is determined by the size of the quantum dot. Larger dots have more closely spaced energy levels in which the electron-hole pair can be trapped. Therefore, electron-hole pairs in larger dots live longer causing larger dots to show a longer lifetime.

To improve fluorescence quantum yield, quantum dots can be made with "shells" of a larger bandgap semiconductor material around them. The improvement is suggested to be due to the reduced access of electron and hole to non-radiative surface recombination pathways in some cases, but also due to reduced auger recombination in others.

Potential Applications

Quantum dots are particularly promising for optical applications due to their high extinction coefficient. They operate like a single electron transistor and show the Coulomb blockade effect. Quantum dots have also been suggested as implementations of qubits for quantum information processing.

Tuning the size of quantum dots is attractive for many potential applications. For instance, larger quantum dots have a greater spectrum-shift towards red compared to smaller dots, and exhibit less pronounced quantum properties. Conversely, the smaller particles allow one to take advantage of more subtle quantum effects.

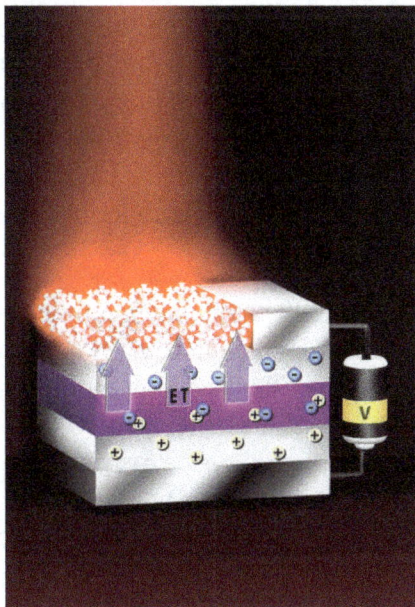

A device that produces visible light, through energy transfer from thin layers of quantum wells to crystals above the layers.

Being zero-dimensional, quantum dots have a sharper density of states than higher-dimensional structures. As a result, they have superior transport and optical properties. They have potential uses in diode lasers, amplifiers, and biological sensors. Quantum dots may be excited within a locally enhanced electromagnetic field produced by gold nanoparticles, which can then be observed from the surface plasmon resonance in the photoluminescent excitation spectrum of (CdSe)ZnS nanocrystals. High-quality quantum dots are well suited for optical encoding and multiplexing applications due to their broad excitation profiles and narrow/symmetric emission spectra. The new generations of quantum dots have far-reaching potential for the study of intracellular processes at the single-molecule level, high-resolution cellular imaging, long-term in vivo observation of cell trafficking, tumor targeting, and diagnostics.

CdSe nanocrystals are efficient triplet photosensitizers. Laser excitation of small CdSe nanoparticles enables the extraction of the excited state energy from the Quantum Dots into bulk solution, thus opening the door to a wide range of potential applications such as photodynamic therapy, photovoltaic devices, molecular electronics, and catalysis.

Computing

Quantum dot technology is potentially relevant to solid-state quantum computation. By applying small voltages to the leads, current through the quantum dot can be controlled and thereby precise measurements of the spin and other properties therein can be made. With several entangled quantum dots, or qubits, plus a way of performing operations, quantum calculations and the computers that would perform them might be possible.

Biology

In modern biological analysis, various kinds of organic dyes are used. However, as technology advances, greater flexibility in these dyes is sought. To this end, quantum dots have quickly filled in the role, being found to be superior to traditional organic dyes on several counts, one of the most immediately obvious being brightness (owing to the high extinction coefficient combined with a comparable quantum yield to fluorescent dyes) as well as their stability (allowing much less photobleaching). It has been estimated that quantum dots are 20 times brighter and 100 times more stable than traditional fluorescent reporters. For single-particle tracking, the irregular blinking of quantum dots is a minor drawback. However, there have been groups which have developed quantum dots which are essentially nonblinking and demonstrated their utility in single molecule tracking experiments.

The use of quantum dots for highly sensitive cellular imaging has seen major advances. The improved photostability of quantum dots, for example, allows the acquisition of many consecutive focal-plane images that can be reconstructed into a high-resolution three-dimensional image. Another application that takes advantage of the extraordinary photostability of quantum dot probes is the real-time tracking of molecules and cells over extended periods of time. Antibodies, streptavidin, peptides, DNA, nucleic acid aptamers, or small-molecule ligands can be used to target quantum dots to specific proteins on cells. Researchers were able to observe quantum dots in lymph nodes of mice for more than 4 months.

Semiconductor quantum dots have also been employed for in vitro imaging of pre-labeled cells. The ability to image single-cell migration in real time is expected to be important to several research areas such as embryogenesis, cancer metastasis, stem cell therapeutics, and lymphocyte immunology.

One application of quantum dots in biology is as donor fluorophores in Förster resonance energy transfer, where the large extinction coefficient and spectral purity of these fluorophores make them superior to molecular fluorophores It is also worth noting that the broad absorbance of QDs allows selective excitation of the QD donor and a minimum excitation of a dye acceptor in FRET-based studies. The applicability of the FRET model, which assumes that the Quantum Dot can be approximated as a point dipole, has recently been demonstrated

The use of quantum dots for tumor targeting under in vivo conditions employ two targeting schemes: active targeting and passive targeting. In the case of active targeting, quantum dots are functionalized with tumor-specific binding sites to selectively bind to tumor cells. Passive targeting uses the enhanced permeation and retention of tumor cells for the delivery of quantum dot probes. Fast-growing tumor cells typically have more permeable membranes than healthy cells, allowing the leakage of small nanoparticles into the cell body. Moreover, tumor cells lack an effective lymphatic drainage system, which leads to subsequent nanoparticle-accumulation.

Quantum dot probes exhibit in vivo toxicity. For example, CdSe nanocrystals are highly toxic to cultured cells under UV illumination, because the particles dissolve, in a process known as photolysis, to release toxic cadmium ions into the culture medium. In the absence of UV irradiation, however, quantum dots with a stable polymer coating have been found to be essentially nontoxic. Hydrogel encapsulation of quantum dots allows for quantum dots to be introduced into a stable aqueous solution, reducing the possibility of cadmium leakage. Then again, only little is known about the excretion process of quantum dots from living organisms.

In another potential application, quantum dots are being investigated as the inorganic fluorophore for intra-operative detection of tumors using fluorescence spectroscopy.

Delivery of undamaged quantum dots to the cell cytoplasm has been a challenge with existing techniques. Vector-based methods have resulted in aggregation and endosomal sequestration of quantum dots while electroporation can damage the semi-conducting particles and aggregate delivered dots in the cytosol. Via cell squeezing, quantum dots can be efficiently delivered without inducing aggregation, trapping material in endosomes, or significant loss of cell viability. Moreover, it has shown that individual quantum dots delivered by this approach are detectable in the cell cytosol, thus illustrating the potential of this technique for single molecule tracking studies.

Photovoltaic Devices

The tunable absorption spectrum and high extinction coefficients of quantum dots make them attractive for light harvesting technologies such as photovoltaics. Quantum dots may be able to increase the efficiency and reduce the cost of today's typical silicon photovoltaic cells. According to an experimental proof from 2004, quantum dots of lead selenide can produce more than one exciton from one high energy photon via the process of carrier multiplication or multiple exciton generation (MEG). This compares favorably to today's photovoltaic cells which can only manage one exciton per high-energy photon, with high kinetic energy carriers losing their energy as heat. Quantum dot photovoltaics would theoretically be cheaper to manufacture, as they can be made "using simple chemical reactions."

Quantum Dot Only Solar Cells

Aromatic self-assembled monolayers (SAMs) (e.g. 4-nitrobenzoic acid) can be used to improve the band alignment at electrodes for better efficiencies. This technique has provided a record power conversion efficiency (PCE) of 10.7%. The SAM is positioned between ZnO-PbS colloidal quantum dot (CQD) film junction to modify band alignment via the dipole moment of the constituent SAM molecule, and the band tuning may be modified via the density, dipole and the orientation of the SAM molecule.

Quantum Dot in Hybrid Solar Cells

Colloidal quantum dots are also used in inorganic/organic hybrid solar cells. These solar cells are attractive because of potentially their low-cost fabrication and relatively high efficiency. Incorporation of metal oxides, such as ZnO, TiO_2, and Nb_2O_5 nanomaterials into organic photovoltaics have been commercialized using full roll-to-roll processing. A 13.2% power conversion efficiency is claimed in Si nanowire/PEDOT:PSS hybrid solar cells.

Quantum Dot With Nanowire in Solar Cells

Another potential use involves capped single-crystal ZnO nanowires with CdSe quantum dots, immersed in mercaptopropionic acid as hole transport medium in order to obtain a QD-sensitized solar cell. The morphology of the nanowires allowed the electrons to have a direct pathway to the photoanode. This form of solar cell exhibits 50-60% internal quantum efficiencies.

Nanowires with quantum dot coatings on silicon nanowires (SiNW) and carbon quantum dots. The use of SiNWs instead of planar silicon enhances the antiflection properties of Si. The SiNW exhibits a light-trapping effect due to light trapping in the SiNW. This use of SiNWs in conjunction with carbon quantum dots resulted in a solar cell that reached 9.10% PCE.

Graphene quantum dots have also been blended with organic electronic materials to improve efficiency and lower cost in photovoltaic devices and organic light emitting diodes (OLEDs) in compared to graphene sheets. These graphene quantum dots were functionalized with organic ligands that experience photoluminescence from UV-Vis absorption.

Light Emitting Devices

Several methods are proposed for using quantum dots to improve existing light-emitting diode (LED) design, including "Quantum Dot Light Emitting Diode" (QD-LED) displays and "Quantum Dot White Light Emitting Diode" (QD-WLED) displays. Because Quantum dots naturally produce monochromatic light, they can be more efficient than light sources which must be color filtered. QD-LEDs can be fabricated on a silicon substrate, which allows them to be integrated onto standard silicon-based integrated circuits or microelectromechanical systems. Quantum dots are valued for displays, because they emit light in very specific gaussian distributions. This can result in a display with visibly more accurate colors. A conventional color liquid crystal display (LCD) is usually backlit by fluorescent lamps (CCFLs) or conventional white LEDs that are color filtered to produce red, green, and blue pixels. An improvement is using conventional blue-emitting LEDs as the light sources and converting part of the emitted light into *pure* green and red light by the appropriate quantum dots placed in front of the blue LED or using a quantum dot infused diffuser sheet in the backlight optical stack. This type of white light as the backlight of an LCD panel allows for the best color gamut at lower cost than a RGB LED combination using three LEDs.

The ability of QDs to precisely convert and tune a spectrum makes them attractive for LCD displays. Previous LCD displays can waste energy converting red-green poor, blue-yellow rich white light into a more balanced lighting. By using QDs, only the necessary colors for ideal images are contained in the screen. The result is a screen that is brighter, clearer, and more energy-efficient. The first commercial application of quantum dots was the Sony XBR X900A series of flat panel televisions released in 2013.

In June 2006, QD Vision announced technical success in making a proof-of-concept quantum dot display and show a bright emission in the visible and near infra-red region of the spectrum. A QD-LED integrated at a scanning microscopy tip was used to demonstrate fluorescence near-field scanning optical microscopy (NSOM) imaging.

Photodetector Devices

Quantum dot photodetectors (QDPs) can be fabricated either via solution-processing, or from conventional single-crystalline semiconductors. Conventional single-crystalline semiconductor QDPs are precluded from integration with flexible organic electronics due to the incompatibility of their growth conditions with the process windows required by organic semiconductors. On the other hand, solution-processed QDPs can be readily integrated with an almost infinite variety of substrates, and also postprocessed atop other integrated circuits. Such colloidal QDPs have potential applications in surveillance, machine vision, industrial inspection, spectroscopy, and fluorescent biomedical imaging.

Photocatalysts

Quantum dots also function as photocatalysts for the light driven chemical conversion of water into hydrogen as a pathway to solar fuel. In photocatalysis, electron hole pairs formed in the dot under band gap excitation drive redox reactions in the surrounding liquid. Generally, the photocatalytic activity of the dots is related to the particle size and its degree of quantum confinement. This is because the band gap determines the chemical energy that is stored in the dot in the excited state. An obstacle for the use of quantum dots in photocatalysis is the presence of surfactants on the surface of the dots. These surfactants (or ligands) interfere with the chemical reactivity of the dots by slowing down mass transfer and electron transfer processes. Also, quantum dots made of metal chalcogenides are chemically unstable under oxidizing conditions and undergo photo corrosion reactions.

Theory

Quantum dots are theoretically described as a point like, or a zero dimensional (0D) entity. Most of their properties depend on the dimensions, shape and materials of which QDs are made. Generally QDs present different thermodynamic properties from the bulk materials of which they are made. One of these effects is the Melting-point depression. Optical properties of spherical metallic QDs are well described by the Mie scattering theory.

Superalloy

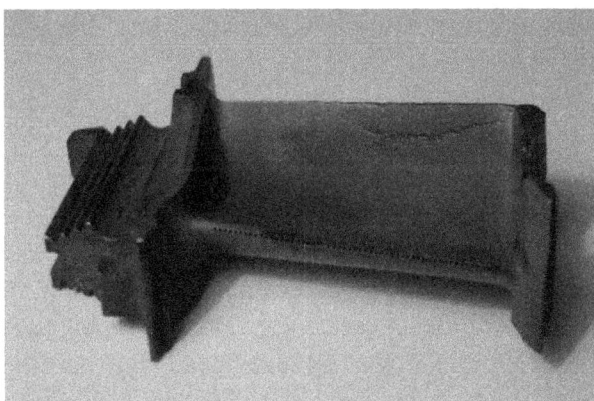

Nickel superalloy jet engine (RB199) turbine blade

A superalloy, or high-performance alloy, is an alloy that exhibits several key characteristics: excellent mechanical strength, resistance to thermal creep deformation, good surface stability, and resistance to corrosion or oxidation. The crystal structure is typically face-centered cubic austenitic. Examples of such alloys are Hastelloy, Inconel, Waspaloy, Rene alloys, Haynes alloys, Incoloy, MP98T, TMS alloys, and CMSX single crystal alloys.

Superalloy development has relied heavily on both chemical and process innovations. Superalloys develop high temperature strength through solid solution strengthening. An important strengthening mechanism is precipitation strengthening which forms secondary phase precipitates such as gamma prime and carbides. Oxidation or corrosion resistance is provided by elements such as aluminium and chromium.

The primary application for such alloys is in turbine engines, both aerospace and marine.

Chemical Development

Because these alloys are intended for high temperature applications (i.e. holding their shape at temperatures near their melting point) their creep and oxidation resistance are of primary importance. Nickel (Ni) based superalloys have emerged as the material of choice for these applications. The properties of these Ni based superalloys can be tailored to a certain extent through the addition of many other elements, both common and exotic, including not only metals, but also metalloids and nonmetals; chromium, iron, cobalt, molybdenum, tungsten, tantalum, aluminium, titanium, zirconium, niobium, rhenium, yttrium, vanadium, carbon, boron or hafnium are some examples of the alloying additions used. Each of these additions has been chosen to serve a particular purpose in optimizing the properties for high temperature application.

Creep resistance is dependent on slowing the speed of dislocation motion within a crystal structure. In modern Ni based superalloys, the γ'-Ni_3(Al,Ti) phase present acts as a barrier to dislocation motion. For this reason, this γ' intermetallic phase, when present in high volume fractions, drastically increases the strength of these alloys due to its ordered nature and high coherency with the γ matrix. The chemical additions of aluminum and titanium promote the creation of the γ' phase. The γ' phase size can be precisely controlled by careful precipitation strengthening heat treatments. Many superalloys are produced using a two-phase heat treatment that creates a dispersion of cuboidal γ' particles known as the primary phase, with a fine dispersion between these known as secondary γ'. In order to improve the oxidation resistance of these alloys, Al, Cr, B, and Y are added. The Al and Cr form oxide layers that passivate the surface and protect the superalloy from further oxidation while B and Y are used to improve the adhesion of this oxide scale to the substrate. Cr, Fe, Co, Mo and Re all preferentially partition to the γ matrix while Al, Ti, Nb, Ta, and V preferentially partition to the γ' precipitates and solid solution strengthen the matrix and precipitates respectively. In addition to solid solution strengthening, if grain boundaries are present, certain elements are chosen for grain boundary strengthening. B and Zr tend to segregate to the grain boundaries which reduces the grain boundary energy and results in better grain boundary cohesion and ductility. Another form of grain boundary strengthening is achieved through the addition of C and a carbide former, such as Cr, Mo, W, Nb, Ta, Ti, or Hf, which drives precipitation of carbides at grain boundaries and thereby reduces grain boundary sliding.

While Ni based superalloys are excellent high temperature materials and have proven very useful, Co based superalloys potentially possess superior hot corrosion, oxidation, and wear resistance as compared to Ni-based superalloys. For this reason, efforts have also been put into developing Co based superalloys over the past several years. Despite that, traditional Co based superalloys have not found widespread usage because they have a lower strength at high temperature than Ni based superalloys. The main reason for this is that they appear to lack the γ' precipitation strengthening that is so important in the high temperature strength of Ni-based superalloys. A 2006 report on metastable γ'-$Co_3(Al,W)$ intermetallic compound with the $L1_2$ structure suggests Co based alloys as alternative to traditional Ni based superalloys. However this class of alloys was reported in a PhD thesis by C.S. Lee in 1971. The two-phase microstructure consists of cuboidal γ' precipitates embedded in a continuous γ matrix and is therefore morphologically identical to the microstructure observed in Ni based superalloys. Like in the Ni-based system, there is a high degree of coherency between the two phases which is one of the main factors resulting in the superior strength at high temperatures. This provides a pathway for the development of a new class of load-bearing Co based superalloys for application in severe environments. In these alloys 'W' is the crucial addition for getting γ' intermetallic compound that makes them much denser (>9.6 gm/cm^3) compared to Ni-based superalloys. Recently a new class of γ - γ' cobalt based superalloys have been developed that are "W" free and have much lower density comparable to nickel based superalloys. In addition to the fact that many of the properties of these new Co based superalloys could be better than those of the more traditional Ni based ones, Co also has a higher melting temperature than Ni. Therefore, if the high temperature strength could be improved, the development of novel Co based superalloys could allow for an increase in jet engine operation temperature resulting in an increased efficiency.

Metallurgy of Superalloys

Ni-based Superalloy Phases

- Gamma (γ): This phase composes the matrix of Ni-based superalloy. It is a solid solution fcc austenitic phase of the alloying elements. Alloying elements found in most commercial Ni-based alloys are, C, Cr, Mo, W, Nb, Fe, Ti, Al, V, and Ta. During the formation of these materials, as the Ni-alloys are cooled from the melt, carbides begin to precipitate, at even lower temperatures γ'phase precipitates.

- Gamma Prime (γ'): This phase constitutes the precipitate used to strengthen the alloy. It is an intermetallic phase based on $Ni_3(Ti,Al)$ which have an ordered FCC $L1_2$ structure. The γ' phase is coherent with the matrix of the superalloy having a lattice parameter that varies by around 0.5%. Ni (Ti,Al) are ordered systems with Ni atoms at the cube faces and either Al or Ti atoms at the cube edges As particles of γ' precipitates aggregate, they decrease their energy states by aligning along the <100> directions forming cuboidal structures. This phase has a window of instability between 600 °C and 850 °C, inside of which γ' will transform into the HCP η phase. For applications at temperatures below 650 °C, the γ'' phase can be utilized for strengthening.

- Gamma Double Prime (γ''): This phase typically possesses the composition of Ni_3Nb or Ni_3V and is used to strengthen Ni-based superalloys at lower temperatures (<650 °C) relative to γ'. The crystal structure of γ'' is body-centered tetragonal (BCT), and the phase precipitates as 60 nm by 10 nm discs with the (001) planes in γ'' parallel to the {001} family in γ. These anisotropic discs form as a result of lattice mismatch between the BCT precipitate and the FCC matrix. This lattice mismatch leads to high coherency strains which, together with order hardening, comprise the primary strengthening mechanisms. The γ'' phase is unstable above approximately 650 °C.

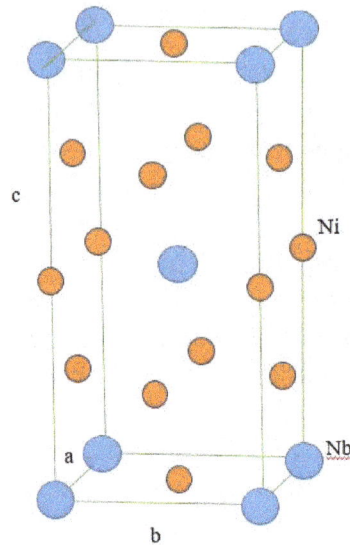

Crystal structure for γ" (Ni$_3$Nb) (Body Centered Tetragonal)

- Carbide Phases: Carbide formation is usually considered deleterious although in Ni-based superalloys they are used to stabilize the structure of the material against deformation at high temperatures. Carbides form at the grain boundaries inhibiting grain boundary motion.

- Topologically Close-Packed (TCP) Phases: The term "TCP Phase" refers to any member of a family of phases (including the σ phase, the χ phase, the μ phase, and the Laves phase) which are not atomically close-packed but possess some close-packed planes with HCP stacking. TCP phases are characterized by their tendency to be highly brittle and deplete the γ matrix of strengthening, solid solution refractory elements (including Cr, Co, W, and Mo). These phases form as a result of kinetics after long periods of time (thousands of hours) at high temperatures (>750 °C).

Co-based Superalloy Phases

- Gamma (γ): Similar to Ni-based superalloys, this is the phase of the superalloy's matrix. While not used commercially to the extent of Ni-based superalloys, alloying elements found in research Co-based alloys are C, Cr, W, Ni, Ti, Al, Ir, and Ta. Chromium is also used in Cobalt based superalloys (occasionally up to 20 wt.%) as it provides oxidation and corrosion resistance, critical for material use in gas turbines.

- Gamma Prime (γ'): Just as in Ni-based super alloys, this phase constitutes the precipitate used to strengthen the alloy. In this case, it is usually close packed with a L1$_2$ structure of Co$_3$Ti or fcc Co$_3$T, though both W and Al have been found to integrate into these cuboidal precipitates quite well. The elements Ta, Nb, and Ti integrate into the γ' phase and are quite effective at stabilizing it at high temperatures. This stabilization is quite important as the lack of stability is one of the key factors that makes Co-based superalloys weaker than their Ni-base cousins at elevated temperatures.

- Carbide Phases: As is common with carbide formation, its appearance in Co-based superalloys does provide precipitation hardening, but does decrease low-temperature ductility.

Microstructure of Superalloys

In pure Ni_3Al phase atoms of aluminium are placed at the vertices of the cubic cell and form the sublattice A. Atoms of nickel are located at centers of the faces and form the sublattice B. The phase is not strictly stoichiometric. There may exist an excess of vacancies in one of the sublattices, which leads to deviations from stoichiometry. Sublattices A and B of the γ'-phase can solute a considerable proportion of other elements. The alloying elements are dissolved in the γ-phase as well. The γ'-phase hardens the alloy through an unusual mechanism called the yield strength anomaly. Dislocations dissociate in the γ'-phase, leading to the formation of an anti-phase boundary. It turns out that at elevated temperature, the free energy associated with the anti-phase boundary (APB) is considerably reduced if it lies on a particular plane, which by coincidence is not a permitted slip plane. One set of partial dislocations bounding the APB cross-slips so that the APB lies on the low-energy plane, and, since this low-energy plane is not a permitted slip plane, the dissociated dislocation is now effectively locked. By this mechanism, the yield strength of γ'-phase Ni_3Al actually *increases* with temperature up to about 1000 °C, giving superalloys their currently unrivalled high-temperature strength.

Initial material selection for blade applications in Gas Turbine engines included alloys like the Nimonic series alloys in the 1940s. The early Nimonic series incorporated γ' $Ni_3(Al,Ti)$ precipitates in a γ matrix, as well as various metal-carbon carbides (e.g. $Cr_{23}C_6$) at the grain boundaries for additional grain boundary strength. Turbine blade components were forged until vacuum induction casting technologies were introduced in the 1950s. This process significantly improved cleanliness, reduced defects, and increased the strength and temperature capability of the material.

Modern superalloys were developed in the 1980s. The first generation superalloys incorporated increased aluminium, titanium, tantalum, and niobium content in order to increase the γ' volume fraction in these alloys. Examples of first generation superalloys include: PWA1480, René N4 and SRR99. Additionally, the volume fraction of the γ' precipitates increased to about 50-70%. With the advent of single crystal, or monocrystal, solidification techniques for superalloys that enable grain boundaries to be entirely eliminated from a casting. Because the material contained no grain boundaries, carbides were unnecessary as grain boundary strengthers and were thus eliminated.

The second and third generation superalloys introduced about 3 and 6 weight per cent Rhenium, for increased temperature capability. Re is a slow diffuser and typically partitions to the γ matrix, decreasing the rate of diffusion (and thereby high temperature creep) and improving high temperature performance and increasing service temperatures by 30 °C and 60 °C in second and third generation superalloys, respectively. Re has also been shown to promote the formation of rafts of the γ' phase (as opposed to cuboidal precipitates). The presence of rafts can decrease creep rate in the power-law regime (controlled by dislocation climb), but can also potentially increase the creep rate if the dominant mechanism is particle shearing. Furthermore, Re tends to promote the formation of brittle TCP phases, which has led to the strategy of reducing Co, W, Mo, and particularly Cr. Younger generations of Ni-based superalloys have significantly reduced Cr content for this reason, however with the reduction in Cr comes a reduction in oxidation resistance. Advanced coating techniques are now used to offset the loss of oxidation resistance accompanying the decreased Cr contents. Examples of second generation superalloys include PWA1484, CMSX-4 and René N5.

Third generation alloys include CMSX-10, and René N6. Fourth, Fifth, and even Sixth generation superalloys have been developed which incorporate Ruthenium additions, making them more expensive still than the prior generation's Re-containing alloys. The effect of Ru on the promotion of TCP phases is not well-determined. Early reports determined that Ru decreased the supersaturation of Re in the matrix and thereby diminished the susceptibility to TCP phase formation. More recent studies have noted the opposite effect. Chen, et al., found that in two alloys differing significantly only in Ru content (USTB-F3 and USTB-F6) that the addition of Ru increased both the partitioning ratio as well as the supersaturation in the γ matrix of Cr and Re, and thereby promoted the formation of TCP phases.

The current trend is to avoid very expensive and very heavy elements. An example is Eglin steel, a budget material with compromised temperature range and chemical resistance. It does not contain rhenium or ruthenium and its nickel content is limited. To reduce fabrication costs, it was chemically designed to melt in a ladle (though with improved properties in a vacuum crucible). Also, conventional welding and casting is possible before heat-treatment. The original purpose was to produce high-performance, inexpensive bomb casings, but the material has proven widely applicable to structural applications, including armor.

Single-crystal Superalloys

Single-crystal superalloys (SX or SC superalloys) are formed as a single crystal using a modified version of the directional solidification technique, so there are no grain boundaries in the material. The mechanical properties of most other alloys depend on the presence of grain boundaries, but at high temperatures, they would participate in creep and must be replaced by other mechanisms. In many such alloys, islands of an ordered intermetallic phase sit in a matrix of disordered phase, all with the same crystalline lattice. This approximates the dislocation-pinning behavior of grain boundaries, without introducing any amorphous solid into the structure.

Single crystal (SX) superalloys have wide application in the high pressure turbine section of aero and industrial gas turbine engines due to the unique combination of properties and performance. Since introduction of single crystal casting technology, SX alloy development has focused on increased temperature capability, and major improvements in alloy performance have been associated with the introduction of new alloying elements, including rhenium (Re) and ruthenium (Ru).

With increasing turbine entry temperature, it is important to gain a fundamental understanding of the physical phenomena occurring during creep deformation of single crystal superalloys under such extreme condition (i.e. high temperature and high stress). The creep deformation behavior of superalloy single crystal is strongly temperature, stress, orientation and alloy dependent. For a single-crystal superalloy, there are 3 different mode of creep deformation under regimes of different temperature and stress: Rafting, Tertiary and Primary. At low temperature (~750 °C), SX alloys exhibits mostly primary creep behavior. Matan et al. concluded that the extent of primary creep deformation depends strongly on the angle between the tensile axis and the <001>/<011> symmetry boundary. At temperature above 850 °C, tertiary creep dominates and promotes strain softening behavior. When temperature exceeds 1000 °C, the rafting effect is prevalent where cubic particles transform into flat shapes under tensile stress The rafts would also form perpendicular to the tensile axis, since γ phase was transported out of the vertical channels and into the horizontal ones. After conducting uniaxial creep deformation of <001> orientated CMSX-4 single crystal superalloy at 1105 °C and 100 MPa, Reed et al. has established that rafting is beneficial to creep life since it delays evolution of creep strain. In addition, rafting would occur quickly and suppress the accumulation of creep strain until a critical strain is reached.

Oxidation in Superalloys

For superalloys operating at high temperatures and exposed to corrosive environments, the oxidation behavior is of paramount concern. Oxidation involves chemical reactions of the alloying elements with oxygen to form new oxide phases, generally at the surface of the metal. If unmitigated, oxidation can degrade the alloy over time in a variety of ways, including:

- sequential oxidation, cracking, and spalling of the surface, leading to erosion of the alloy over time.

- embrittlement of the surface through the introduction of oxide phases, promoting crack formation and fatigue failure

- depletion of key alloying elements, affecting the mechanical properties of the superalloy and possibly compromising its performance.

The primary strategy used to limit these deleterious processes is called selective oxidation. Simply, the alloy is designed such that the ratio of alloying elements promotes formation of a specific oxide phase that can then act as a barrier to further oxidation. Most commonly, aluminum and chromium are used in this role, because they form relatively thin and continuous oxide layers of alumina (Al_2O_3) and chromia (Cr_2O_3), respectively. Furthermore, they possess low oxygen diffusivities, effectively halting further oxidation beneath this layer. In the ideal case, oxidation proceeds through 2 stages. First, transient oxidation involves the conversion of various elements, especially the majority elements (e.g. nickel or cobalt). Transient oxidation proceeds until the selective oxidation of the sacrificial element forms a complete barrier layer.

The protective effect of selective oxidation can be undermined by numerous mechanisms. The continuity of the thin sacrificial oxide layer can be compromised by mechanical disruption due to stress or may be disrupted as a result of the kinetics of oxidation (e.g. if diffusion of oxygen is too fast). If the layer is not continuous, its effectiveness as a diffusion barrier to oxygen is significantly reduced. The stability of the oxide layer is also strongly influenced by the presence of other minority elements. For example, the addition of boron, silicon, and yttrium to superalloys promotes oxide layer adhesion, reducing spalling and maintaining the integrity of the protective oxide layer.

It should be noted that oxidation is only the most basic form of chemical degradation superalloys may experience. More complex corrosion processes are common when operating environments include salts and sulfur compounds, or under chemical conditions that change dramatically over time. These issues and those of basic oxidation are often also addressed through thin coatings.

Superalloy Processing

The historical developments in superalloy processing have brought about considerable increases in superalloy operating temperatures. Superalloys were originally iron based and cold wrought prior to the 1940s. In the 1940s investment casting of cobalt base alloys significantly raised operating temperatures. The development of vacuum melting in the 1950s allowed for very fine control of the chemical composition of superalloys and reduction in contamination and in turn led to a revolution in processing techniques such as directional solidification of alloys and single crystal superalloys.

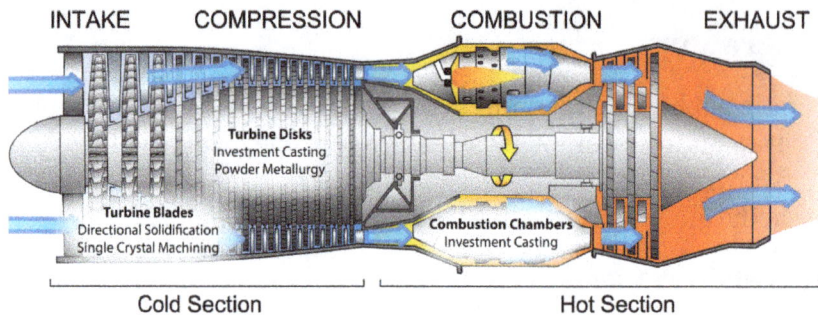

Processing methods used to generate various portions of a jet turbine engine.

There are many forms of superalloy present within a gas turbine engine, and processing methods vary widely depending on the necessary properties of each specific part.

Casting and Forging

Casting and forging are traditional metallurgical processing techniques that can be used to generate both polycrystalline and monocrystalline products. Polycrystalline casts tend to have higher fracture resistance, while monocrystalline casts have higher creep resistance.

Jet turbine engines employ both poly and mono crystalline components to take advantage of their individual strengths. The disks of the high pressure turbine, which are near the central hub of the engine are polycrystalline, The turbine blades, which extend radially into the engine housing, experience a much greater centripetal force, necessitating creep resistance. As a result, turbine blades are typically monocrystalline or polycrystalline with a preferred crystal orientation.

Investment Casting

Investment casting is a metallurgical processing technique in which a wax form is fabricated and used as a template for a ceramic mold. Briefly, a ceramic mold is poured around the wax form, the wax form is melted out of the ceramic mold, and molten metal is poured into the void left by the wax. This leads to a metal form in the same shape as the original wax form. Investment casting leads to a polycrystalline final product, as nucleation and growth of crystal grains occurs at numerous locations throughout the solid matrix. Generally, the polycrystalline product has no preferred grain orientation.

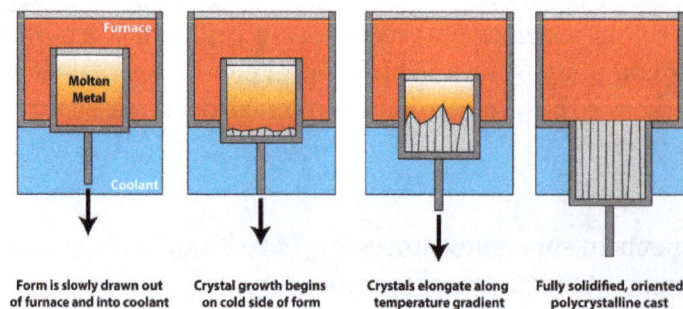

Schematic of directional solidification.

Directional Solidification

Directional solidification uses a thermal gradient to promote nucleation of metal grains on a low temperature surface, as well as to promote their growth along the temperature gradient. This leads to grains elongated along the temperature gradient, and significantly greater creep resistance parallel to the long grain direction. In polycrystalline turbine blades, directional solidification is used to orient the grains parallel to the centripetal force.

Single Crystal Growth

Single crystal growth starts with a seed crystal which is used to template growth of a larger crystal. The overall process is lengthy, and additional processing via machining is necessary after the single crystal is grown.

Powder Metallurgy

Powder metallurgy is a class of modern processing techniques in which metals are first converted into a powder form, and then formed into the desired shape by heating below the melting point. This is in contrast to casting, which occurs with molten metal. Superalloy manufacturing often employs powder metallurgy because of its material efficiency - typically much less waste metal must be machined away from the final product - and its ability to facilitate mechanical alloying. Mechanical alloying is a process by which reinforcing particles are incorporated into the superalloy matrix material by repeated fracture and welding.

Sintering and Hot Isostatic Pressing

Sintering and hot isostatic pressing are processing techniques used to densify materials from a loosely packed "green body" into a solid object with physically merged grains. Sintering occurs below the melting point, and causes adjacent particles to merge at their boundaries, leading to a strong bond between them. In hot isostatic pressing, a sintered material is placed in a pressure vessel and compressed from all directions (isostatically) in an inert atmosphere to affect densification.

Additive Manufacturing

Selective laser sintering is an additive manufacturing procedure used to create intricately detailed forms from a CAD file. In CAD, a shape is designed and the converted into slices. These slices are sent to a laser writer to print the final product. In brief, a bed of metal powder is prepared, and the first slice of the CAD design is formed in the powder bed by a high energy laser sintering the particles together. After this first slice is generated, the powder bed moves downwards, and a new batch of metal powder is rolled over the top of the slice. The second layer is then sintered with the laser, and the process is repeated until all the slices in the CAD file have been processed.

Coating of Superalloys

In modern gas turbine, the turbine entry temperature (~1750K) has exceeded the incipient melting temperature of superalloys (~1600K), with the help of surface engineering. Under such extreme working condition, the qualification of coating becomes vital.

Different Types of Coating

Historically, three "generations" of coatings have been developed: diffusion coatings, overlay coatings and thermal barrier coatings. Diffusion coatings, mainly constituted with aluminide or platinum-aluminide, is still the most common form of surface protection. To further enhance resistance to corrosion and oxidation, MCrAlX-based overlay coatings (M=Ni or Co, X=Y, Hf, Si) are deposited to surface of superalloys. Compared to diffusion coatings, overlay coatings are less dependent on the composition of the substrate, but also more expensive, since they must be carried out by air or vacuum plasma spraying (APS/VPS) or else electron beam physical vapour deposition (EB-PVD). Thermal barrier coatings provide by far the best enhancement in working temperature and coating life. It is estimated that modern TBC of thickness 300 μm, if used in conjunction with a hollow component and cooling air, has the potential to lower metal surface temperatures by a few hundred degrees.

Thermal Barrier Coatings

Thermal barrier coatings (TBCs) are used extensively on the surface of superalloy in both commercial and military gas turbine engines to increase component life and engine performance. A coating of about 1-200 μm can reduce the temperature at the superalloy surface by up to 200K. TBCs are really a system of coatings consisting of a bond coat, a thermally grown oxide (TGO), and a thermally insulating ceramic top coat. In most applications, the bond coat is either a MCrAlY (where M=Ni or NiCo) or a Pt modified aluminide coating. A dense bond coat is required to provide protection of the superalloy substrate from oxidation and hot corrosion attack and to form an adherent, slow growing TGO on its surface. The TGO is formed by oxidation of the aluminum that is contained in the bond coat. The current (first generation) thermal insulation layer is composed of 7wt % yttria stabilized zirconia (7YSZ) with a typical thickness of 100-300 μm. Yttria stabilized zirconia is used due to its low thermal conductivity (2.6W/mK for fully dense material), relatively high coefficient of thermal expansion, and good high temperature stability. The electron beam directed vapor deposition (EB-DVD) process used to apply the TBC to turbine airfoils produces a columnar microstructure with several levels of porosity. The porosity between the columns is critical to providing strain tolerance (via a very low in-plane modulus), as it would otherwise spall on thermal cycling due to thermal expansion mismatch with the superalloy substrate. The porosity within the columns reduces the thermal conductivity of the coating.

Bond Coat

The bond coat adheres the thermal barrier coating to the superalloy substrate. Additionally, the bond coat provides oxidation protection and functions as a diffusion barrier against the motion of substrate atoms towards the environment. There are five major types of bond coats, the aluminides, the platinum-aluminides, MCrAlY, cobalt-cermets, and nickel-chromium. For the aluminide bond coatings, the final composition and structure of the coating depends on the composition of the substrate. Aluminides also lack ductility below 750 °C, and exhibit a limited by thermomechanical fatigue strength. The Pt-aluminides are very similar to the aluminide bond coats except for a layer of Pt (5-10 μm) deposited on the blade. The Pt is believed to aid in oxide adhesion and contributes to hot corrosion. The cost of Pt plating is justified by the increased blade life span. The MCrAlY is the latest generation of bond coat and does not strongly interact with the substrate. Normally applied by plasma spraying, MCrAlY coatings are secondary aluminum oxide formers. This means that the coatings form an outer layer of chromium oxide (chromia), and a secondary aluminum oxide (alumina) layer underneath. These oxide formations occur at high temperatures in the range of those that superalloys usually encounter. The chromia provides oxidation and hot-corrosion resistance. The alumina controls oxidation mechanisms by limiting oxide growth by self-passivating. The yttrium enhances the oxide adherence to the substrate, and limits the growth of grain boundaries (which can lead to flaking of the coating). Investigation indicates that addition of rhenium and tantalum increases oxidation resistance. Cobalt-cermet based coatings consisting of materials such as tungsten carbide/cobalt can be used due to excellent resistance to abrasion, corrosion, erosion, and heat. These cermet coatings perform well in situations where temperature and oxidation damage are significant concerns, such as boilers. One of the unique advantages of cobalt cermet coatings is a minimal loss of coating mass over time, due to the strength of carbides within the mixture. Overall, cermet coatings are useful in situations where mechanical demands are equal to chemical demands for superalloys. Nickel-chromium coatings are used most frequently in boilers fed by fossil fuels, electric furnaces, and waste incineration furnaces, where the danger of oxidizing agents and corrosive compounds in the vapor must be dealt with. The specific method of spray-coating depends on the composition of the coatings. Nickel-chromium coatings that also contain iron or aluminum perform much better (in terms of corrosion resistance) when they are sprayed and laser glazed, while pure nickel-chromium coatings perform better when thermally sprayed exclusively.

Process Methods of Coating

Superalloy products that are subjected to high working temperatures and corrosive atmosphere (such as high pressure turbine region of jet engines) are coated with various kinds of coating. Several kinds of coating process are applied: pack cementation process, gas phase coating (both are a type of chemical vapor deposition (CVD)), thermal spraying, and physical vapor deposition. In most cases, after the coating process near-surface regions of parts are enriched with aluminium, the matrix of the coating being nickel aluminide.

Pack Cementation Process

The pack cementation process is carried out at lower temperatures, about 750 °C. The parts are loaded into boxes that contain a mixture of powders: active coating material, containing aluminum, activator (chloride or fluoride), and thermal ballast, like aluminum oxide. At high temperatures the gaseous aluminum chloride is transferred to the surface of the part and diffuses inside (mostly inward diffusion). After the end of the process the so-called "green coating" is produced, which is too thin and brittle for direct use. A subsequent diffusion heat treatment (several hours at temperatures about 1080 °C) leads to further inward diffusion and formation of the desired coating.

Thermal Spraying

Thermal spraying is a process of applying coatings by heating a feedstock of precursor material and spraying it on a surface. Different specific techniques are used depending on desired particle size, coat thickness, spray speed, desired area, etc. The coatings applied by thermal spraying of any kind, however, rely on adhesion to the surface. As a result, the surface of the superalloy must be cleaned and prepared, usually polished, before application of the thermal coating.

Plasma Spraying

Of the various thermal spray methods, one of the more ideal and commonly used techniques for coating superalloys is plasma spraying. This is due to the versatility of usable coatings, and the high-temperature performance of plasma-sprayed coatings. Plasma spraying can accommodate a very wide range of materials, much more so than other techniques. As long as the difference between melting and decomposition temperatures is greater than 300 Kelvin, a material can be melted and applied as a coating via plasma spraying.

Gas Phase Coating

This process is carried out at higher temperatures, about 1080 °C. The coating material is usually loaded onto special trays without physical contact with the parts to be coated. The coating mixture contains active coating material and activator, but usually does not contain thermal ballast. As in the pack cementation process, the gaseous aluminium chloride (or fluoride) is transferred to the surface of the part. However, in this case the diffusion is outwards. This kind of coating also requires diffusion heat treatment.

Failure Mechanisms in Thermal Barrier Coating Systems

Failure of thermal barrier coating usually manifests as delamination, which arises from the temperature gradient during thermal cycling between ambient temperature and working conditions coupled with the difference in thermal expansion coefficient of the substrate and coating. It is rare for the coating to fail completely – some pieces of it remain intact, and significant scatter is observed in the time to failure if testing is repeated under identical conditions. There are various degradation mechanisms for thermal barrier coating, and some or all of these must operate before failure finally occurs:

- Oxidation at the interface of thermal barrier coating and underlying bond coat;

- The depletion of aluminum in bond coat due to oxidation and diffusion with substrate;

- Thermal stresses from mismatch in thermal expansion coefficient and growth stress due to the formation of thermally grown oxide layer;

- Imperfections near thermally grown oxide layer;

- Various other complicating factors during engine operation.

Additionally, TBC life is very dependent upon the combination of materials (substrate, bond coat, ceramic) and processes (EB-PVD, plasma spraying) used.

Applications

Nickel-based superalloys are used in load-bearing structures to the highest homologous temperature of any common alloy system (Tm = 0.9, or 90% of their melting point). Among the most demanding applications for a structural material are those in the hot sections of turbine engines. The preeminence of superalloys is reflected in the fact that they currently comprise over 50% of the weight of advanced aircraft engines. The widespread use of superalloys in turbine engines coupled with the fact that the thermodynamic efficiency of turbine engines is increased with increasing turbine inlet temperatures has, in part, provided the motivation for increasing the maximum-use temperature of superalloys. In fact, during the past 30 years turbine airfoil temperature capability has increased on average by about 4 °F (2.2 °C) per year. Two major factors which have made this increase possible are

1. Advanced processing techniques, which improved alloy cleanliness (thus improving reliability) and/or enabled the production of tailored microstructures such as directionally solidified or single-crystal material.

2. Alloy development resulting in higher-use-temperature materials primarily through the additions of refractory elements such as Re, W, Ta, and Mo.

About 60% of the use-temperature increases have occurred due to advanced cooling concepts; 40% have resulted from material improvements. State-of-the-art turbine blade surface temperatures are near 2,100 °F (1,150 °C); the most severe combinations of stress and temperature corresponds to an average bulk metal temperature approaching 1,830 °F (1,000 °C).

Although superalloys retain significant strength to temperatures near 1,800 °F (980 °C), they tend to be susceptible to environmental attack because of the presence of reactive alloying elements (which provide their high-temperature strength). Surface attack includes oxidation, hot corrosion, and thermal fatigue. In the most demanding applications, such as turbine blade and vanes, superalloys are often coated to improve environmental resistance.

Research and Development of New Superalloys

The availability of superalloys during past decades has led to a steady increase in the turbine entry temperatures and the trend is expected to continue. Sandia National Laboratories is studying a new method for making superalloys, known as radiolysis. It introduces an entirely new area of research into creating alloys and superalloys through nanoparticle synthesis. This process holds promise as a universal method of nanoparticle formation. By developing an understanding of the basic material science behind these nanoparticle formations, there is speculation that it might be possible to expand research into other aspects of superalloys.

There may be considerable disadvantages in making alloys by this method. About half of the use of superalloys is in applications where the service temperature is close to the melting temperature of the alloy. It is common therefore to use single crystals. The above method produces polycrystalline alloys, which suffer from an unacceptable level of creep.

Future paradigm in alloy development focus on reduction of weight, improving oxidation and corrosion resistance while maintaining the strength of the alloy. Furthermore, with the increasing demand for turbine blade for power generation, another focus of alloy design is to reduce the cost of super alloys.

References

-   Engheta, Nader; Richard W. Ziolkowski (June 2006). Metamaterials: Physics and Engineering Explorations. Wiley & Sons. pp. xv, 3–30, 37, 143–150, 215–234, 240–256. ISBN 978-0-471-76102-0.

- Mourachkine, A. (2004). Room-Temperature Superconductivity. Cambridge International Science Publishing. arXiv:cond-mat/0606187 . ISBN 1-904602-27-4.

- "Thermal conductivity" in Lide, D. R., ed. (2005). CRC Handbook of Chemistry and Physics (86th ed.). Boca Raton (FL): CRC Press. ISBN 0-8493-0486-5. Section 12, p. 227

- Zouhdi, Saïd; Ari Sihvola; Alexey P. Vinogradov (December 2008). Metamaterials and Plasmonics: Fundamentals, Modelling, Applications. New York: Springer-Verlag. pp. 3–10, Chap. 3, 106. ISBN 978-1-4020-9406-4.

- Eleftheriades, George V.; Keith G. Balmain (2005). Negative-refraction metamaterials: fundamental principles and applications. Wiley, John & Sons. p. 340. ISBN 978-0-471-60146-3.

- Engheta, Nader; Richard W. Ziolkowski (June 2006). Metamaterials: physics and engineering explorations (added this reference on 2009-12-14.). Wiley & Sons. pp. 211–221. ISBN 978-0-471-76102-0.

- Capolino, Filippo (October 2009). Applications of Metamaterials. Taylor & Francis, Inc. pp. 29–1, 25–14, 22–1. ISBN 978-1-4200-5423-1. Retrieved 2009-10-01.

- McCarthy, Wil (2003). Hacking Matter: Levitating Chairs, Quantum Mirages, and the Infinite Weirdness of Programmable Atoms. New York: Basic Books. ISBN 0-465-04428-X.

- Rothman, D.H.; Zaleski, S. (2004) [1997]. Lattice Gas Cellular Automata. Cambridge University Press. ISBN 9780521607605.

Chemical Bonding: An Overview

The bond between atoms and chemical compounds is known as chemical bonding. Ionic bonding, covalent bond, metallic bonding and electronegativity are other aspects of chemical bonding. This section is an overview of the subject matter incorporating all the major aspects of chemical bonding.

Chemical Bond

A chemical bond is a lasting attraction between atoms that enables the formation of chemical compounds. The bond may result from the electrostatic force of attraction between atoms with opposite charges, or through the sharing of electrons as in the covalent bonds. The strength of chemical bonds varies considerably; there are "strong bonds" such as covalent or ionic bonds and "weak bonds" such as Dipole-dipole interaction, the London dispersion force and hydrogen bonding.

Since opposite charges attract via a simple electromagnetic force, the negatively charged electrons that are orbiting the nucleus and the positively charged protons in the nucleus attract each other. An electron positioned between two nuclei will be attracted to both of them, and the nuclei will be attracted toward electrons in this position. This attraction constitutes the chemical bond. Due to the matter wave nature of electrons and their smaller mass, they must occupy a much larger amount of volume compared with the nuclei, and this volume occupied by the electrons keeps the atomic nuclei relatively far apart, as compared with the size of the nuclei themselves. This phenomenon limits the distance between nuclei and atoms in a bond.

In general, strong chemical bonding is associated with the sharing or transfer of electrons between the participating atoms. The atoms in molecules, crystals, metals and diatomic gases—indeed most of the physical environment around us—are held together by chemical bonds, which dictate the structure and the bulk properties of matter.

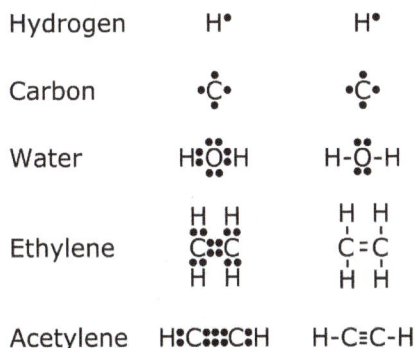

Examples of Lewis dot-style representations of chemical bonds between carbon (C), hydrogen (H), and oxygen (O). Lewis dot diagrams were an early attempt to describe chemical bonding and are still widely used today.

All bonds can be explained by quantum theory, but, in practice, simplification rules allow chemists to predict the strength, directionality, and polarity of bonds. The octet rule and VSEPR theory are two examples. More sophisticated theories are valence bond theory which includes orbital hybridization and resonance, and molecular orbital theory which includes linear combination of atomic orbitals and ligand field theory. Electrostatics are used to describe bond polarities and the effects they have on chemical substances.

Overview of Main Types of Chemical Bonds

A chemical bond is an attraction between atoms. This attraction may be seen as the result of different behaviors of the outermost or valence electrons of atoms. These behaviors merge into each other seamlessly in various circumstances, so that there is no clear line to be drawn between them. However it remains useful and customary to differentiate between different types of bond, which result in different properties of condensed matter.

In the simplest view of a covalent bond, one or more electrons (often a pair of electrons) are drawn into the space between the two atomic nuclei. Energy is released by bond formation. This is not as a reduction in potential energy, because the attraction of the two electrons to the two protons is offset by the electron-electron and proton-proton repulsions. Instead, the release of energy (and hence stability of the bond) arises from the reduction in kinetic energy due to the electrons being in a more spatially distributed (i.e. longer de Broglie wavelength) orbital compared with each electron being confined closer to its respective nucleus. These bonds exist between two particular identifiable atoms and have a direction in space, allowing them to be shown as single connecting lines between atoms in drawings, or modeled as sticks between spheres in models.

In a polar covalent bond, one or more electrons are unequally shared between two nuclei. Covalent bonds often result in the formation of small collections of better-connected atoms called molecules, which in solids and liquids are bound to other molecules by forces that are often much weaker than the covalent bonds that hold the molecules internally together. Such weak intermolecular bonds give organic molecular substances, such as waxes and oils, their soft bulk character, and their low melting points (in liquids, molecules must cease most structured or oriented contact with each other). When covalent bonds link long chains of atoms in large molecules, however (as in polymers such as nylon), or when covalent bonds extend in networks through solids that are not composed of discrete molecules (such as diamond or quartz or the silicate minerals in many types of rock) then the structures that result may be both strong and tough, at least in the direction oriented correctly with networks of covalent bonds. Also, the melting points of such covalent polymers and networks increase greatly.

In a simplified view of an *ionic* bond, the bonding electron is not shared at all, but transferred. In this type of bond, the outer atomic orbital of one atom has a vacancy which allows the addition of one or more electrons. These newly added electrons potentially occupy a lower energy-state (effectively closer to more nuclear charge) than they experience in a different atom. Thus, one nucleus offers a more tightly bound position to an electron than does another nucleus, with the result that one atom may transfer an electron to the other. This transfer causes one atom to assume a net positive charge, and the other to assume a net negative charge. The *bond* then results from electrostatic attraction between atoms and the atoms become positive or negatively charged ions. Ionic bonds may be seen as extreme examples of polarization in covalent bonds. Often, such bonds have no particular orientation in space, since they result from equal electrostatic attraction of each ion to all ions around them. Ionic bonds are strong (and thus ionic substances require high temperatures to melt) but also brittle, since the forces between ions are short-range and do not easily bridge cracks and fractures. This type of bond gives rise to the physical characteristics of crystals of classic mineral salts, such as table salt.

A less often mentioned type of bonding is *metallic* bonding. In this type of bonding, each atom in a metal donates one or more electrons to a "sea" of electrons that reside between many metal atoms. In this sea, each electron is free (by virtue of its wave nature) to be associated with a great many atoms at once. The bond results because the metal atoms become somewhat positively charged due to loss of their electrons while the electrons remain attracted to many atoms, without being part of any given atom. Metallic bonding may be seen as an extreme example of delocalization of electrons over a large system of covalent bonds, in which every atom participates. This type of bonding is often very strong (resulting in the tensile strength of metals). However, metallic bonding is more collective in nature than other types, and so they allow metal crystals to more easily deform, because they are composed of atoms attracted to each other, but not in any particularly-oriented ways. This results in the malleability of metals. The sea of electrons in metallic bonding causes the characteristically good electrical and thermal conductivity of metals, and also their "shiny" reflection of most frequencies of white light.

Early speculations about the nature of the chemical bond, from as early as the 12th century, supposed that certain types of chemical species were joined by a type of chemical affinity. In 1704, Sir Isaac Newton famously outlined his atomic bonding theory, in "Query 31" of his *Opticks*, whereby atoms attach to each other by some "force". Specifically, after acknowledging the various popular theories in vogue at the time, of how atoms were reasoned to attach to each other, i.e. "hooked atoms", "glued together by rest", or "stuck together by conspiring motions", Newton states that he would rather infer from their cohesion, that "particles attract one another by some force, which in immediate contact is exceedingly strong, at small distances performs the chemical operations, and reaches not far from the particles with any sensible effect."

In 1819, on the heels of the invention of the voltaic pile, Jöns Jakob Berzelius developed a theory of chemical combination stressing the electronegative and electropositive characters of the combining atoms. By the mid 19th century, Edward Frankland, F.A. Kekulé, A.S. Couper, Alexander Butlerov, and Hermann Kolbe, building on the theory of radicals, developed the theory of valency, originally called "combining power", in which compounds were joined owing to an attraction of positive and negative poles. In 1916, chemist Gilbert N. Lewis developed the concept of the electron-pair bond, in which two atoms may share one to six electrons, thus forming the single electron bond, a single bond, a double bond, or a triple bond; in Lewis's own words, "An electron may form a part of the shell of two different atoms and cannot be said to belong to either one exclusively."

That same year, Walther Kossel put forward a theory similar to Lewis' only his model assumed complete transfers of electrons between atoms, and was thus a model of ionic bonding. Both Lewis and Kossel structured their bonding models on that of Abegg's rule (1904).

In 1927, the first mathematically complete quantum description of a simple chemical bond, i.e. that produced by one electron in the hydrogen molecular ion, H_2^+, was derived by the Danish physicist Oyvind Burrau. This work showed that the quantum approach to chemical bonds could be fundamentally and quantitatively correct, but the mathematical methods used could not be extended to molecules containing more than one electron. A more practical, albeit less quantitative, approach was put forward in the same year by Walter Heitler and Fritz London. The Heitler-London method forms the basis of what is now called valence bond theory. In 1929, the linear combination of atomic orbitals molecular orbital method (LCAO) approximation was introduced by Sir John Lennard-Jones, who also suggested methods to derive electronic structures of molecules of F_2 (fluorine) and O_2 (oxygen) molecules, from basic quantum principles. This molecular orbital theory represented a covalent bond as an orbital formed by combining the quantum mechanical Schrödinger atomic orbitals which had been hypothesized for electrons in single atoms. The equations for bonding electrons in multi-electron atoms could not be solved to mathematical perfection (i.e., *analytically*), but approximations for them still gave many good qualitative predictions and results. Most quantitative calculations in modern quantum chemistry use either valence bond or molecular orbital theory as a starting point, although a third approach, density functional theory, has become increasingly popular in recent years.

In 1933, H. H. James and A. S. Coolidge carried out a calculation on the dihydrogen molecule that, unlike all previous calculation which used functions only of the distance of the electron from the atomic nucleus, used functions which also explicitly added the distance between the two electrons. With up to 13 adjustable parameters they obtained a result very close to the experimental result for the dissociation energy. Later extensions have used up to 54 parameters and gave excellent agreement with experiments. This calculation convinced the scientific community that quantum theory could give agreement with experiment. However this approach has none of the physical pictures of the valence bond and molecular orbital theories and is difficult to extend to larger molecules.

Bonds in Chemical Formulas

Because atoms and molecules are three-dimensional, it is difficult to use a single method to indicate orbitals and bonds. In molecular formulas the chemical bonds (binding orbitals) between atoms are indicated in different ways depending on the type of discussion. Sometimes, some details are neglected. For example, in organic chemistry one is sometimes concerned only with the functional group of the molecule. Thus, the molecular formula of ethanol may be written in conformational form, three-dimensional form, full two-dimensional form (indicating every bond with no three-dimensional directions), compressed two-dimensional form (CH_3-CH_2-OH), by separating the functional group from another part of the molecule (C_2H_5OH), or by its atomic constituents (C_2H_6O), according to what is discussed. Sometimes, even the non-bonding valence shell electrons (with the two-dimensional approximate directions) are marked, e.g. for elemental carbon ˙C. Some chemists may also mark the respective orbitals, e.g. the hypothetical ethene^{-4} anion ($\backslash/C=C\backslash$ $^{-4}$) indicating the possibility of bond formation.

Strong Chemical Bonds

Typical bond lengths in pm and bond energies in kJ/mol. Bond lengths can be converted to Å by division by 100 (1 Å = 100 pm). Data taken from University of Waterloo.		
Bond	**Length (pm)**	**Energy (kJ/mol)**

H — Hydrogen		
H–H	74	436
H–O	96	366
H–F	92	568
H–Cl	127	432
C — Carbon		
C–H	109	413
C–C	154	348
C–C=	151	
=C–C≡	147	
=C–C=	148	
C=C	134	614
C≡C	120	839
C–N	147	308
C–O	143	360
C–F	134	488
C–Cl	177	330
N — Nitrogen		
N–H	101	391
N–N	145	170
N≡N	110	945
O — Oxygen		
O–O	148	145
O=O	121	498
F, Cl, Br, I — Halogens		
F–F	142	158
Cl–Cl	199	243
Br–H	141	366
Br–Br	228	193
I–H	161	298
I–I	267	151

Strong chemical bonds are the *intramolecular* forces which hold atoms together in molecules. A strong chemical bond is formed from the transfer or sharing of electrons between atomic centers and relies on the electrostatic attraction between the protons in nuclei and the electrons in the orbitals.

The types of strong bond differ due to the difference in electronegativity of the constituent elements. A large difference in electronegativity leads to more polar (ionic) character in the bond.

Ionic Bonding

Ionic bonding is a type of electrostatic interaction between atoms which have a large electronegativity difference. There is no precise value that distinguishes ionic from covalent bonding, but a difference of electronegativity of over 1.7 is likely to be ionic, and a difference of less than 1.7 is likely to be covalent. Ionic bonding leads to separate positive and negative ions. Ionic charges are commonly between −3e to +3e. Ionic bonding commonly occurs in metal salts such as sodium chloride (table salt). A typical feature of ionic bonds is that the species form into ionic crystals, in which no ion is specifically paired with any single other ion, in a specific directional bond. Rather, each species of ion is surrounded by ions of the opposite charge, and the spacing between it and each of the oppositely charged ions near it, is the same for all surrounding atoms of the same type. It is thus no longer possible to associate an ion with any specific other single ionized atom near it. This is a situation unlike that in covalent crystals, where covalent bonds between specific atoms are still discernible from the shorter distances between them, as measured via such techniques as X-ray diffraction.

Ionic crystals may contain a mixture of covalent and ionic species, as for example salts of complex acids, such as sodium cyanide, NaCN. X-ray diffraction shows that in NaCN, for example, the bonds between sodium cations (Na^+) and the cyanide anions (CN^-) are *ionic*, with no sodium ion associated with any particular cyanide. However, the bonds between C and N atoms in cyanide are of the *covalent* type, making each of the carbon and nitrogen associated with *just one* of its opposite type, to which it is physically much closer than it is to other carbons or nitrogens in a sodium cyanide crystal.

When such crystals are melted into liquids, the ionic bonds are broken first because they are non-directional and allow the charged species to move freely. Similarly, when such salts dissolve into water, the ionic bonds are typically broken by the interaction with water, but the covalent bonds continue to hold. For example, in solution, the cyanide ions, still bound together as single CN^- ions, move independently through the solution, as do sodium ions, as Na^+. In water, charged ions move apart because each of them are more strongly attracted to a number of water molecules, than to each other. The attraction between ions and water molecules in such solutions is due to a type of weak dipole-dipole type chemical bond. In melted ionic compounds, the ions continue to be attracted to each other, but not in any ordered or crystalline way.

Covalent Bond

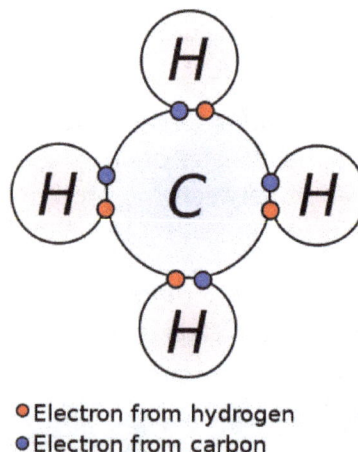

● Electron from hydrogen
● Electron from carbon

Nonpolar covalent bonds in methane (CH_4). The Lewis structure shows electrons shared between C and H atoms.

Covalent bonding is a common type of bonding, in which two atoms share two valence electrons, one from each of the atoms. In nonpolar covalent bonds, the electronegativity difference between the bonded atoms is small, typically 0 to 0.3. Bonds within most organic compounds are described as covalent. The figure shows methane (CH_4), in which each hydrogen forms a covalent bond with the carbon.

Molecules which are formed primarily from non-polar covalent bonds are often immiscible in water or other polar solvents, but much more soluble in non-polar solvents such as hexane.

A polar covalent bond is a covalent bond with a significant ionic character. This means that the two shared electrons are closer to one of the atoms than the other, creating an imbalance of charge. Such bonds occur between two atoms with moderately different electronegativities and give rise to dipole-dipole interactions. The electronegativity difference between the two atoms in these bonds is 0.3 to 1.7.

Single and Multiple Bonds

A single bond between two atoms corresponds to the sharing of one pair of electrons. The electron density of these two bonding electrons is concentrated in the region between the two atoms, which is the defining quality of a sigma bond.

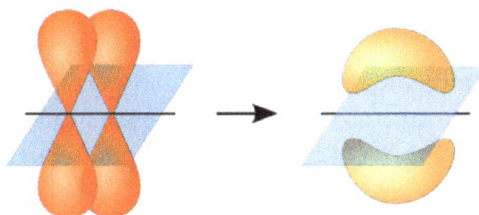

Two p-orbitals forming a pi-bond.

A double bond between two atoms is formed by the sharing of two pairs of electrons, one in a sigma bond and one in a pi bond, with electron density concentrated on two opposite sides of the internuclear axis. A triple bond consists of three shared electron pairs, forming one sigma and two pi bonds.

Quadruple and higher bonds are very rare and occur only between certain transition metal atoms.

Coordinate Covalent Bond (Dipolar Bond)

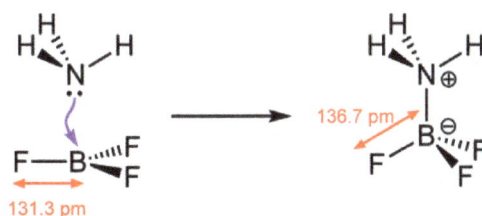

Adduct of ammonia and boron trifluoride

A coordinate covalent bond is a covalent bond in which the two shared bonding electrons are from the same one of the atoms involved in the bond. For example, boron trifluoride (BF_3) and ammonia (NH_3) from an adduct or coordination complex $F_3B \leftarrow NH_3$ with a B—N bond in which a lone pair of electrons on N is shared with an empty atomic orbital on B. BF_3 with an empty orbital is described as an electron pair acceptor or Lewis acid, while NH_3 with a lone pair which can be shared is described as an electron-pair donor or Lewis base. The electrons are shared roughly equally between the atoms in contrast to ionic bonding. Such bonding is shown by an arrow pointing to the Lewis acid.

Transition metal complexes are generally bound by coordinate covalent bonds. For example, the ion Ag^+ reacts as a Lewis acid with two molecules of the Lewis base NH_3 to form the complex ion $Ag(NH_3)_2^+$, which has two $Ag \leftarrow N$ coordinate covalent bonds.

Metallic Bonding

In metallic bonding, bonding electrons are delocalized over a lattice of atoms. By contrast, in ionic compounds, the locations of the binding electrons and their charges are static. The freely-moving or delocalization of bonding electrons leads to classical metallic properties such as luster (surface light reflectivity), electrical and thermal conductivity, ductility, and high tensile strength.

Intermolecular Bonding

There are four basic types of bonds that can be formed between two or more (otherwise non-associated) molecules, ions or atoms. Intermolecular forces cause molecules to be attracted or repulsed by each other. Often, these define some of the physical characteristics (such as the melting point) of a substance.

- A large difference in electronegativity between two bonded atoms will cause a permanent charge separation, or dipole, in a molecule or ion. Two or more molecules or ions with permanent dipoles can interact within dipole-dipole interactions. The bonding electrons in a molecule or ion will, on average, be closer to the more electronegative atom more frequently than the less electronegative one, giving rise to partial charges on each atom, and causing electrostatic forces between molecules or ions.

- A hydrogen bond is effectively a strong example of an interaction between two permanent dipoles. The large difference in electronegativities between hydrogen and any of fluorine, nitrogen and oxygen, coupled with their lone pairs of electrons cause strong electrostatic forces between molecules. Hydrogen bonds are responsible for the high boiling points of water and ammonia with respect to their heavier analogues.

- The London dispersion force arises due to instantaneous dipoles in neighbouring atoms. As the negative charge of the electron is not uniform around the whole atom, there is always a charge imbalance. This small charge will induce a corresponding dipole in a nearby molecule; causing an attraction between the two. The electron then moves to another part of the electron cloud and the attraction is broken.

- A cation–pi interaction occurs between a pi bond and a cation.

Theories of Chemical Bonding

In the (unrealistic) limit of "pure" ionic bonding, electrons are perfectly localized on one of the two atoms in the bond. Such bonds can be understood by classical physics. The forces between the atoms are characterized by isotropic continuum electrostatic potentials. Their magnitude is in simple proportion to the charge difference.

Covalent bonds are better understood by valence bond theory or molecular orbital theory. The properties of the atoms involved can be understood using concepts such as oxidation number. The electron density within a bond is not assigned to individual atoms, but is instead delocalized between atoms. In valence bond theory, the two electrons on the two atoms are coupled together with the bond strength depending on the overlap between them. In molecular orbital theory, the linear combination of atomic orbitals (LCAO) helps describe the delocalized molecular orbital structures and energies based on the atomic orbitals of the atoms they came from. Unlike pure ionic bonds, covalent bonds may have directed anisotropic properties. These may have their own names, such as sigma bond and pi bond.

In the general case, atoms form bonds that are intermediate between ionic and covalent, depending on the relative electronegativity of the atoms involved. This type of bond is sometimes called polar covalent.

Ionic Bonding

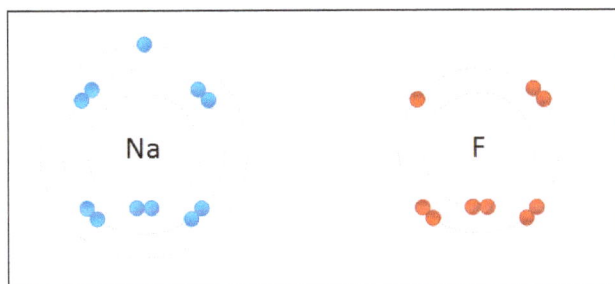

Sodium and fluorine undergoing a redox reaction to form sodium fluoride. Sodium loses its outer electron to give it a stable electron configuration, and this electron enters the fluorine atom exothermically. The oppositely charged ions – typically a great many of them – are then attracted to each other to form a solid.

Ionic bonding is a type of chemical bond that involves the electrostatic attraction between oppositely charged ions, and is the primary interaction occurring in ionic compounds. The ions are atoms that have gained one or more electrons (known as anions, which are negatively charged) and atoms that have lost one or more electrons (known as cations, which are positively charged). This transfer of electrons is known as electrovalence in contrast to covalence. In the simplest case, the cation is a metal atom and the anion is a nonmetal atom, but these ions can be of a more complex nature, e.g. molecular ions like NH_4^+ or SO_4^{2-}. In simpler words, an ionic bond is the transfer of electrons from a metal to a non-metal in order for both atoms to obtain a full valence shell.

It is important to recognize that *clean* ionic bonding – in which one atom or molecule completely share an electron from another – cannot exist: all ionic compounds have some degree of covalent bonding, or electron sharing. Thus, the term "ionic bonding" is given when the ionic character is greater than the covalent character—that is, a bond in which a large electronegativity difference exists between the two atoms, causing the bonding to be more polar (ionic) than in covalent bonding where electrons are shared more equally. Bonds with partially ionic and partially covalent character are called polar covalent bonds.

Ionic compounds conduct electricity when molten or in solution, typically as a solid. Ionic compounds generally have a high melting point, depending on the charge of the ions they consist of. The higher the charges the stronger the cohesive forces and the higher the melting point. They also tend to be soluble in water. Here, the opposite trend roughly holds: the weaker the cohesive forces, the greater the solubility.

Overview

Atoms that have an almost full or almost empty valence shells tend to be very reactive. Atoms that are strongly electronegative (as is the case with halogens) often only have one or two empty orbitals in their valence shell, and frequently bond with other molecules or gain electrons to form anions. Atoms that are weakly electronegative (such as alkali metals) have relatively few valence electrons that can easily be shared with atoms that are strongly electronegative. As a result, weakly electronegative atoms tend to distort their electrons cloud and form cations.

Formation

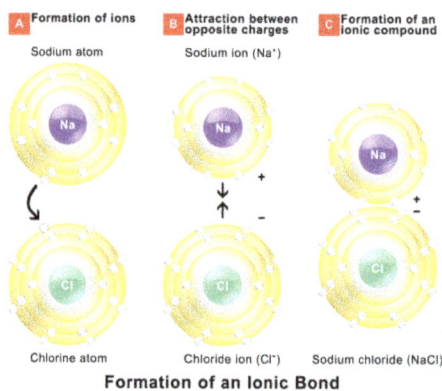

Formation of an Ionic Bond

Ionic bonds in sodium chloride

Ionic bonding can result from a redox reaction when atoms of an element (usually metal), whose ionization energy is low, give some of their electrons to achieve a stable electron configuration. In doing so, cations are formed. The atom of another element (usually nonmetal), whose electron affinity is positive, then accepts the electron(s), again to attain a stable electron configuration, and after accepting electron(s) the atom becomes an anion. Typically, the stable electron configuration is one of the noble gases for elements in the s-block and the p-block, and particular stable electron configurations for d-block and f-block elements. The electrostatic attraction between the anions and cations leads to the formation of a solid with a crystallographic lattice in which the ions are stacked in an alternating fashion. In such a lattice, it is usually not possible to distinguish discrete molecular units, so that the compounds formed are not molecular in nature. However, the ions themselves can be complex and form molecular ions like the acetate anion or the ammonium cation.

For example, common table salt is sodium chloride. When sodium (Na) and chlorine (Cl) are combined, the sodium atoms each lose an electron, forming cations (Na^+), and the chlorine atoms each gain an electron to form anions (Cl^-). These ions are then attracted to each other in a 1:1 ratio to form sodium chloride (NaCl).

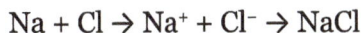

$$Na + Cl \rightarrow Na^+ + Cl^- \rightarrow NaCl$$

However, to maintain charge neutrality, strict ratios between anions and cations are observed so that ionic compounds, in general, obey the rules of stoichiometry despite not being molecular compounds. For compounds that are transitional to the alloys and possess mixed ionic and metallic bonding, this may not be the case anymore. Many sulfides, e.g., do form non-stoichiometric compounds.

Many ionic compounds are referred to as salts as they can also be formed by the neutralization reaction of an Arrhenius base like NaOH with an Arrhenius acid like HCl

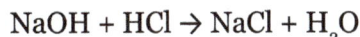

$$NaOH + HCl \rightarrow NaCl + H_2O$$

The salt NaCl is then said to consist of the acid rest Cl^- and the base rest Na^+.

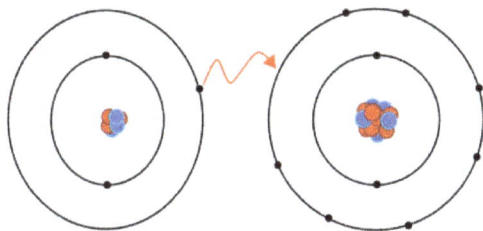

Representation of ionic bonding between lithium and fluorine to form lithium fluoride. Lithium has a low ionization energy and readily gives up its lone valence electron to a fluorine atom, which has a positive electron affinity and accepts the electron that was donated by the lithium atom. The end-result is that lithium is isoelectronic with helium and fluorine is isoelectronic with neon. Electrostatic interaction occurs between the two resulting ions, but typically aggregation is not limited to two of them. Instead, aggregation into a whole lattice held together by ionic bonding is the result.

The removal of electrons from the cation is endothermic, raising the system's overall energy. There may also be energy changes associated with breaking of existing bonds or the addition of more than one electron to form anions. However, the action of the anion's accepting the cation's valence electrons and the subsequent attraction of the ions to each other releases (lattice) energy and, thus, lowers the overall energy of the system.

Ionic bonding will occur only if the overall energy change for the reaction is favorable. In general, the reaction is exothermic, but, e.g., the formation of mercuric oxide (HgO) is endothermic. The charge of the resulting ions is a major factor in the strength of ionic bonding, e.g. a salt C^+A^- is held together by electrostatic forces roughly four times weaker than $C^{2+}A^{2-}$ according to Coulombs law, where C and A represent a generic cation and anion respectively. Of course the sizes of the ions and the particular packing of the lattice are ignored in this simple argument.

Structures

Ionic compounds in the solid state form lattice structures. The two principal factors in determining the form of the lattice are the relative charges of the ions and their relative sizes. Some structures are adopted by a number of compounds; for example, the structure of the rock salt sodium chloride is also adopted by many alkali halides, and binary oxides such as MgO. Pauling's rules provide guidelines for predicting and rationalizing the crystal structures of ionic crystals

Bond Strength

For a solid crystalline ionic compound the enthalpy change in forming the solid from gaseous ions is termed the lattice energy. The experimental value for the lattice energy can be determined using the Born-Haber cycle. It can also be calculated (predicted) using the Born-Landé equation as the sum of the electrostatic potential energy, calculated by summing interactions between cations and anions, and a short-range repulsive potential energy term. The electrostatic potential can be expressed in terms of the inter-ionic separation and a constant (Madelung constant) that takes account of the geometry of the crystal. The further away from the nucleus the weaker the shield. The Born-Landé equation gives a reasonable fit to the lattice energy of, e.g., sodium chloride, where the calculated (predicted) value is −756 kJ/mol, which compares to −787 kJ/mol using the Born-Haber cycle.

Polarization Effects

Ions in crystal lattices of purely ionic compounds are spherical; however, if the positive ion is small and/or highly charged, it will distort the electron cloud of the negative ion, an effect summarised in Fajans' rules. This polarization of the negative ion leads to a build-up of extra charge density between the two nuclei, i.e., to partial covalency. Larger negative ions are more easily polarized, but the effect is usually important only when positive ions with charges of 3+ (e.g., Al^{3+}) are involved. However, 2+ ions (Be^{2+}) or even 1+ (Li^{+}) show some polarizing power because their sizes are so small (e.g., LiI is ionic but has some covalent bonding present). Note that this is not the ionic polarization effect that refers to displacement of ions in the lattice due to the application of an electric field.

Comparison with Covalent Bonding

In ionic bonding, the atoms are bound by attraction of opposite ions, whereas, in covalent bonding, atoms are bound by sharing electrons to attain stable electron configurations. In covalent bonding, the molecular geometry around each atom is determined by valence shell electron pair repulsion VSEPR rules, whereas, in ionic materials, the geometry follows maximum packing rules. One could say that covalent bonding is more *directional* in the sense that the energy penalty for not adhering to the optimum bond angles is large, whereas ionic bonding has no such penalty. There are no shared electron pairs to repel each other, the ions should simply be packed as efficiently as possible. This often leads to much higher coordination numbers. In NaCl, each ion has 6 bonds and all bond angles are 90 degrees. In CsCl the coordination number is 8. By comparison carbon typically has a maximum of four bonds.

Purely ionic bonding cannot exist, as the proximity of the entities involved in the bonding allows some degree of sharing electron density between them. Therefore, all ionic bonding has some covalent character. Thus, bonding is considered ionic where the ionic character is greater than the covalent character. The larger the difference in electronegativity between the two types of atoms involved in the bonding, the more ionic (polar) it is. Bonds with partially ionic and partially covalent character are called polar covalent bonds. For example, Na–Cl and Mg–O interactions have a few percent covalency, while Si–O bonds are usually ~50% ionic and ~50% covalent. Pauling estimated that an electronegativity difference of 1.7 (on the Pauling scale) corresponds to 50% ionic character, so that a difference greater than 50% corresponds to a bond which is predominantly ionic. Ionic character in covalent bonds can be directly measured for atoms having quadrupolar nuclei (^{2}H, ^{14}N, ^{81,79}Br, ^{35,37}Cl or ^{127}I). These nuclei are generally objects of NQR nuclear quadrupole resonance and NMR nuclear magnetic resonance studies. Interactions between the nuclear quadrupole moments Q and the electric field gradients (EFG) are characterized via the nuclear quadrupole coupling constants $QCC = e^2q_{zz}Q/h$ where the eq_{zz} term corresponds to the principal component of the EFG tensor and e is the elementary charge. In turn, the electric field gradient opens the way to description of bonding modes in molecules when the QCC values are accurately determined by NMR or NQR methods.

In general, when ionic bonding occurs in the solid (or liquid) state, it is not possible to talk about a single "ionic bond" between two individual atoms, because the cohesive forces that keep the lattice together are of a more collective nature. This is quite different in the case of covalent bonding, where we can often speak of a distinct bond localized between two particular atoms. However, even if ionic bonding is combined with some covalency, the result is *not* necessarily discrete bonds of a localized character. In such cases, the resulting bonding often requires description in terms of a band structure consisting of gigantic molecular orbitals spanning the entire crystal. Thus, the bonding in the solid often retains its collective rather than localized nature. When the difference in electronegativity is decreased, the bonding may then lead to a semiconductor, a semimetal or eventually a metallic conductor with metallic bonding.

Covalent Bond

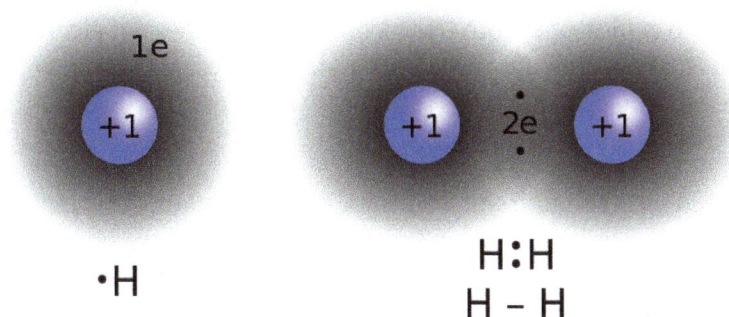

A covalent bond forming H_2 (right) where two hydrogen atoms share the two electrons

A covalent bond, also called a molecular bond, is a chemical bond that involves the sharing of electron pairs between atoms. These electron pairs are known as shared pairs or bonding pairs, and the stable balance of attractive and repulsive forces between atoms, when they share electrons, is known as covalent bonding. For many molecules, the sharing of electrons allows each atom to attain the equivalent of a full outer shell, corresponding to a stable electronic configuration.

Covalent bonding includes many kinds of interactions, including σ-bonding, π-bonding, metal-to-metal bonding, agostic interactions, bent bonds, and three-center two-electron bonds. The term *covalent bond* dates from 1939. The prefix *co-* means *jointly, associated in action, partnered to a lesser degree,* etc.; thus a "co-valent bond", in essence, means that the atoms share "valence", such as is discussed in valence bond theory.

In the molecule H2, the hydrogen atoms share the two electrons via covalent bonding. Covalency is greatest between atoms of similar electronegativities. Thus, covalent bonding does not necessarily require that the two atoms be of the same elements, only that they be of comparable electronegativity. Covalent bonding that entails sharing of electrons over more than two atoms is said to be delocalized.

History

The term *covalence* in regard to bonding was first used in 1919 by Irving Langmuir in a *Journal of the American Chemical Society* article entitled "The Arrangement of Electrons in Atoms and Molecules". Langmuir wrote that "we shall denote by the term *covalence* the number of pairs of electrons that a given atom shares with its neighbors."

The idea of covalent bonding can be traced several years before 1919 to Gilbert N. Lewis, who in 1916 described the sharing of electron pairs between atoms. He introduced the *Lewis notation* or *electron dot notation* or *Lewis dot structure*, in which valence electrons (those in the outer shell) are represented as dots around the atomic symbols. Pairs of electrons located between atoms represent covalent bonds. Multiple pairs represent multiple bonds, such as double bonds and triple bonds. An alternative form of representation, not shown here, has bond-forming electron pairs represented as solid lines.

Lewis proposed that an atom forms enough covalent bonds to form a full (or closed) outer electron shell. In the methane diagram shown here, the carbon atom has a valence of four and is, therefore, surrounded by eight electrons (the octet rule), four from the carbon itself and four from the hydrogens bonded to it. Each hydrogen has a valence of one and is surrounded by two electrons (a duet rule) – its own one electron plus one from the carbon. The numbers of electrons correspond to full shells in the quantum theory of the atom; the outer shell of a carbon atom is the $n = 2$ shell, which can hold eight electrons, whereas the outer (and only) shell of a hydrogen atom is the $n = 1$ shell, which can hold only two.

While the idea of shared electron pairs provides an effective qualitative picture of covalent bonding, quantum mechanics is needed to understand the nature of these bonds and predict the structures and properties of simple molecules. Walter Heitler and Fritz London are credited with the first successful quantum mechanical explanation of a chemical bond (molecular hydrogen) in 1927. Their work was based on the valence bond model, which assumes that a chemical bond is formed when there is good overlap between the atomic orbitals of participating atoms.

Types of Covalent Bonds

Atomic orbitals (except for s orbitals) have specific directional properties leading to different types of covalent bonds. Sigma (σ) bonds are the strongest covalent bonds and are due to head-on overlapping of orbitals on two different atoms. A single bond is usually a σ bond. Pi (π) bonds are weaker and are due to lateral overlap between p (or d) orbitals. A double bond between two given atoms consists of one σ and one π bond, and a triple bond is one σ and two π bonds.

Covalent bonds are also affected by the electronegativity of the connected atoms which determines the chemical polarity of the bond. Two atoms with equal electronegativity will make nonpolar covalent bonds such as H–H. An unequal relationship creates a polar covalent bond such as with H–Cl.

Covalent Structures

There are several types of structures for covalent substances, including individual molecules, molecular structures, macromolecular structures and giant covalent structures. Individual molecules have strong bonds that hold the atoms together, but there are negligible forces of attraction between molecules. Such covalent substances are usually gases, for example, HCl, SO_2, CO_2, and CH_4. In molecular structures, there are weak forces of attraction. Such covalent substances are low-boiling-temperature liquids (such as ethanol), and low-melting-temperature solids (such as iodine and solid CO_2). Macromolecular structures have large numbers of atoms linked by covalent bonds in chains, including synthetic polymers such as polyethylene and nylon, and biopolymers such as proteins and starch. Network covalent structures (or giant covalent structures) contain large numbers of atoms linked in sheets (such as graphite), or 3-dimensional structures (such as diamond and quartz). These substances have high melting and boiling points, are frequently brittle, and tend to have high electrical resistivity. Elements that have high electronegativity, and the ability to form three or four electron pair bonds, often form such large macromolecular structures.

One- and Three-electron Bonds

Bonds with one or three electrons can be found in radical species, which have an odd number of electrons. The simplest example of a 1-electron bond is found in the dihydrogen cation, H+2. One-electron bonds often have about half the bond energy of a 2-electron bond, and are therefore called "half bonds". However, there are exceptions: in the case of dilithium, the bond is actually stronger for the 1-electron Li+2 than for the 2-electron Li_2. This exception can be explained in terms of hybridization and inner-shell effects.

Comparison of the electronic structure of the three-electron bond to the conventional covalent bond.

The simplest example of three-electron bonding can be found in the helium dimer cation, He+2. It is considered a "half bond" because it consists of only one shared electron (rather than two); in molecular orbital terms, the third electron is in an anti-bonding orbital which cancels out half of the bond formed by the other two electrons. Another example of a molecule containing a 3-electron bond, in addition to two 2-electron bonds, is nitric oxide, NO. The oxygen molecule, O_2 can also be regarded as having two 3-electron bonds and one 2-electron bond, which accounts for its paramagnetism and its formal bond order of 2. Chlorine dioxide and its heavier analogues bromine dioxide and iodine dioxide also contain three-electron bonds.

Molecules with odd-electron bonds are usually highly reactive. These types of bond are only stable between atoms with similar electronegativities.

Resonance

There are situations whereby a single Lewis structure is insufficient to explain the electron configuration in a molecule, hence a superposition of structures are needed. The same two atoms in such molecules can be bonded differently in different structures (a single bond in one, a double bond in another, or even none at all), resulting in a non-integer bond order. The nitrate ion is one such example with three equivalent structures. The bond between the nitrogen and each oxygen is a double bond in one structure and a single bond in the other two, so that the average bond order for each N–O interaction is $2 + 1 + 1/3 = 4/3$.

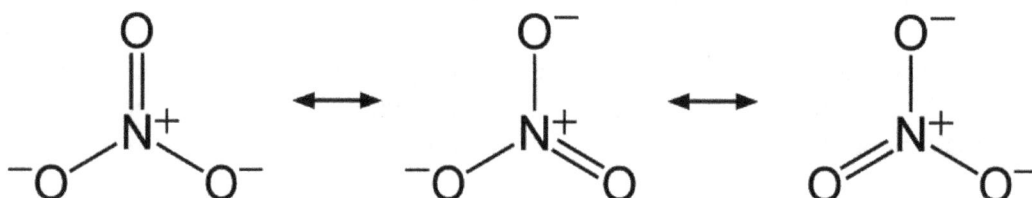

Aromaticity

In organic chemistry, when a molecule with a planar ring obeys Hückel's rule, where the number of π electrons fit the formula $4n + 2$ (where n is an integer), it attains extra stability and symmetry. In benzene, the prototypical aromatic compound, there are 6 π bonding electrons ($n = 1$, $4n + 2 = 6$). These occupy three delocalized π molecular orbitals (molecular orbital theory) or form conjugate π bonds in two resonance structures that linearly combine (valence bond theory), creating a regular hexagon exhibiting a greater stabilization than the hypothetical 1,3,5-cyclohexatriene.

In the case of heterocyclic aromatics and substituted benzenes, the electronegativity differences between different parts of the ring may dominate the chemical behaviour of aromatic ring bonds, which otherwise are equivalent.

Hypervalence

Certain molecules such as xenon difluoride and sulfur hexafluoride have higher co-ordination numbers than would be possible due to strictly covalent bonding according to the octet rule. This is explained by the three-center four-electron bond ("3c–4e") model which interprets the molecular wavefunction in terms of non-bonding highest occupied molecular orbitals in molecular orbital theory and ionic-covalent resonance in valence bond theory.

Electron-deficiency

In three-center two-electron bonds ("3c–2e") three atoms share two electrons in bonding. This type of bonding occurs in electron deficient compounds like diborane. Each such bond (2 per molecule in diborane) contains a pair of electrons which connect the boron atoms to each other in a banana shape, with a proton (nucleus of a hydrogen atom) in the middle of the bond, sharing electrons with both boron atoms. In certain cluster compounds, so-called four-center two-electron bonds also have been postulated.

Quantum Mechanical Description

After the development of quantum mechanics, two basic theories were proposed to provide a quantum description of chemical bonding: valence bond (VB) theory and molecular orbital (MO) theory. A more recent quantum description is given in terms of atomic contributions to the electronic density of states.

Metallic Bonding

Metallic bonding arises from the electrostatic attractive force between conduction electrons (in the form of an electron cloud of delocalized electrons) and positively charged metal ions. It may be described as the sharing of *free* electrons among a lattice of positively charged ions (cations). Metallic bonding accounts for many physical properties of metals, such as strength, ductility, thermal and electrical resistivity and conductivity, opacity, and luster.

Metallic bonding is not the only type of chemical bonding a metal can exhibit, even as a pure substance. For example, elemental gallium consists of covalently-bound pairs of atoms in both liquid and solid state—these pairs form a crystal lattice with metallic bonding between them. Another example of a metal–metal covalent bond is mercurous ion ($Hg2+2$).

History

As chemistry developed into a science it became clear that metals formed the large majority of the periodic table of the elements and great progress was made in the description of the salts that can be formed in reactions with acids. With the advent of electrochemistry it became clear that metals generally go into solution as positively charged ions and the oxidation reactions of the metals became well understood in the electrochemical series. A picture emerged of metals as positive ions held together by an ocean of negative electrons.

With the advent of quantum mechanics this picture was given more formal interpretation in the form of the free electron model and its further extension, the nearly free electron model. In both of these models the electrons are seen as a gas traveling through the lattice of the solid with an energy that is essentially isotropic in that it depends on the square of the magnitude, *not* the direction of the momentum vector k. In three-dimensional k-space, the set of points of the highest filled levels (the Fermi surface) should therefore be a sphere. In the nearly free correction of the model, box-like Brillouin zones are added to k-space by the periodic potential experienced from the (ionic) lattice, thus mildly breaking the isotropy.

The advent of X-ray diffraction and thermal analysis made it possible to study the structure of crystalline solids, including metals and their alloys, and the construction of phase diagrams became accessible. Despite all this progress the nature of intermetallic compounds and alloys largely remained a mystery and their study was often empirical. Chemists generally steered away from anything that did not seem to follow Dalton's laws of multiple proportions and the problem was considered the domain of a different science, metallurgy.

The almost-free electron model was eagerly taken up by some researchers in this field, notably Hume-Rothery, in an attempt to explain why certain intermetallic alloys with certain compositions would form and others would not. Initially his attempts were quite successful. His idea was to add electrons to inflate the spherical Fermi-balloon inside the series of Brillouin-boxes and determine when a certain box would be full. This indeed predicted a fairly large number of observed alloy compositions. Unfortunately, as soon as cyclotron resonance became available and the shape of the balloon could be determined, it was found that the assumption that the balloon was spherical did not hold at all, except perhaps in the case of caesium. This reduced many of the conclusions to examples of how a model can sometimes give a whole series of correct predictions, yet still be wrong.

The free-electron debacle showed researchers that the model assuming that the ions were in a sea of free electrons needed modification, and so a number of quantum mechanical models such as band structure calculations based on molecular orbitals or the density functional theory were developed. In these models, one either departs from the atomic orbitals of neutral atoms that share their electrons or (in the case of density functional theory) departs from the total electron density. The free-electron picture has, nevertheless, remained a dominant one in education.

The electronic band structure model became a major focus not only for the study of metals but even more so for the study of semiconductors. Together with the electronic states, the vibrational states were also shown to form bands. Rudolf Peierls showed that, in the case of a one-dimensional row of metallic atoms, say hydrogen, an instability had to arise that would lead to the breakup of such a chain into individual molecules. This sparked an interest in the general question: When is collective metallic bonding stable and when will a more localized form of bonding take its place? Much research went into the study of clustering of metal atoms.

As powerful as the concept of the band structure proved to be in the description of metallic bonding, it does have a drawback. It remains a one-electron approximation to a multitudinous many-body problem. In other words, the energy states of each electron are described as if all the other electrons simply form a homogeneous background. Researchers like Mott and Hubbard realized that this was perhaps appropriate for strongly delocalized s- and p-electrons but for d-electrons, and even more for f-electrons the interaction with electrons (and atomic displacements) in the local environment may become stronger than the delocalization that leads to broad bands. Thus, the transition from localized unpaired electrons to itinerant ones partaking in metallic bonding became more comprehensible.

The Nature of Metallic Bonding

The combination of two phenomena gives rise to metallic bonding: delocalization of electrons and the availability of a far larger number of delocalized energy states than of delocalized electrons. The latter could be called electron deficiency.

In 2D

Graphene is an example of two-dimensional metallic bonding. Its metallic bonds are similar to aromatic bonding in benzene, naphthalene, anthracene, ovalene, and so on.

In 3D

Metal aromaticity in metal clusters is another example of delocalization, this time often in three-dimensional entities. Metals take the delocalization principle to its extreme and one could say that a crystal of a metal represents a single molecule over which all conduction electrons are delocalized in all three dimensions. This means that inside the metal one can generally not distinguish molecules, so that the metallic bonding is neither intra- nor intermolecular. 'Nonmolecular' would perhaps be a better term. Metallic bonding is mostly non-polar, because even in alloys there is little difference among the electronegativities of the atoms participating in the bonding interaction (and, in pure elemental metals, none at all). Thus, metallic bonding is an extremely delocalized communal form of covalent bonding. In a sense, metallic bonding is not a 'new' type of bonding at all, therefore, and it describes the bonding only as present in a *chunk* of condensed matter, be it crystalline solid, liquid, or even glass. Metallic vapors by contrast are often atomic (Hg) or at times contain molecules like Na_2 held together by a more conventional covalent bond. This is why it is not correct to speak of a single 'metallic bond'.

The delocalization is most pronounced for s- and p-electrons. For caesium it is so strong that the electrons are virtually free from the caesium atoms to form a gas constrained only by the surface of the metal. For caesium, therefore, the picture of Cs^+ ions held together by a negatively charged electron gas is not too inaccurate. For other elements the electrons are less free, in that they still experience the potential of the metal atoms, sometimes quite strongly. They require a more intricate quantum mechanical treatment (e.g., tight binding) in which the atoms are viewed as neutral, much like the carbon atoms in benzene. For d- and especially f-electrons the delocalization is not strong at all and this explains why these electrons are able to continue behaving as unpaired electrons that retain their spin, adding interesting magnetic properties to these metals.

Electron Deficiency and Mobility

Metal atoms contain few electrons in their valence shells relative to their periods or energy levels. They are electron deficient elements and the communal sharing does not change that. There remain far more available energy states than there are shared electrons. Both requirements for conductivity are therefore fulfilled: strong delocalization and partly filled energy bands. Such electrons can therefore easily change from one energy state into a slightly different one. Thus, not only do they become delocalized, forming a sea of electrons permeating the lattice, but they are also able to migrate through the lattice when an external electrical field is imposed, leading to electrical conductivity. Without the field, there are electrons moving equally in all directions. Under the field, some will adjust their state slightly, adopting a different wave vector. As a consequence, there will be more moving one way than the other and a net current will result.

The freedom of conduction electrons to migrate also give metal atoms, or layers of them, the capacity to slide past each other. Locally, bonds can easily be broken and replaced by new ones after the deformation. This process does not affect the communal metallic bonding very much. This gives rise to metals' typical characteristic phenomena of malleability and ductility. This is particularly true for pure elements. In the presence of dissolved impurities, the defects in the lattice that function as cleavage points may get blocked and the material becomes harder. Gold, for example, is very soft in pure form (24-karat), which is why alloys of 18-karat or lower are preferred in jewelry.

Metals are typically also good conductors of heat, but the conduction electrons only contribute partly to this phenomenon. Collective (i.e., delocalized) vibrations of the atoms known as phonons that travel through the solid as a wave, contribute strongly.

However, the latter also holds for a substance like diamond. It conducts heat quite well but *not* electricity. The latter is *not* a consequence of the fact that delocalization is absent in diamond, but simply that carbon is not electron deficient. The electron deficiency is an important point in distinguishing metallic from more conventional covalent bonding. Thus, we should amend the expression given above into: *Metallic bonding is an extremely delocalized communal form of electron deficient covalent bonding.*

Metallic Radius

Metallic radius is defined as one-half of the distance between the two adjacent metal ions in the metallic lattice. This radius depends on the nature of the atom as well as its environment—specifically, on the coordination number (CN), which in turn depends on the temperature and applied pressure.

When comparing periodic trends in the size of atoms it is often desirable to apply so-called Goldschmidt correction, which converts the radii to the values the atoms would have if they were 12-coordinated. Since metallic radii are always biggest for the highest coordination number, correction for less dense coordinations involves multiplying by x, where $0 < x < 1$. Specifically, for CN = 4, x = 0.88; for CN = 6, x = 0.96, and for CN = 8, x = 0.97. The correction is named after Victor Goldschmidt who obtained the numerical values quoted above.

The radii follow general periodic trends: they decrease across the period due to increase in the effective nuclear charge, which is not offset by the increased number of valence electrons. The radii also increase down the group due to increase in principal quantum number. Between rows 3 and 4, the lanthanide contraction is observed – there is very little increase of the radius down the group due to the presence of poorly shielding f orbitals.

Strength of the Bond

The atoms in metals have a strong attractive force between them. Much energy is required to overcome it. Therefore, metals often have high boiling points, with tungsten (5828 K) being extremely high. A remarkable exception are the elements of the zinc group: Zn, Cd, and Hg. Their electron configuration ends in $...ns^2$ and this comes to resemble a noble gas configuration like that of helium more and more when going down in the periodic table because the energy distance to the empty np orbitals becomes larger. These metals are therefore relatively volatile, and are avoided in ultra-high vacuum systems.

Otherwise, metallic bonding can be very strong, even in molten metals, such as Gallium. Even though gallium will melt from the heat of one's hand just above room temperature, its boiling point is not far from that of copper. Molten gallium is therefore a very nonvolatile liquid thanks to its strong metallic bonding.

The strong bonding of metals in the liquid form demonstrates that the energy of a metallic bond is not a strong function of the direction of the metallic bond; this lack of bond directionality is a direct consequence of electron delocalization, and is best understood in contrast to the directional bonding of covalent bonds. The energy of a metallic bond is thus mostly a function of the amount of electrons which surround the metallic atom, as exemplified by the Embedded atom model. This typically results in metals assuming relatively simple, close-packed crystal structures, such as FCC, BCC, and HCP.

Given high enough cooling rates and appropriate alloy composition, metallic bonding can occur even in glasses with an amorphous structure.

Much biochemistry is mediated by the weak interaction of metal ions and biomolecules. Such interactions and their associated conformational change has been measured using dual polarisation interferometry.

Solubility and Compound Formation

Metals are insoluble in water or organic solvents unless they undergo a reaction with them. Typically this is an oxidation reaction that robs the metal atoms of their itinerant electrons, destroying the metallic bonding. However metals are often readily soluble in each other while retaining the metallic character of their bonding. Gold, for example, dissolves easily in mercury, even at room temperature. Even in solid metals, the solubility can be extensive. If the structures of the two metals are the same, there can even be complete solid solubility, as in the case of electrum, the alloys of silver and gold. At times, however, two metals will form alloys with different structures than either of the two parents. One could call these materials metal compounds, but, because materials with metallic bonding are typically not molecular, Dalton's law of integral proportions is not valid and often a range of stoichiometric ratios can be achieved. It is better to abandon such concepts as 'pure substance' or 'solute' is such cases and speak of phases instead. The study of such phases has traditionally been more the domain of metallurgy than of chemistry, although the two fields overlap considerably.

Localization and Clustering: from Bonding to Bonds

The metallic bonding in complicated compounds does not necessarily involve all constituent elements equally. It is quite possible to have an element or more that do not partake at all. One could picture the conduction electrons flowing around them like a river around an island or a big rock. It is possible to observe which elements do partake, e.g., by looking at the core levels in an X-ray photoelectron spectroscopy (XPS) spectrum. If an element partakes, its peaks tend to be skewed.

Some intermetallic materials e.g. do exhibit metal clusters, reminiscent of molecules and these compounds are more a topic of chemistry than of metallurgy. The formation of the clusters could be seen as a way to 'condense out' (localize) the electron deficient bonding into bonds of a more localized nature. Hydrogen is an extreme example of this form of condensation. At high pressures it is a metal. The core of the planet Jupiter could be said to be held together by a combination of metallic bonding and high pressure induced by gravity. At lower pressures however the bonding becomes entirely localized into a regular covalent bond. The localization is so complete that the (more familiar) H_2 gas results. A similar argument holds for an element like boron. Though it is electron deficient compared to carbon, it does not form a metal. Instead it has a number of complicated structures in which icosahedral B_{12} clusters dominate. Charge density waves are a related phenomenon.

As these phenomena involve the movement of the atoms towards or away from each other, they can be interpreted as the coupling between the electronic and the vibrational states (i.e. the phonons) of the material. A different such electron-phonon interaction is thought to cause a very different result at low temperatures, that of superconductivity. Rather than blocking the mobility of the charge carriers by forming electron pairs in localized bonds, Cooper-pairs are formed that no longer experience any resistance to their mobility.

Optical Properties

The presence of an ocean of mobile charge carriers has profound effects on the optical properties of metals. They can only be understood by considering the electrons as a *collective* rather than considering the states of individual electrons involved in more conventional covalent bonds.

Light consists of a combination of an electrical and a magnetic field. The electrical field is usually able to excite an elastic response from the electrons involved in the metallic bonding. The result is that photons are not able to penetrate very far into the metal and are typically reflected. They bounce off, although some may also be absorbed. This holds equally for all photons of the visible spectrum, which is why metals are often silvery white or grayish with the characteristic specular reflection of metallic luster. The balance between reflection and absorption determines how white or how gray they are, although surface tarnish can obscure such observations. Silver, a very good metal with high conductivity is one of the whitest.

Notable exceptions are reddish copper and yellowish gold. The reason for their color is that there is an upper limit to the frequency of the light that metallic electrons can readily respond to, the plasmon frequency. At the plasmon frequency, the frequency-dependent dielectric function of the free electron gas goes from negative (reflecting) to positive (transmitting); higher frequency photons are not reflected at the surface, and do not contribute to the color of the metal. There are some materials like indium tin oxide (ITO) that are metallic conductors (actually degenerate semiconductors) for which this threshold is in the infrared, which is why they are transparent in the visible, but good mirrors in the IR.

For silver the limiting frequency is in the far UV, but for copper and gold it is closer to the visible. This explains the colors of these two metals. At the surface of a metal resonance effects known as surface plasmons can result. They are collective oscillations of the conduction electrons like a ripple in the electronic ocean. However, even if photons have enough energy they usually do not have enough momentum to set the ripple in motion. Therefore, plasmons are hard to excite on a bulk metal. This is why gold and copper still look like lustrous metals albeit with a dash of color. However, in colloidal gold the metallic bonding is confined to a tiny metallic particle, preventing the oscillation wave of the plasmon from 'running away'. The momentum selection rule is therefore broken, and the plasmon resonance causes an extremely intense absorption in the green with a resulting beautiful purple-red color. Such colors are orders of magnitude more intense than ordinary absorptions seen in dyes and the like that involve individual electrons and their energy states.

Electronegativity

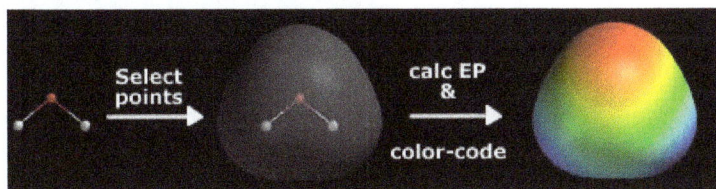

Electrostatic potential map of a water molecule, where the oxygen atom has a more negative charge (red) than the positive (blue) hydrogen atoms

Electronegativity, symbol χ, is a chemical property that describes the tendency of an atom or a functional group to attract electrons (or electron density) towards itself. An atom's electronegativity is affected by both its atomic number and the distance at which its valence electrons reside from the charged nucleus. The higher the associated electronegativity number, the more an element or compound attracts electrons towards it.

The term "electronegativity" was introduced by Jöns Jacob Berzelius in 1811, though the concept was known even before that and was studied by many chemists including Avogadro. In spite of its long history, an accurate scale of electronegativity had to wait till 1932, when Linus Pauling proposed an electronegativity scale, which depends on bond energies, as a development of valence bond theory. It has been shown to correlate with a number of other chemical properties. Electronegativity cannot be directly measured and must be calculated from other atomic or molecular properties. Several methods of calculation have been proposed, and although there may be small differences in the numerical values of the electronegativity, all methods show the same periodic trends between elements.

The most commonly used method of calculation is that originally proposed by Linus Pauling. This gives a dimensionless quantity, commonly referred to as the Pauling scale, on a relative scale running from around 0.7 to 3.98 (hydrogen = 2.20). When other methods of calculation are used, it is conventional (although not obligatory) to quote the results on a scale that covers the same range of numerical values: this is known as an electronegativity in Pauling units.

As it is usually calculated, electronegativity is not a property of an atom alone, but rather a property of an atom in a molecule. Properties of a free atom include ionization energy and electron affinity. It is to be expected that the electronegativity of an element will vary with its chemical environment, but it is usually considered to be a transferable property, that is to say that similar values will be valid in a variety of situations.

On the most basic level, electronegativity is determined by factors like the nuclear charge (the more protons an atom has, the more "pull" it will have on electrons) and the number/location of other electrons present in the atomic shells (the more electrons an atom has, the farther from the nucleus the valence electrons will be, and as a result the less positive charge they will experience—both because of their increased distance from the nucleus, and because the other electrons in the lower energy core orbitals will act to shield the valence electrons from the positively charged nucleus).

The opposite of electronegativity is electropositivity: a measure of an element's ability to donate electrons.

Caesium is the least electronegative element in the periodic table (=0.79), while fluorine is most electronegative (=3.98). (Francium and caesium were originally both assigned 0.7; caesium's value was later refined to 0.79, but no experimental data allows a similar refinement for francium. However, francium's ionization energy is known to be slightly higher than caesium's, in accordance with the relativistic stabilization of the 7s orbital, and this in turn implies that francium is in fact more electronegative than caesium.)

Electronegativities of the Elements

Periodic table of electronegativity by Pauling scale

→ Atomic radius decreases → Ionization energy increases → Electronegativity increases →

	1	2	3	4	5	6	7	8	9	10	11	12	13	14	15	16	17	18
Group →																		
↓ Period																		
1	H 2.20																	He
2	Li 0.98	Be 1.57											B 2.04	C 2.55	N 3.04	O 3.44	F 3.98	Ne
3	Na 0.93	Mg 1.31											Al 1.61	Si 1.90	P 2.19	S 2.58	Cl 3.16	Ar
4	K 0.82	Ca 1.00	Sc 1.36	Ti 1.54	V 1.63	Cr 1.66	Mn 1.55	Fe 1.83	Co 1.88	Ni 1.91	Cu 1.90	Zn 1.65	Ga 1.81	Ge 2.01	As 2.18	Se 2.55	Br 2.96	Kr 3.00
5	Rb 0.82	Sr 0.95	Y 1.22	Zr 1.33	Nb 1.6	Mo 2.16	Tc 1.9	Ru 2.2	Rh 2.28	Pd 2.20	Ag 1.93	Cd 1.69	In 1.78	Sn 1.96	Sb 2.05	Te 2.1	I 2.66	Xe 2.60
6	Cs 0.79	Ba 0.89	*	Hf 1.3	Ta 1.5	W 2.36	Re 1.9	Os 2.2	Ir 2.20	Pt 2.28	Au 2.54	Hg 2.00	Tl 1.62	Pb 1.87	Bi 2.02	Po 2.0	At 2.2	Rn 2.2
7	Fr 0.7[en 1]	Ra 0.9	**	Rf	Db	Sg	Bh	Hs	Mt	Ds	Rg	Cn	Uut	Fl	Uup	Lv	Uus	Uuo
		*	La 1.1	Ce 1.12	Pr 1.13	Nd 1.14	Pm 1.13	Sm 1.17	Eu 1.2	Gd 1.2	Tb 1.1	Dy 1.22	Ho 1.23	Er 1.24	Tm 1.25	Yb 1.1	Lu 1.27	
		**	Ac 1.1	Th 1.3	Pa 1.5	U 1.38	Np 1.36	Pu 1.28	Am 1.13	Cm 1.28	Bk 1.3	Cf 1.3	Es 1.3	Fm 1.3	Md 1.3	No 1.3	Lr	

Values are given for the elements in their most common and stable oxidation states.

1. Electronegativity of francium was chosen by Pauling as 0.7, close to that of caesium (also assessed 0.7 at that point). The base value of hydrogen was later increased by 0.10 and caesium's electronegativity was later refined to 0.79; however, no refinements have been made for francium as no experiment has been conducted and the old value was kept. However, francium is expected and, to a small extent, observed to be less electropositive than caesium.

Methods of Calculation

Pauling Electronegativity

Pauling first proposed the concept of electronegativity in 1932 as an explanation of the fact that the covalent bond between two different atoms (A–B) is stronger than would be expected by taking the average of the strengths of the A–A and B–B bonds. According to valence bond theory, of which Pauling was a notable proponent, this "additional stabilization" of the heteronuclear bond is due to the contribution of ionic canonical forms to the bonding.

The difference in electronegativity between atoms A and B is given by:

$$|\chi_A - \chi_B| = (eV)^{-1/2} \sqrt{E_d(AB) - [E_d(AA) + E_d(BB)]/2}$$

where the dissociation energies, E_d, of the A–B, A–A and B–B bonds are expressed in electron-volts, the factor $(eV)^{-\frac{1}{2}}$ being included to ensure a dimensionless result. Hence, the difference in Pauling electronegativity between hydrogen and bromine is 0.73 (dissociation energies: H–Br, 3.79 eV; H–H, 4.52 eV; Br–Br 2.00 eV)

As only differences in electronegativity are defined, it is necessary to choose an arbitrary reference point in order to construct a scale. Hydrogen was chosen as the reference, as it forms covalent bonds with a large variety of elements: its electronegativity was fixed first at 2.1, later revised to 2.20. It is also necessary to decide which of the two elements is the more electronegative (equivalent to choosing one of the two possible signs for the square root). This is usually done using "chemical intuition": in the above example, hydrogen bromide dissolves in water to form H^+ and Br^- ions, so it may be assumed that bromine is more electronegative than hydrogen. However, in principle, since the same electronegativities should be obtained for any two bonding compounds, the data are in fact overdetermined, and the signs are unique once a reference point is fixed (usually, for H or F).

To calculate Pauling electronegativity for an element, it is necessary to have data on the dissociation energies of at least two types of covalent bond formed by that element. A. L. Allred updated Pauling's original values in 1961 to take account of the greater availability of thermodynamic data, and it is these "revised Pauling" values of the electronegativity that are most often used.

The essential point of Pauling electronegativity is that there is an underlying, quite accurate, semi-empirical formula for dissociation energies, namely:

$$E_d(AB) = [E_d(AA) + E_d(BB)]/2 + (\chi_A - \chi_B)^2 eV$$

or sometimes, a more accurate fit

$$E_d(AB) = \sqrt{E_d(AA)E_d(BB)} + 1.3(\chi_A - \chi_B)^2 eV$$

This is an approximate equation, but holds with good accuracy. Pauling obtained it by noting that a bond can be approximately represented as a quantum mechanical superposition of a covalent bond and two ionic bond-states. The covalent energy of a bond is approximately, by quantum mechanical calculations, the geometric mean of the two energies of covalent bonds of the same molecules, and there is an additional energy that comes from ionic factors, i.e. polar character of the bond.

The geometric mean is approximately equal to the arithmetic mean - which is applied in the first formula above - when the energies are of the similar value, e.g., except for the highly electropositive elements, where there is a larger difference of two dissociation energies; the geometric mean is more accurate and almost always gives a positive excess energy, due to ionic bonding. The square root of this excess energy, Pauling notes, is approximately additive, and hence one can introduce the electronegativity. Thus, it is this semi-empirical formula for bond energy that underlies Pauling electronegativity concept.

The formulas are approximate, but this rough approximation is in fact relatively good and gives the right intuition, with the notion of polarity of the bond and some theoretical grounding in quantum mechanics. The electronegativities are then determined to best fit the data.

In more complex compounds, there is additional error since electronegativity depends on the molecular environment of an atom. Also, the energy estimate can be only used for single, not for multiple bonds. The energy of formation of a molecule containing only single bonds then can be approximated from an electronegativity table, and depends on the constituents and sum of squares of differences of electronegativities of all pairs of bonded atoms. Such a formula for estimating energy typically has relative error of order of 10%, but can be used to get a rough qualitative idea and understanding of a molecule.

Mulliken Electronegativity

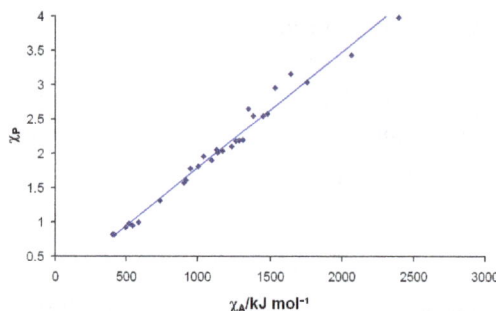

The correlation between Mulliken electronegativities (x-axis, in kJ/mol) and Pauling electronegativities (y-axis).

Robert S. Mulliken proposed that the arithmetic mean of the first ionization energy (E_i) and the electron affinity (E_{ea}) should be a measure of the tendency of an atom to attract electrons. As this definition is not dependent on an arbitrary relative scale, it has also been termed absolute electronegativity, with the units of kilojoules per mole or electronvolts.

$$\chi = (E_i + E_{ea}) / 2$$

However, it is more usual to use a linear transformation to transform these absolute values into values that resemble the more familiar Pauling values. For ionization energies and electron affinities in electronvolts,

$$\chi = 0.187(E_i + E_{ea}) + 0.17$$

and for energies in kilojoules per mole,

$$\chi = (1.97 \times 10^{-3})(E_i + E_{ea}) + 0.19.$$

The Mulliken electronegativity can only be calculated for an element for which the electron affinity is known, fifty-seven elements as of 2006. The Mulliken electronegativity of an atom is sometimes said to be the negative of the chemical potential. By inserting the energetic definitions of the ionization potential and electron affinity into the Mulliken electronegativity, it is possible to show that the Mulliken chemical potential is a finite difference approximation of the electronic energy with respect to the number of electrons., i.e.,

$$\mu(\text{Mulliken}) = -\chi(\text{Mulliken}) = -(E_i + E_{ea})/2$$

Allred–Rochow Electronegativity

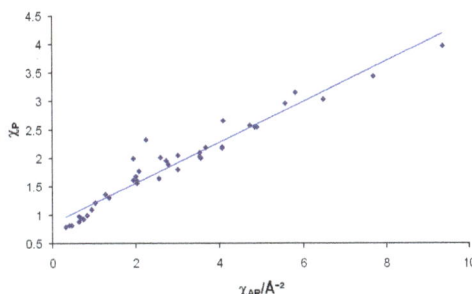

The correlation between Allred–Rochow electronegativities (x-axis, in Å^{-2}) and Pauling electronegativities (y-axis).

A. Louis Allred and Eugene G. Rochow considered that electronegativity should be related to the charge experienced by an electron on the "surface" of an atom: The higher the charge per unit area of atomic surface the greater the tendency of that atom to attract electrons. The effective nuclear charge, Z_{eff}, experienced by valence electrons can be estimated using Slater's rules, while the surface area of an atom in a molecule can be taken to be proportional to the square of the covalent radius, r_{cov}. When r_{cov} is expressed in picometres,

$$\chi = 3590\frac{Z_{eff}}{r_{cov}^2} + 0.744$$

Sanderson Electronegativity Equalization

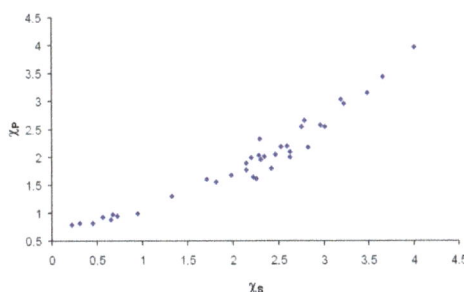

The correlation between Sanderson electronegativities (x-axis, arbitrary units) and Pauling electronegativities (y-axis).

R.T. Sanderson has also noted the relationship between Mulliken electronegativity and atomic size, and has proposed a method of calculation based on the reciprocal of the atomic volume.

With a knowledge of bond lengths, Sanderson's model allows the estimation of bond energies in a wide range of compounds. Sanderson's model has also been used to calculate molecular geometry, s-electrons energy, NMR spin-spin constants and other parameters for organic compounds. This work underlies the concept of electronegativity equalization, which suggests that electrons distribute themselves around a molecule to minimize or to equalize the Mulliken electronegativity. This behavior is analogous to the equalization of chemical potential in macroscopic thermodynamics.

Allen Electronegativity

Perhaps the simplest definition of electronegativity is that of Leland C. Allen, who has proposed that it is related to the average energy of the valence electrons in a free atom,

$$\chi = \frac{n_s \varepsilon_s + n_p \varepsilon_p}{n_s + n_p}$$

where $\varepsilon_{s,p}$ are the one-electron energies of s- and p-electrons in the free atom and $n_{s,p}$ are the number of s- and p-electrons in the valence shell. It is usual to apply a scaling factor, 1.75×10^{-3} for energies expressed in kilojoules per mole or 0.169 for energies measured in electronvolts, to give values that are numerically similar to Pauling electronegativities.

The one-electron energies can be determined directly from spectroscopic data, and so electronegativities calculated by this method are sometimes referred to as spectroscopic electronegativities. The necessary data are available for almost all elements, and this method allows the estimation of electronegativities for elements that cannot be treated by the other methods, e.g. francium, which has an Allen electronegativity of 0.67. However, it is not clear what should be considered to be valence electrons for the d- and f-block elements, which leads to an ambiguity for their electronegativities calculated by the Allen method.

In this scale neon has the highest electronegativity of all elements, followed by fluorine, helium, and oxygen.

Correlation of Electronegativity With Other Properties

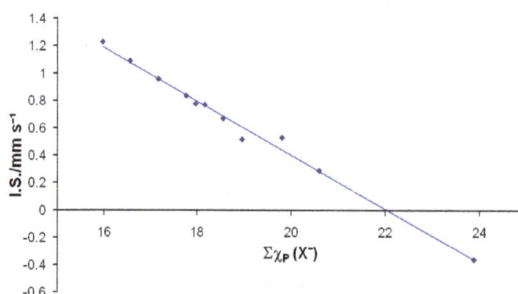

The variation of the isomer shift (y-axis, in mm/s) of $[SnX_6]^{2-}$ anions, as measured by [119]Sn Mössbauer spectroscopy, against the sum of the Pauling electronegativities of the halide substituents (x-axis).

The wide variety of methods of calculation of electronegativities, which all give results that correlate well with one another, is one indication of the number of chemical properties which might be affected by electronegativity. The most obvious application of electronegativities is in the discussion of bond polarity, for which the concept was introduced by Pauling. In general, the greater the difference in electronegativity between two atoms the more polar the bond that will be formed between them, with the atom having the higher electronegativity being at the negative end of the dipole. Pauling proposed an equation to relate "ionic character" of a bond to the difference in electronegativity of the two atoms, although this has fallen somewhat into disuse.

Several correlations have been shown between infrared stretching frequencies of certain bonds and the electronegativities of the atoms involved: however, this is not surprising as such stretching frequencies depend in part on bond strength, which enters into the calculation of Pauling electronegativities. More convincing are the correlations between electronegativity and chemical shifts in NMR spectroscopy or isomer shifts in Mössbauer spectroscopy. Both these measurements depend on the s-electron density at the nucleus, and so are a good indication that the different measures of electronegativity really are describing "the ability of an atom in a molecule to attract electrons to itself".

Trends in Electronegativity

Periodic Trends

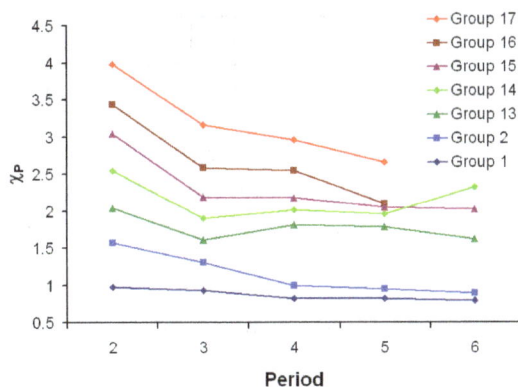

The variation of Pauling electronegativity (y-axis) as one descends the main groups of the periodic table from the second period to the sixth period

In general, electronegativity increases on passing from left to right along a period, and decreases on descending a group. Hence, fluorine is the most electronegative of the elements (not counting noble gases), whereas caesium is the least electronegative, at least of those elements for which substantial data is available. This would lead one to believe that caesium fluoride is the compound whose bonding features the most ionic character.

There are some exceptions to this general rule. Gallium and germanium have higher electronegativities than aluminium and silicon, respectively, because of the d-block contraction. Elements of the fourth period immediately after the first row of the transition metals have unusually small atomic radii because the 3d-electrons are not effective at shielding the increased nuclear charge, and smaller atomic size correlates with higher electronegativity. The anomalously high electronegativity of lead, in particular when compared to thallium and bismuth, appears to be an artifact of data selection (and data availability)—methods of calculation other than the Pauling method show the normal periodic trends for these elements.

Variation of Electronegativity With Oxidation Number

In inorganic chemistry it is common to consider a single value of the electronegativity to be valid for most "normal" situations. While this approach has the advantage of simplicity, it is clear that the electronegativity of an element is *not* an invariable atomic property and, in particular, increases with the oxidation state of the element.

Allred used the Pauling method to calculate separate electronegativities for different oxidation states of the handful of elements (including tin and lead) for which sufficient data was available. However, for most elements, there are not enough different covalent compounds for which bond dissociation energies are known to make this approach feasible. This is particularly true of the transition elements, where quoted electronegativity values are usually, of necessity, averages over several different oxidation states and where trends in electronegativity are harder to see as a result.

Acid	Formula	Chlorine oxidation state	pK_a
Hypochlorous acid	HClO	+1	+7.5
Chlorous acid	$HClO_2$	+3	+2.0
Chloric acid	$HClO_3$	+5	−1.0
Perchloric acid	$HClO_4$	+7	−10

The chemical effects of this increase in electronegativity can be seen both in the structures of oxides and halides and in the acidity of oxides and oxoacids. Hence CrO_3 and Mn_2O_7 are acidic oxides with low melting points, while Cr_2O_3 is amphoteric and Mn_2O_3 is a completely basic oxide.

The effect can also be clearly seen in the dissociation constants of the oxoacids of chlorine. The effect is much larger than could be explained by the negative charge being shared among a larger number of oxygen atoms, which would lead to a difference in pK_a of $\log_{10}(¼) = −0.6$ between hypochlorous acid and perchloric acid. As the oxidation state of the central chlorine atom increases, more electron density is drawn from the oxygen atoms onto the chlorine, reducing the partial negative charge on the oxygen atoms and increasing the acidity.

Group Electronegativity

In organic chemistry, electronegativity is associated more with different functional groups than with individual atoms. The terms group electronegativity and substituent electronegativity are used synonymously. However, it is common to distinguish between the inductive effect and the resonance effect, which might be described as σ- and π-electronegativities, respectively. There are a number of linear free-energy relationships that have been used to quantify these effects, of which the Hammett equation is the best known. Kabachnik parameters are group electronegativities for use in organophosphorus chemistry.

Electropositivity

Electropositivity is a measure of an element's ability to donate electrons, and therefore form positive ions; thus, it is opposed to electronegativity.

Mainly, this is an attribute of metals, meaning that, in general, the greater the metallic character of an element the greater the electropositivity. Therefore, the alkali metals are most electropositive of all. This is because they have a single electron in their outer shell and, as this is relatively far from the nucleus of the atom, it is easily lost; in other words, these metals have low ionization energies.

While electronegativity increases along periods in the periodic table, and decreases down groups, electropositivity *decreases* along periods (from left to right) and *increases* down groups.

Intermolecular Force

Intermolecular forces (IMFs) are forces of attraction or repulsion which act between neighboring particles (atoms, molecules, or ions). They are weak compared to the intramolecular forces, the forces which keep a molecule together. For example the covalent bond, involving the sharing of electron pairs between atoms is much stronger than the forces present between the neighboring molecules. They are an essential part of force fields frequently used in molecular mechanics.

The investigation of intermolecular forces starts from macroscopic observations which point out the existence and action of forces at a molecular level. These observations include non-ideal-gas thermodynamic behavior reflected by virial coefficients, vapor pressure, viscosity, superficial tension and absorption data.

The first reference to the nature of microscopic forces is found in Alexis Clairaut's work *Theorie de la Figure de la Terre*. Other scientists who have contributed to the investigation of microscopic forces include: Laplace, Gauss, Maxwell and Boltzmann.

Attractive intermolecular forces are considered by the following types:

- Ion-induced dipole forces

- Ion-dipole forces

- van der Waals forces (Keesom force, Debye force, and London dispersion force)

Information on intermolecular force is obtained by macroscopic measurements of properties like viscosity, PVT data. The link to microscopic aspects is given by virial coefficients and Lennard-Jones potentials.

Ion-dipole and Ion-induced Dipole Forces

Ion-dipole and ion-induced dipole forces are similar to dipole-dipole and induced-dipole interactions but involve ions, instead of only polar and non-polar molecules. Ion-dipole and ion-induced dipole forces are stronger than dipole-dipole interactions because the charge of any ion is much greater than the charge of a dipole moment. Ion-dipole bonding is stronger than hydrogen bonding.

An ion-dipole force consists of an ion and a polar molecule interacting. They align so that the positive and negative groups are next to one another, allowing for maximum attraction.

An ion-induced dipole force consists of an ion and a non-polar molecule interacting. Like a dipole-induced dipole force, the charge of the ion causes distortion of the electron cloud on the non-polar molecule.

Hydrogen Bonding

A hydrogen bond is the attraction between the lone pair of an electronegative atom and a hydrogen atom that is bonded to either nitrogen, oxygen, or fluorine. The hydrogen bond is often described as a strong electrostatic dipole-dipole interaction. However, it also has some features of covalent bonding: it is directional, stronger than a van der Waals interaction, produces interatomic distances shorter than the sum of van der Waals radius, and usually involves a limited number of interaction partners, which can be interpreted as a kind of valence.

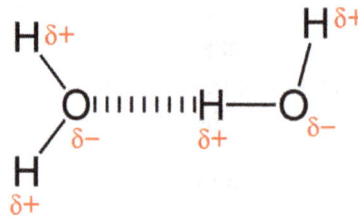

Intermolecular hydrogen bonding is responsible for the high boiling point of water (100 °C) compared to the other group 16 hydrides, which have no hydrogen bonds. Intramolecular hydrogen oxygen bonding is partly responsible for the secondary, tertiary, and quaternary structures of proteins and nucleic acids. It also plays an important role in the structure of polymers, both synthetic and natural.

Van Der Waals Forces

The van der Waals forces arise from interaction between uncharged atoms or molecules, leading not only to such phenomena as the cohesion of condensed phases and physical adsorption of gases, but also to a universal force of attraction between macroscopic bodies.

Keesom (Permanent-permanent Dipoles) Interaction

The first contribution to van der Waals forces is due to electrostatic interactions between charges (in molecular ions), dipoles (for polar molecules), quadrupoles (all molecules with symmetry lower than cubic), and permanent multipoles. It is referred to as Keesom interactions(named after Willem Hendrik Keesom). These forces originate from the attraction between permanent dipoles (dipolar molecules) and are temperature dependent.

They consist of attractive interactions between dipoles that are ensemble averaged over different rotational orientations of the dipoles. It is assumed that the molecules are constantly rotating and never get locked into place. This is a good assumption, but at some point molecules do get locked into place. The energy of a Keesom interaction depends on the inverse sixth power of the distance, unlike the interaction energy of two spatially fixed dipoles, which depends on the inverse third power of the distance. The Keesom interaction can only occur among molecules that possess permanent dipole moments a.k.a. two polar molecules. Also Keesom interactions are very weak van der Waals interactions and do not occur in aqueous solutions that contain electrolytes. The angle averaged interaction is given by the following equation:

$$\frac{-m_1^2 m_2^2}{24\pi^2 \varepsilon_o^2 \varepsilon_r^2 k_b T r^6} = V$$

Where m = charge per length, ε_o = permitivity of free space, ε_r = dielectric constant of surrounding material, T = temperature, k_b = Boltzmann constant, and r = distance between molecules.

Debye (Permanent-induced Dipoles) Force

The second contribution is the induction (also known as polarization) or Debye force, arising from interactions between rotating permanent dipoles and from the polarizability of atoms and molecules (induced dipoles). These induced dipoles occur when one molecule with a permanent dipole repels another molecule's electrons. A molecule with permanent dipole can induce a dipole in a similar neighboring molecule and cause mutual attraction. Debye forces cannot occur between atoms. The forces between induced and permanent dipoles are not as temperature dependent as Keesom interactions because the induced dipole is free to shift and rotate around the non-polar molecule. The Debye induction effects and Keesom orientation effects are referred to as polar interactions.

The induced dipole forces appear from the induction (also known as polarization), which is the attractive interaction between a permanent multipole on one molecule with an induced (by the former di/multi-pole) multipole on another. This interaction is called the Debye force, named after Peter J.W. Debye.

One example of an induction-interaction between permanent dipole and induced dipole is the interaction between HCl and Ar. In this system, Ar experiences a dipole as its electrons are attracted (to the H side of HCl) or repelled (from the Cl side) by HCl. The angle averaged interaction is given by the following equation.

$$\frac{-m_1^2 \alpha_2}{16\pi^2 \varepsilon_o^2 \varepsilon_r^2 r^6} = V$$

Where α = polarizability

This kind of interaction can be expected between any polar molecule and non-polar/symmetrical molecule. The induction-interaction force is far weaker than dipole-dipole interaction, but stronger than the London dispersion force.

London Dispersion Force (Dipole-induced Dipoles Interaction)

The third and dominant contribution is the dispersion or London force (fluctuating dipole-induced dipole), which arises due to the non-zero instantaneous dipole moments of all atoms and molecules. Such polarization can be induced either by a polar molecule or by the repulsion of negatively charged electron clouds in non-polar molecules. Thus, London interactions are caused by random fluctuations of electron density in an electron cloud. An atom with a large number of electrons will have a greater associated London force than an atom with fewer electrons. The dispersion (London) force is the most important component because all materials are polarizable, whereas Keesom and Debye forces require permanent dipoles. The London interaction is universal and is present in atom-atom interactions as well. For various reasons, London interactions (dispersion) have been considered relevant for interactions between macroscopic bodies in condensed systems. Hamaker developed the theory of van der Waals between macroscopic bodies in 1937 and showed that the additivity of these interactions renders them considerably more long-range.

Relative Strength of Forces

Bond type	Dissociation energy (kcal/mol)
Ionic Lattice Energy	250–4000
Covalent Bond Energy	30–260
Hydrogen Bonds	1–12 (about 5 in water)
Dipole–Dipole	0.5–2
London Dispersion Forces	<1 to 15 (estimated from the enthalpies of vaporization of hydrocarbons)

Note: this comparison is only approximate – the actual relative strengths will vary depending on the molecules involved. Ionic and covalent bonding will always be stronger than intermolecular forces in any given substance.

Effect on the Behavior of Gases

Intermolecular forces are repulsive at short distances and attractive at long distances. In a gas, the repulsive force chiefly has the effect of keeping two molecules from occupying the same volume. This gives a real gas a tendency to occupy a larger volume than an ideal gas at the same temperature and pressure. The attractive force draws molecules closer together and gives a real gas a tendency to occupy a smaller volume than an ideal gas. Which interaction is more important depends on temperature and pressure.

In a gas, the distances between molecules are generally large, so intermolecular forces have only a small effect. The attractive force is not overcome by the repulsive force, but by the thermal energy of the molecules. Temperature is the measure of thermal energy, so increasing temperature reduces the influence of the attractive force. In contrast, the influence of the repulsive force is essentially unaffected by temperature.

When a gas is compressed to increase its density, the influence of the attractive force increases. If the gas is made sufficiently dense, the attractions can become large enough to overcome the tendency of thermal motion to cause the molecules to spread out. Then the gas can condense to form a solid or liquid (i.e., a condensed phase). Lower temperature favors the formation of a condensed phase. In a condensed phase, there is very nearly a balance between the attractive and repulsive forces.

Quantum Mechanical Theories

Intermolecular forces observed between atoms and molecules can be described phenomenologically as occurring between permanent and instantaneous dipoles, as outlined above. Alternatively, one may seek a fundamental, unifying theory that is able to explain the various types of interactions such as hydrogen bonding, van der Waals forces and dipole-dipole interactions. Typically, this is done by applying the ideas of quantum mechanics to molecules, and Rayleigh–Schrödinger perturbation theory has been especially effective in this regard. When applied to existing quantum chemistry methods, such a quantum mechanical explanation of intermolecular interactions, this provides an array of approximate methods that can be used to analyze intermolecular interactions.

References

- Stranks, D. R.; Heffernan, M. L.; Lee Dow, K. C.; McTigue, P. T.; Withers, G. R. A. (1970). Chemistry: A structural view. Carlton, Vic.: Melbourne University Press. p. 184. ISBN 0-522-83988-6.

- Campbell, Neil A.; Williamson, Brad; Heyden, Robin J. (2006). Biology: Exploring Life. Boston, MA: Pearson Prentice Hall. ISBN 0-13-250882-6. Retrieved 2012-02-05.[better source needed]

- March, Jerry (1992). Advanced Organic Chemistry: Reactions, Mechanisms, and Structure. John Wiley & Sons. ISBN 0-471-60180-2.

- Atkins, Peter; Loretta Jones (1997). Chemistry: Molecules, Matter and Change. New York: W. H. Freeman & Co. pp. 294–295. ISBN 0-7167-3107-X.

- SW Rick; SJ Stuart (2002). "Electronegativity equalization models". In Kenny B. Lipkowitz; Donald B. Boyd. Reviews in computational chemistry. Wiley. p. 106. ISBN 0-471-21576-7.

- Robert G. Parr; Weitao Yang (1994). Density-functional theory of atoms and molecules. Oxford University Press. p. 91. ISBN 0-19-509276-7.

- Margenau, H. and Kestner, N. (1969) Theory of intermolecular forces, International Series of Monographs in Natural Philosophy, Pergamon Press, ISBN 1483119289

- Majer, V. and Svoboda, V. (1985) Enthalpies of Vaporization of Organic Compounds, Blackwell Scientific Publications, Oxford. ISBN 0632015292

Carbon Nanotubes: An Integrated Study

Carbon nanotubes are allotropes of carbon. They have unusual properties and are very valuable for electronics, optics and nanotechnology. Topics such as optical properties of carbon nanotubes, synthesis of carbon nanotubes, carbon nanotube chemistry and selective chemistry of single-walled nanotubes have been explicated in the section.

Carbon Nanotube

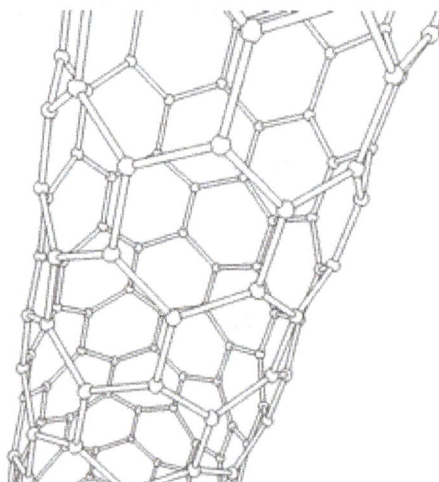

Rotating single-walled zigzag carbon nanotube

Carbon nanotubes (CNTs) are allotropes of carbon with a cylindrical nanostructure. These cylindrical carbon molecules have unusual properties, which are valuable for nanotechnology, electronics, optics and other fields of materials science and technology. Owing to the material's exceptional strength and stiffness, nanotubes have been constructed with length-to-diameter ratio of up to 132,000,000:1, significantly larger than for any other material.

In addition, owing to their extraordinary thermal conductivity, mechanical, and electrical properties, carbon nanotubes find applications as additives to various structural materials. For instance, nanotubes form a tiny portion of the material(s) in some (primarily carbon fiber) baseball bats, golf clubs, car parts or damascus steel.

Nanotubes are members of the fullerene structural family. Their name is derived from their long, hollow structure with the walls formed by one-atom-thick sheets of carbon, called graphene. These sheets are rolled at specific and discrete ("chiral") angles, and the combination of the rolling angle and radius decides the nanotube properties; for example, whether the individual nanotube shell is a metal or semiconductor. Nanotubes are categorized as single-walled nanotubes (SWNTs) and multi-walled nanotubes (MWNTs). Individual nanotubes naturally align themselves into "ropes" held together by van der Waals forces, more specifically, pi-stacking.

Applied quantum chemistry, specifically, orbital hybridization best describes chemical bonding in nanotubes. The chemical bonding of nanotubes is composed entirely of sp^2 bonds, similar to those of graphite. These bonds, which are stronger than the sp^3 bonds found in alkanes and diamond, provide nanotubes with their unique strength.

Types of Carbon Nanotubes and Related Structures

Terminology

There is no consensus on some terms describing carbon nanotubes in scientific literature: both "-wall" and "-walled" are being used in combination with "single", "double", "triple" or "multi", and the letter C is often omitted in the abbreviation; for example, multi-walled carbon nanotube (MWNT).

Most single-walled nanotubes (SWNTs) have a diameter of close to 1 nanometer, and can be many millions of times longer. The structure of a SWNT can be conceptualized by wrapping a one-atom-thick layer of graphite called graphene into a seamless cylinder. The way the graphene sheet is wrapped is represented by a pair of indices (n,m). The integers n and m denote the number of unit vectors along two directions in the honeycomb crystal lattice of graphene. If $m = 0$, the nanotubes are called zigzag nanotubes, and if $n = m$, the nanotubes are called armchair nanotubes. Otherwise, they are called chiral. The diameter of an ideal nanotube can be calculated from its (n,m) indices as follows

$$d = \frac{a}{\pi}\sqrt{(n^2 + nm + m^2)} = 78.3\sqrt{((n+m)^2 - nm)}\text{pm},$$

where a = 0.246 nm.

SWNTs are an important variety of carbon nanotube because most of their properties change significantly with the (n,m) values, and this dependence is non-monotonic. In particular, their band gap can vary from zero to about 2 eV and their electrical conductivity can show metallic or semi-conducting behavior. Single-walled nanotubes are likely candidates for miniaturizing electronics. The most basic building block of these systems is the electric wire, and SWNTs with diameters of an order of a nanometer can be excellent conductors. One useful application of SWNTs is in the development of the first intermolecular field-effect transistors (FET). The first intermolecular logic gate using SWCNT FETs was made in 2001. A logic gate requires both a p-FET and an n-FET. Because SWNTs are p-FETs when exposed to oxygen and n-FETs otherwise, it is possible to expose half of an SWNT to oxygen and protect the other half from it. The resulting SWNT acts as a *not* logic gate with both p and n-type FETs in the same molecule.

Prices for single-walled nanotubes declined from around $1500 per gram as of 2000 to retail prices of around $50 per gram of as-produced 40–60% by weight SWNTs as of March 2010. As of 2016 the retail price of as-produced 75% by weight SWNTs were $2 per gram, cheap enough for widespread use. SWNTs are forecast to make a large impact in electronics applications by 2020 according to the *The Global Market for Carbon Nanotubes* report.

Multi-Walled

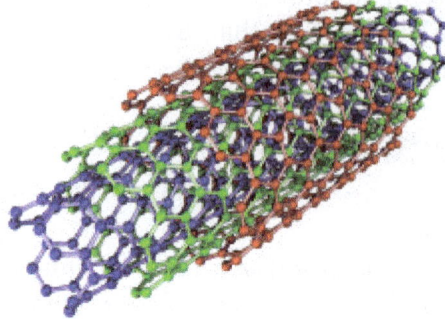

Triple-walled armchair carbon nanotube

Multi-walled nanotubes (MWNTs) consist of multiple rolled layers (concentric tubes) of graphene. There are two models that can be used to describe the structures of multi-walled nanotubes. In the *Russian Doll* model, sheets of graphite are arranged in concentric cylinders, e.g., a (0,8) single-walled nanotube (SWNT) within a larger (0,17) single-walled nanotube. In the *Parchment* model, a single sheet of graphite is rolled in around itself, resembling a scroll of parchment or a rolled newspaper. The interlayer distance in multi-walled nanotubes is close to the distance between graphene layers in graphite, approximately 3.4 Å. The Russian Doll structure is observed more commonly. Its individual shells can be described as SWNTs, which can be metallic or semiconducting. Because of statistical probability and restrictions on the relative diameters of the individual tubes, one of the shells, and thus the whole MWNT, is usually a zero-gap metal.

Double-walled carbon nanotubes (DWNTs) form a special class of nanotubes because their morphology and properties are similar to those of SWNTs but they are more resistant to chemicals. This is especially important when it is necessary to graft chemical functions to the surface of the nanotubes (functionalization) to add properties to the CNT. Covalent functionalization of SWNTs will break some C=C double bonds, leaving "holes" in the structure on the nanotube, and thus modifying both its mechanical and electrical properties. In the case of DWNTs, only the outer wall is modified. DWNT synthesis on the gram-scale was first proposed in 2003 by the CCVD technique, from the selective reduction of oxide solutions in methane and hydrogen.

The telescopic motion ability of inner shells and their unique mechanical properties will permit the use of multi-walled nanotubes as main movable arms in coming nanomechanical devices. Retraction force that occurs to telescopic motion caused by the Lennard-Jones interaction between shells and its value is about 1.5 nN.

Torus

In theory, a nanotorus is a carbon nanotube bent into a torus (doughnut shape). Nanotori are predicted to have many unique properties, such as magnetic moments 1000 times larger than previously expected for certain specific radii. Properties such as magnetic moment, thermal stability, etc. vary widely depending on radius of the torus and radius of the tube.

Nanobud

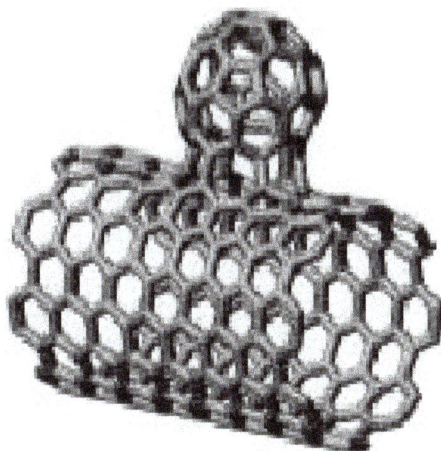

A stable nanobud structure

Carbon nanobuds are a newly created material combining two previously discovered allotropes of carbon: carbon nanotubes and fullerenes. In this new material, fullerene-like "buds" are covalently bonded to the outer sidewalls of the underlying carbon nanotube. This hybrid material has useful properties of both fullerenes and carbon nanotubes. In particular, they have been found to be exceptionally good field emitters. In composite materials, the attached fullerene molecules may function as molecular anchors preventing slipping of the nanotubes, thus improving the composite's mechanical properties.

Three-dimensional Carbon Nanotube Architectures

3D carbon scaffolds

Recently, several studies have highlighted the prospect of using carbon nanotubes as building blocks to fabricate three-dimensional macroscopic (>100 nm in all three dimensions) all-carbon devices. Lalwani et al. have reported a novel radical initiated thermal crosslinking method to fabricate macroscopic, free-standing, porous, all-carbon scaffolds using single- and multi-walled carbon nanotubes as building blocks. These scaffolds possess macro-, micro-, and nano- structured pores and the porosity can be tailored for specific applications. These 3D all-carbon scaffolds/ architectures may be used for the fabrication of the next generation of energy storage, supercapacitors, field emission transistors, high-performance catalysis, photovoltaics, and biomedical devices and implants.

Graphenated Carbon Nanotubes (G-CNTs)

SEM series of graphenated CNTs with varying foliate density

Graphenated CNTs are a relatively new hybrid that combines graphitic foliates grown along the sidewalls of multiwalled or bamboo style CNTs. Yu *et al.* reported on "chemically bonded graphene leaves" growing along the sidewalls of CNTs. Stoner *et al.* described these structures as "graphenated CNTs" and reported in their use for enhanced supercapacitor performance. Hsu *et al.* further reported on similar structures formed on carbon fiber paper, also for use in supercapacitor applications. Pham *et al.* also reported a similar structure, namely "graphene-carbon nanotube hybrids", grown directly onto carbon fiber paper to form an integrated, binder free, high surface area conductive catalyst support for Proton Exchange Membrane Fuel Cells electrode applications with enhanced performance and durability. The foliate density can vary as a function of deposition conditions (e.g. temperature and time) with their structure ranging from few layers of graphene (< 10) to thicker, more graphite-like.

The fundamental advantage of an integrated graphene-CNT structure is the high surface area three-dimensional framework of the CNTs coupled with the high edge density of graphene. Graphene edges provide significantly higher charge density and reactivity than the basal plane, but they are difficult to arrange in a three-dimensional, high volume-density geometry. CNTs are readily aligned in a high density geometry (i.e., a vertically aligned forest) but lack high charge density surfaces—the sidewalls of the CNTs are similar to the basal plane of graphene and exhibit low charge density except where edge defects exist. Depositing a high density of graphene foliates along the length of aligned CNTs can significantly increase the total charge capacity per unit of nominal area as compared to other carbon nanostructures.

Nitrogen-doped Carbon Nanotubes

Nitrogen doped carbon nanotubes (N-CNTs) can be produced through five main methods, chemical vapor deposition, high-temperature and high-pressure reactions, gas-solid reaction of amorphous carbon with NH_3 at high temperature, solid reaction, and solvothermal synthesis.

N-CNTs can also be prepared by a CVD method of pyrolyzing melamine under Ar at elevated temperatures of 800–980 °C. However synthesis by CVD of melamine results in the formation of bamboo-structured CNTs. XPS spectra of grown N-CNTs reveal nitrogen in five main components, pyridinic nitrogen, pyrrolic nitrogen, quaternary nitrogen, and nitrogen oxides. Furthermore, synthesis temperature affects the type of nitrogen configuration.

Nitrogen doping plays a pivotal role in lithium storage, as it creates defects in the CNT walls allowing for Li ions to diffuse into interwall space. It also increases capacity by providing more favorable bind of N-doped sites. N-CNTs are also much more reactive to metal oxide nanoparticle deposition which can further enhance storage capacity, especially in anode materials for Li-ion batteries. However boron-doped nanotubes have been shown to make batteries with triple capacity.

Peapod

A carbon peapod is a novel hybrid carbon material which traps fullerene inside a carbon nanotube.

It can possess interesting magnetic properties with heating and irradiation. It can also be applied as an oscillator during theoretical investigations and predictions.

Cup-stacked Carbon Nanotubes

Cup-stacked carbon nanotubes (CSCNTs) differ from other quasi-1D carbon structures, which normally behave as quasi-metallic conductors of electrons. CSCNTs exhibit semiconducting behaviors due to the stacking microstructure of graphene layers.

Extreme Carbon Nanotubes

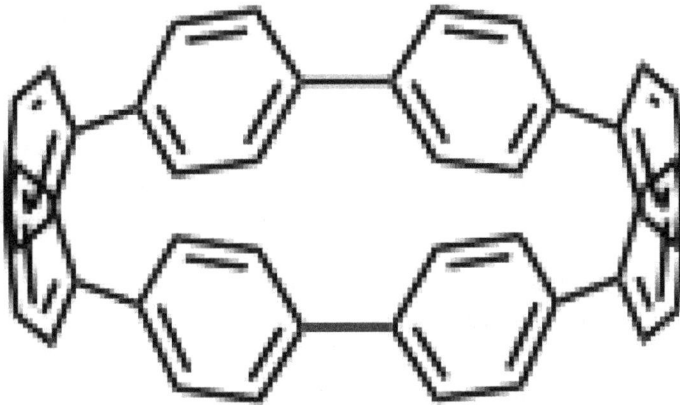

Cycloparaphenylene

The observation of the *longest* carbon nanotubes grown so far are over 1/2 m (550 mm long) was reported in 2013. These nanotubes were grown on Si substrates using an improved chemical vapor deposition (CVD) method and represent electrically uniform arrays of single-walled carbon nanotubes.

The *shortest* carbon nanotube is the organic compound cycloparaphenylene, which was synthesized in 2008.

The *thinnest* carbon nanotube is the armchair (2,2) CNT with a diameter of 0.3 nm. This nanotube was grown inside a multi-walled carbon nanotube. Assigning of carbon nanotube type was done by a combination of high-resolution transmission electron microscopy (HRTEM), Raman spectroscopy and density functional theory (DFT) calculations.

The *thinnest freestanding* single-walled carbon nanotube is about 0.43 nm in diameter. Researchers suggested that it can be either (5,1) or (4,2) SWCNT, but the exact type of carbon nanotube remains questionable. (3,3), (4,3) and (5,1) carbon nanotubes (all about 0.4 nm in diameter) were unambiguously identified using aberration-corrected high-resolution transmission electron microscopy inside double-walled CNTs.

The *highest density* of CNTs was achieved in 2013, grown on a conductive titanium-coated copper surface that was coated with co-catalysts cobalt and molybdenum at lower than typical temperatures of 450 °C. The tubes averaged a height of 380 nm and a mass density of 1.6 g cm^{-3}. The material showed ohmic conductivity (lowest resistance ~22 kΩ).

Properties

Strength

Carbon nanotubes are the strongest and stiffest materials yet discovered in terms of tensile strength and elastic modulus respectively. This strength results from the covalent sp^2 bonds formed between the individual carbon atoms. In 2000, a multi-walled carbon nanotube was tested to have a tensile strength of 63 gigapascals (9,100,000 psi). (For illustration, this translates into the ability to endure tension of a weight equivalent to 6,422 kilograms-force (62,980 N; 14,160 lbf) on a cable with cross-section of 1 square millimetre (0.0016 sq in).) Further studies, such as one conducted in 2008, revealed that individual CNT shells have strengths of up to ~100 gigapascals (15,000,000 psi), which is in agreement with quantum/atomistic models. Since carbon nanotubes have a low density for a solid of 1.3 to 1.4 g/cm^3, its specific strength of up to 48,000 kN·m·kg^{-1} is the best of known materials, compared to high-carbon steel's 154 kN·m·kg^{-1}.

Under excessive tensile strain, the tubes will undergo plastic deformation, which means the deformation is permanent. This deformation begins at strains of approximately 5% and can increase the maximum strain the tubes undergo before fracture by releasing strain energy.

Although the strength of individual CNT shells is extremely high, weak shear interactions between adjacent shells and tubes lead to significant reduction in the effective strength of multi-walled carbon nanotubes and carbon nanotube bundles down to only a few GPa. This limitation has been recently addressed by applying high-energy electron irradiation, which crosslinks inner shells and tubes, and effectively increases the strength of these materials to ~60 GPa for multi-walled carbon nanotubes and ~17 GPa for double-walled carbon nanotube bundles.

CNTs are not nearly as strong under compression. Because of their hollow structure and high aspect ratio, they tend to undergo buckling when placed under compressive, torsional, or bending stress.

Comparison of mechanical properties			
Material	**Young's modulus (TPa)**	**Tensile strength (GPa)**	**Elongation at break (%)**
SWNT[E]	~1 (from 1 to 5)	13–53	16
Armchair SWNT[T]	0.94	126.2	23.1
Zigzag SWNT[T]	0.94	94.5	15.6–17.5
Chiral SWNT	0.92		
MWNT[E]	0.2–0.8–0.95	11–63–150	
Stainless steel[E]	0.186–0.214	0.38–1.55	15–50
Kevlar–29&149[E]	0.06–0.18	3.6–3.8	~2

[E]Experimental observation; [T]Theoretical prediction

The above discussion referred to axial properties of the nanotube, whereas simple geometrical considerations suggest that carbon nanotubes should be much softer in the radial direction than along the tube axis. Indeed, TEM observation of radial elasticity suggested that even the van der Waals forces can deform two adjacent nanotubes. Nanoindentation experiments, performed by several groups on multiwalled carbon nanotubes and tapping/contact mode atomic force microscope measurements performed on single-walled carbon nanotubes, indicated a Young's modulus of the order of several GPa, confirming that CNTs are indeed rather soft in the radial direction.

Hardness

Standard single-walled carbon nanotubes can withstand a pressure up to 25 GPa without [plastic/permanent] deformation. They then undergo a transformation to superhard phase nanotubes. Maximum pressures measured using current experimental techniques are around 55 GPa. However, these new superhard phase nanotubes collapse at an even higher, albeit unknown, pressure.

The bulk modulus of superhard phase nanotubes is 462 to 546 GPa, even higher than that of diamond (420 GPa for single diamond crystal).

Wettability

The surface wettability of CNT is of importance for its applications in various settings. Although the intrinsic contact angle of graphite is around 90°, the contact angles of most as-synthesized CNT arrays are over 160°, exhibiting a superhydrophobic property. By applying a voltage as low as 1.3V, the extreme water repellant surface can be switched to a superhydrophilic one.

Kinetic Properties

Multi-walled nanotubes are multiple concentric nanotubes precisely nested within one another. These exhibit a striking telescoping property whereby an inner nanotube core may slide, almost without friction, within its outer nanotube shell, thus creating an atomically perfect linear or rotational bearing. This is one of the first true examples of molecular nanotechnology, the precise positioning of atoms to create useful machines. Already, this property has been utilized to create the world's smallest rotational motor. Future applications such as a gigahertz mechanical oscillator are also envisioned.

Electrical Properties

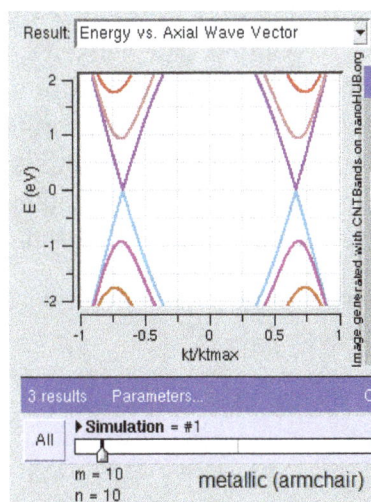

Band structures computed using tight binding approximation for (6,0) CNT (zigzag, metallic), (10,2) CNT (semiconducting) and (10,10) CNT (armchair, metallic).

Unlike graphene, which is a two-dimensional semimetal, carbon nanotubes are either metallic or semiconducting along the tubular axis. For a given (n,m) nanotube, if $n = m$, the nanotube is metallic; if $n - m$ is a multiple of 3, then the nanotube is semiconducting with a very small band gap, otherwise the nanotube is a moderate semiconductor. Thus all armchair $(n = m)$ nanotubes are metallic, and nanotubes $(6,4)$, $(9,1)$, etc. are semiconducting. Carbon nanotubes are not semi-metallic because the degenerate point (that point where the π [bonding] band meets the π^* [anti-bonding] band, at which the energy goes to zero) is slightly shifted away from the K point in the Brillouin zone due to the curvature of the tube surface, casing hybridization between the σ^* and π^* anti-bonding bands, modifying the band dispersion.

The rule regarding metallic versus semiconductor behavior has exceptions, because curvature effects in small diameter tubes can strongly influence electrical properties. Thus, a $(5,0)$ SWCNT that should be semiconducting in fact is metallic according to the calculations. Likewise, zigzag and chiral SWCNTs with small diameters that should be metallic have a finite gap (armchair nanotubes remain metallic). In theory, metallic nanotubes can carry an electric current density of 4×10^9 A/cm^2, which is more than 1,000 times greater than those of metals such as copper, where for copper interconnects current densities are limited by electromigration. Carbon nanotubes are thus being explored as conductivity enhancing components in composite materials and many groups are attempting to commercialize highly conducting electrical wire assembled from individual carbon nanotubes. There are significant challenges to be overcome, however, such as the much more resistive nanotube-to-nanotube junctions and impurities, all of which lower the electrical conductivity of the macroscopic nanotube wires by orders of magnitude, as compared to the conductivity of the individual nanotubes.

Because of its nanoscale cross-section, electrons propagate only along the tube's axis. As a result, carbon nanotubes are frequently referred to as one-dimensional conductors. The maximum electrical conductance of a single-walled carbon nanotube is $2G_0$, where $G_0 = 2e^2/h$ is the conductance of a single ballistic quantum channel.

Due to the role of the π-electron system in determining the electronic properties of graphene, doping in carbon nanotubes differs from that of bulk crystalline semiconductors from the same group of the periodic table (e.g. silicon). Graphitic substitution of carbon atoms in the nanotube wall by boron or nitrogen dopants leads to p-type and n-type behavior, respectively, as would be expected in silicon. However, some non-substitutional (intercalated or adsorbed) dopants introduced into a carbon nanotube, such as alkali metals as well as electron-rich metallocenes, result in n-type conduction because they donate electrons to the π-electron system of the nanotube. By contrast, π-electron acceptors such as $FeCl_3$ or electron-deficient metallocenes function as p-type dopants since they draw π-electrons away from the top of the valence band.

Intrinsic superconductivity has been reported, although other experiments found no evidence of this, leaving the claim a subject of debate.

Optical Properties

Thermal Properties

All nanotubes are expected to be very good thermal conductors along the tube, exhibiting a property known as "ballistic conduction", but good insulators lateral to the tube axis. Measurements show that a SWNT has a room-temperature thermal conductivity along its axis of about 3500 $W \cdot m^{-1} \cdot K^{-1}$; compare this to copper, a metal well known for its good thermal conductivity, which transmits 385 $W \cdot m^{-1} \cdot K^{-1}$. A SWNT has a room-temperature thermal conductivity across its axis (in the radial direction) of about 1.52 $W \cdot m^{-1} \cdot K^{-1}$, which is about as thermally conductive as soil. The temperature stability of carbon nanotubes is estimated to be up to 2800 °C in vacuum and about 750 °C in air.

Defects

As with any material, the existence of a crystallographic defect affects the material properties. Defects can occur in the form of atomic vacancies. High levels of such defects can lower the tensile strength by up to 85%. An important example is the Stone Wales defect, which creates a pentagon and heptagon pair by rearrangement of the bonds. Because of the very small structure of CNTs, the tensile strength of the tube is dependent on its weakest segment in a similar manner to a chain, where the strength of the weakest link becomes the maximum strength of the chain.

Crystallographic defects also affect the tube's electrical properties. A common result is lowered conductivity through the defective region of the tube. A defect in armchair-type tubes (which can conduct electricity) can cause the surrounding region to become semiconducting, and single monatomic vacancies induce magnetic properties.

Crystallographic defects strongly affect the tube's thermal properties. Such defects lead to phonon scattering, which in turn increases the relaxation rate of the phonons. This reduces the mean free path and reduces the thermal conductivity of nanotube structures. Phonon transport simulations indicate that substitutional defects such as nitrogen or boron will primarily lead to scattering of high-frequency optical phonons. However, larger-scale defects such as Stone Wales defects cause phonon scattering over a wide range of frequencies, leading to a greater reduction in thermal conductivity.

Safety and Health

The toxicity of carbon nanotubes has been an important question in nanotechnology. As of 2007, such research had just begun. The data is still fragmentary and subject to criticism. Preliminary results highlight the difficulties in evaluating the toxicity of this heterogeneous material. Parameters such as structure, size distribution, surface area, surface chemistry, surface charge, and agglomeration state as well as purity of the samples, have considerable impact on the reactivity of carbon nanotubes. However, available data clearly show that, under some conditions, nanotubes can cross membrane barriers, which suggests that, if raw materials reach the organs, they can induce harmful effects such as inflammatory and fibrotic reactions.

Synthesis

Techniques have been developed to produce nanotubes in sizable quantities, including arc discharge, laser ablation, high-pressure carbon monoxide disproportionation, and chemical vapor deposition (CVD). Most of these processes take place in a vacuum or with process gases. CVD growth of CNTs can occur in vacuum or at atmospheric pressure. Large quantities of nanotubes can be synthesized by these methods; advances in catalysis and continuous growth are making CNTs more commercially viable.

Chemical Modification

Carbon nanotubes can be functionalized to attain desired properties that can be used in a wide variety of applications. The two main methods of carbon nanotube functionalization are covalent and non-covalent modifications. Because of their hydrophobic nature, carbon nanotubes tend to agglomerate hindering their dispersion is solvents or viscous polymer melts. The resulting nanotube bundles or aggregates reduce the mechanical performance of the final composite. The surface of the carbon nanotubes can be modified to reduce the hydrophobicity and improve interfacial adhesion to a bulk polymer through chemical attachment.

Applications

Current

Current use and application of nanotubes has mostly been limited to the use of bulk nanotubes, which is a mass of rather unorganized fragments of nanotubes. Bulk nanotube materials may never achieve a tensile strength similar to that of individual tubes, but such composites may, nevertheless, yield strengths sufficient for many applications. Bulk carbon nanotubes have already been used as composite fibers in polymers to improve the mechanical, thermal and electrical properties of the bulk product.

- Easton-Bell Sports, Inc. have been in partnership with Zyvex Performance Materials, using CNT technology in a number of their bicycle components—including flat and riser handlebars, cranks, forks, seatposts, stems and aero bars.

- Zyvex Technologies has also built a 54' maritime vessel, the Piranha Unmanned Surface Vessel, as a technology demonstrator for what is possible using CNT technology. CNTs help improve the structural performance of the vessel, resulting in a lightweight 8,000 lb boat that can carry a payload of 15,000 lb over a range of 2,500 miles.

- Amroy Europe Oy manufactures Hybtonite carbon nanoepoxy resins where carbon nanotubes have been chemically activated to bond to epoxy, resulting in a composite material that is 20% to 30% stronger than other composite materials. It has been used for wind turbines, marine paints and variety of sports gear such as skis, ice hockey sticks, baseball bats, hunting arrows, and surfboards.

Other current applications include:

- tips for atomic force microscope probes

- in tissue engineering, carbon nanotubes can act as scaffolding for bone growth

There is also ongoing research in using carbon nanotubes as a scaffold for diverse microfabrication techniques.

Potential

The strength and flexibility of carbon nanotubes makes them of potential use in controlling other nanoscale structures, which suggests they will have an important role in nanotechnology engineering. The highest tensile strength of an individual multi-walled carbon nanotube has been tested to be 63 GPa. Carbon nanotubes were found in Damascus steel from the 17th century, possibly helping to account for the legendary strength of the swords made of it. Recently, several studies have highlighted the prospect of using carbon nanotubes as building blocks to fabricate three-dimensional macroscopic (>1mm in all three dimensions) all-carbon devices. Lalwani et al. have reported a novel radical initiated thermal crosslinking method to fabricated macroscopic, free-standing, porous, all-carbon scaffolds using single- and multi-walled carbon nanotubes as building blocks. These scaffolds possess macro-, micro-, and nano- structured pores and the porosity can be tailored for specific applications. These 3D all-carbon scaffolds/architectures maybe used for the fabrication of the next generation of energy storage, supercapacitors, field emission transistors, high-performance catalysis, photovoltaics, and biomedical devices and implants.

Discovery

The true identity of the discoverers of carbon nanotubes is a subject of some controversy. For years, scientists assumed that Sumio Iijima of NEC had discovered carbon nanotubes in 1991. He published a paper describing his discovery which initiated a flurry of excitement and could be credited by inspiring the many scientists now studying applications of carbon nanotubes. Though Iijima has been given much of the credit for discovering carbon nanotubes, it turns out that the timeline of carbon nanotubes goes back much further than 1991. In 1952 L. V. Radushkevich and V. M. Lukyanovich published clear images of 50 nanometer diameter tubes made of carbon in the Soviet *Journal of Physical Chemistry*. This discovery was largely unnoticed, as the article was published in Russian, and Western scientists' access to Soviet press was limited during the Cold War. Before they came to be known as carbon nanotubes, in 1976, Morinobu Endo of CNRS observed hollow tubes of rolled up graphite sheets synthesised by a chemical vapour-growth technique. The first specimens observed would later come to be known as single-walled carbon nanotubes (SWNTs). The three scientists have been the first ones to show images of a nanotube with a solitary graphene wall.

Endo, in his early review of vapor-phase-grown carbon fibers (VPCF), also reminded us that he had observed a hollow tube, linearly extended with parallel carbon layer faces near the fiber core. This appears to be the observation of multi-walled carbon nanotubes at the center of the fiber. The mass-produced MWCNTs today are strongly related to the VPGCF developed by Endo. In fact, they call it the "Endo-process", out of respect for his early work and patents.

In 1979, John Abrahamson presented evidence of carbon nanotubes at the 14th Biennial Conference of Carbon at Pennsylvania State University. The conference paper described carbon nanotubes as carbon fibers that were produced on carbon anodes during arc discharge. A characterization of these fibers was given as well as hypotheses for their growth in a nitrogen atmosphere at low pressures.

In 1981, a group of Soviet scientists published the results of chemical and structural characterization of carbon nanoparticles produced by a thermocatalytical disproportionation of carbon monoxide. Using TEM images and XRD patterns, the authors suggested that their "carbon multi-layer tubular crystals" were formed by rolling graphene layers into cylinders. They speculated that by rolling graphene layers into a cylinder, many different arrangements of graphene hexagonal nets are possible. They suggested two possibilities of such arrangements: circular arrangement (armchair nanotube) and a spiral, helical arrangement (chiral tube).

In 1987, Howard G. Tennent of Hyperion Catalysis was issued a U.S. patent for the production of "cylindrical discrete carbon fibrils" with a "constant diameter between about 3.5 and about 70 nanometers..., length 10^2 times the diameter, and an outer region of multiple essentially continuous layers of ordered carbon atoms and a distinct inner core...."

Iijima's discovery of multi-walled carbon nanotubes in the insoluble material of arc-burned graphite rods in 1991 and Mintmire, Dunlap, and White's independent prediction that if single-walled carbon nanotubes could be made, then they would exhibit remarkable conducting properties helped create the initial buzz that is now associated with carbon nanotubes. Nanotube research accelerated greatly following the independent discoveries by Bethune at IBM and Iijima at NEC of *single-walled* carbon nanotubes and methods to specifically produce them by adding transition-metal catalysts to the carbon in an arc discharge. The arc discharge technique was well-known to produce the famed Buckminster fullerene on a preparative scale, and these results appeared to extend the run of accidental discoveries relating to fullerenes. The discovery of nanotubes remains a contentious issue. Many believe that Iijima's report in 1991 is of particular importance because it brought carbon nanotubes into the awareness of the scientific community as a whole.

Optical Properties of Carbon Nanotubes

Within materials science, the optical properties of carbon nanotubes refer specifically to the absorption, photoluminescence (fluorescence), and Raman spectroscopy of carbon nanotubes. Spectroscopic methods offer the possibility of quick and non-destructive characterization of relatively large amounts of carbon nanotubes. There is a strong demand for such characterization from the industrial point of view: numerous parameters of the nanotube synthesis can be changed, intentionally or unintentionally, to alter the nanotube quality. As shown below, optical absorption, photoluminescence and Raman spectroscopies allow quick and reliable characterization of this "nanotube quality" in terms of non-tubular carbon content, structure (chirality) of the produced nanotubes, and structural defects. Those features determine nearly any other properties such as optical, mechanical, and electrical properties.

Carbon nanotubes are unique "one-dimensional systems" which can be envisioned as rolled single sheets of graphite (or more precisely graphene). This rolling can be done at different angles and curvatures resulting in different nanotube properties. The diameter typically varies in the range 0.4–40 nm (i.e. "only" ~100 times), but the length can vary ~10,000 times, reaching 55.5 cm. The nanotube aspect ratio, or the length-to-diameter ratio, can be as high as 132,000,000:1, which is unequalled by any other material. Consequently, all the properties of the carbon nanotubes relative to those of typical semiconductors are extremely anisotropic (directionally dependent) and tunable.

Whereas mechanical, electrical and electrochemical (supercapacitor) properties of the carbon nanotubes are well established and have immediate applications, the practical use of optical properties is yet unclear. The aforementioned tunability of properties is potentially useful in optics and photonics. In particular, light-emitting diodes (LEDs) and photo-detectors based on a single nanotube have been produced in the lab. Their unique feature is not the efficiency, which is yet relatively low, but the narrow selectivity in the wavelength of emission and detection of light and the possibility of its fine tuning through the nanotube structure. In addition, bolometer and optoelectronic memory devices have been realised on ensembles of single-walled carbon nanotubes.

Terminology

This section uses the following abbreviations:

- Carbon nanotube (CNT)

- Single wall carbon nanotube (SWCNT)

- Multiwall carbon nanotube (MWCNT)

However, C is often omitted in scientific literature, so NT, SWNT and MWNT are more commonly used. Also, "wall" is often exchanged with "walled".

A single-wall carbon nanotube can be imagined as graphene sheet rolled at a certain "chiral" angle with respect to a plane perpendicular to the tube's long axis. Consequently, SWCNT can be defined by its diameter and chiral angle. The chiral angle can range from 0 to 30 degrees.

However, more conveniently, a pair of indices (n, m) is used instead. The indices refer to equally long unit vectors at 60° angles to each other across a single 6-member carbon ring. Taking the origin as carbon number 1, the a_1 unit vector may be considered the line drawn from carbon 1 to carbon 3, and the a_2 unit vector is then the line drawn from carbon 1 to carbon 5. To visualize a CNT with indices (n, m), draw n a_1 unit vectors across the graphene sheet, then draw m a_2 unit vectors at a 60° angle to the a_1 vectors, then add the vectors together. The line representing the sum of the vectors will define the circumference of the CNT along the plane perpendicular to its long axis, connecting one end to the other. In the diagram at right, C_h is a (4, 2) vector: the sum of 4 unit vectors from the origin directly to the right, then 2 unit vectors at a 60° angle down and to the right.

Tubes having $n = m$ (chiral angle = 30°) are called "armchair" and those with $m = 0$ (chiral angle = 0°) «zigzag». Those indices uniquely determine whether CNT is a metal, semimetal or semiconductor, as well as its band gap: when $|m - n| = 3k$ (k is integer), the tube is metallic; but if $|m - n| = 3k \pm 1$, the tube is semiconducting. The nanotube diameter d is related to m and n as

$$d = \frac{a}{\pi}\sqrt{(n^2 + nm + m^2)}.$$

In this equation, $a = 0.246$ nm is the magnitude of either unit vector a_1 or a_2.

The situation in multi-wall CNTs is complicated as their properties are determined by contribution of all individual shells; those shells have different structures, and, because of the synthesis, are usually more defective than SWCNTs. Therefore, optical properties of MWCNTs will not be considered here.

Van Hove Singularities

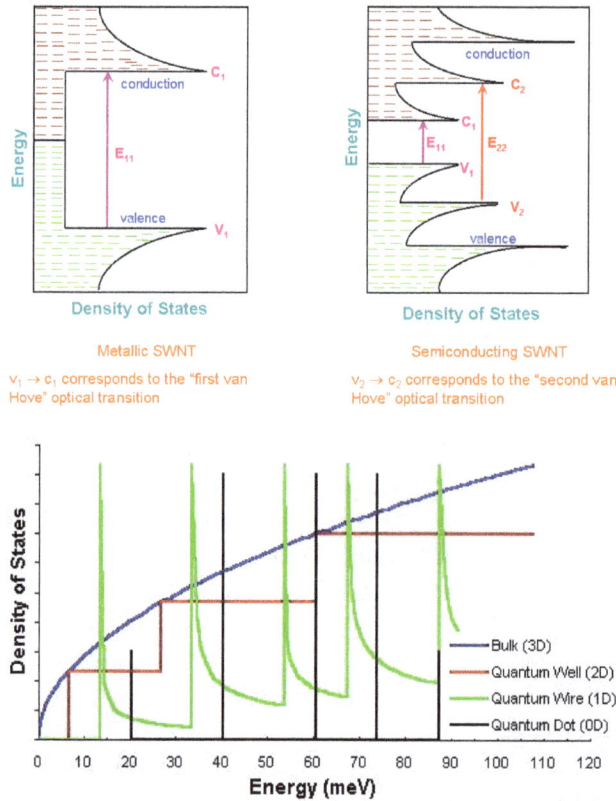

Metallic SWNT

$v_1 \rightarrow c_1$ corresponds to the "first van Hove" optical transition

Semiconducting SWNT

$v_2 \rightarrow c_2$ corresponds to the "second van Hove" optical transition

A bulk 3D material (blue) has continuous DOS, but a 1D wire (green) has Van Hove singularities.

Optical properties of carbon nanotubes derive from electronic transitions within one-dimensional density of states (DOS). A typical feature of one-dimensional crystals is that their DOS is not a continuous function of energy, but it descends gradually and then increases in a discontinuous spike. In contrast, three-dimensional materials have continuous DOS. The sharp peaks found in one-dimensional materials are called Van Hove singularities.

Van Hove singularities result in the following remarkable optical properties of carbon nanotubes:

- Optical transitions occur between the $v_1 - c_1$, $v_2 - c_2$, etc., states of semiconducting or metallic nanotubes and are traditionally labeled as S_{11}, S_{22}, M_{11}, etc., or, if the "conductivity" of the tube is unknown or unimportant, as E_{11}, E_{22}, etc. Crossover transitions $c_1 - v_2$, $c_2 - v_1$, etc., are dipole-forbidden and thus are extremely weak, but they were possibly observed using cross-polarized optical geometry.

- The energies between the Van Hove singularities depend on the nanotube structure. Thus by varying this structure, one can tune the optoelectronic properties of carbon nanotube. Such fine tuning has been experimentally demonstrated using UV illumination of polymer-dispersed CNTs.

- Optical transitions are rather sharp (~10 meV) and strong. Consequently, it is relatively easy to selectively excite nanotubes having certain (n, m) indices, as well as to detect optical signals from individual nanotubes.

Kataura Plot

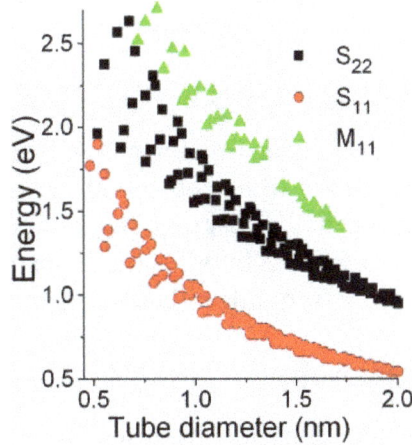

In this Kataura plot, the energy of an electronic transition decreases as the diameter of the nanotube increases.

The band structure of carbon nanotubes having certain (n, m) indexes can be easily calculated. A theoretical graph based on this calculations was designed in 1999 by Hiromichi Kataura to rationalize experimental findings. A Kataura plot relates the nanotube diameter and its bandgap energies for all nanotubes in a diameter range. The oscillating shape of every branch of the Kataura plot reflects the intrinsic strong dependence of the SWCNT properties on the (n, m) index rather than on its diameter. For example, $(10, 1)$ and $(8, 3)$ tubes have almost the same diameter, but very different properties: the former is a metal, but the latter is a semiconductor.

Optical Absorption

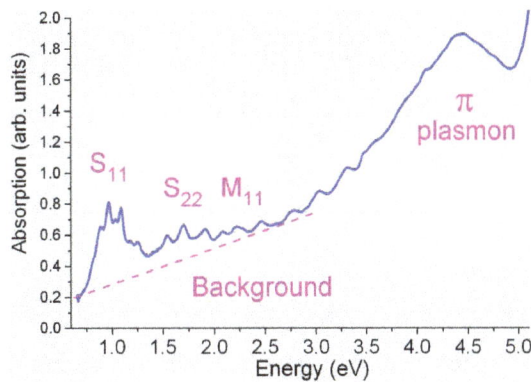

Optical absorption spectrum from dispersed single-wall carbon nanotubes

Optical absorption in carbon nanotubes differs from absorption in conventional 3D materials by presence of sharp peaks (1D nanotubes) instead of an absorption threshold followed by an absorption increase (most 3D solids). Absorption in nanotubes originates from electronic transitions from the v_2 to c_2 (energy E_{22}) or v_1 to c_1 (E_{11}) levels, etc. The transitions are relatively sharp and can be used to identify nanotube types. Note that the sharpness deteriorates with increasing energy, and that many nanotubes have very similar E_{22} or E_{11} energies, and thus significant overlap occurs in absorption spectra. This overlap is avoided in photoluminescence mapping measurements, which instead of a combination of overlapped transitions identifies individual (E_{22}, E_{11}) pairs.

Interactions between nanotubes, such as bundling, broaden optical lines. While bundling strongly affects photoluminescence, it has much weaker effect on optical absorption and Raman scattering. Consequently, sample preparation for the latter two techniques is relatively simple.

Optical absorption is routinely used to quantify quality of the carbon nanotube powders.

The spectrum is analyzed in terms of intensities of nanotube-related peaks, background and pi-carbon peak; the latter two mostly originate from non-nanotube carbon in contaminated samples. However, it has been recently shown that by aggregating nearly single chirality semiconducting nanotubes into closely packed Van der Waals bundles the absorption background can be attributed to free carrier transition originating from intertube charge transfer.

Carbon Nanotubes as a Black Body

An ideal black body should have emissivity or absorbance of 1.0, which is difficult to attain in practice, especially in a wide spectral range. Vertically aligned "forests" of single-wall carbon nanotubes can have absorbances of 0.98–0.99 from the far-ultraviolet (200 nm) to far-infrared (200 µm) wavelengths.

These SWNT forests (buckypaper) were grown by the super-growth CVD method to about 10 µm height. Two factors could contribute to strong light absorption by these structures: (i) a distribution of CNT chiralities resulted in various bandgaps for individual CNTs. Thus a compound material was formed with broadband absorption. (ii) Light might be trapped in those forests due to multiple reflections.

Reflectance measurements			
	UV-to-near IR	**Near-to-mid IR**	**Mid-to-far IR**
Wavelength, µm	0.2-2	2–20	25–200
Incident angle, °	8	5	10
Reflection	Hemispherical-directional	Hemispherical-directional	Specular
Reference	White reflectance standard	Gold mirror	Aluminum mirror
Average reflectance	0.0160	0.0097	0.0017
Standard deviation	0.0048	0.0041	0.0027

Luminescence

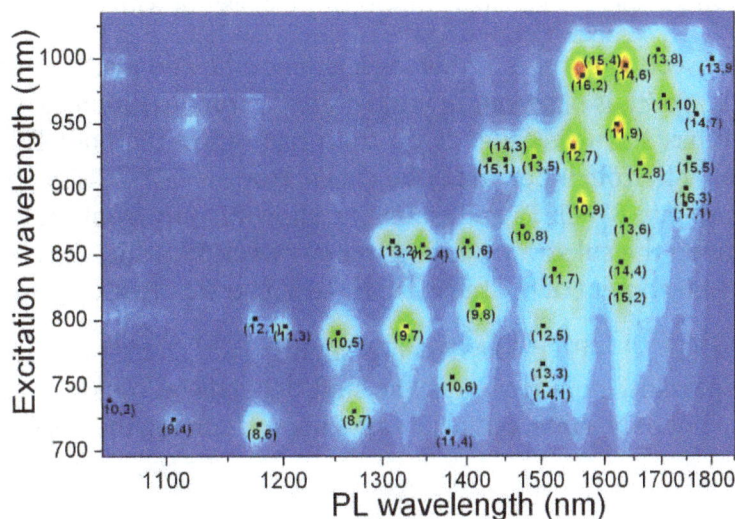

Photoluminescence map from single-wall carbon nanotubes. (n, m) indexes identify certain semiconducting nanotubes. Note that PL measurements do not detect nanotubes with $n = m$ or $m = 0$.

Photoluminescence (Fluorescence)

Semiconducting single-walled carbon nanotubes emit near-infrared light upon photoexcitation, described interchangeably as fluorescence or photoluminescence (PL). The excitation of PL usually occurs as follows: an electron in a nanotube absorbs excitation light via S_{22} transition, creating an electron-hole pair (exciton). Both electron and hole rapidly relax (via phonon-assisted processes) from c_2 to c_1 and from v_2 to v_1 states, respectively. Then they recombine through a $c_1 - v_1$ transition resulting in light emission.

No excitonic luminescence can be produced in metallic tubes. Their electrons can be excited, thus resulting in optical absorption, but the holes are immediately filled by other electrons out of the many available in the metal. Therefore, no excitons are produced.

Salient Properties

- Photoluminescence from SWCNT, as well as optical absorption and Raman scattering, is linearly polarized along the tube axis. This allows monitoring of the SWCNTs orientation without direct microscopic observation.

- PL is quick: relaxation typically occurs within 100 picoseconds.

- PL efficiency was first found to be low (~0.01%), but later studies measured much higher quantum yields. By improving the structural quality and isolation of nanotubes, emission efficiency increased. A quantum yield of 1% was reported in nanotubes sorted by diameter and length through gradient centrifugation, and it was further increased to 20% by optimizing the procedure of isolating individual nanotubes in solution.

- The spectral range of PL is rather wide. Emission wavelength can vary between 0.8 and 2.1 micrometers depending on the nanotube structure.

- Excitons are apparently delocalized over several nanotubes in single chirality bundles as the photoluminescence spectrum displays a splitting consistent with intertube exciton tunneling.

- Interaction between nanotubes or between a nanotube and another material may quench or increase PL. No PL is observed in multi-walled carbon nanotubes. PL from double-wall carbon nanotubes strongly depends on the preparation method: CVD grown DWCNTs show emission both from inner and outer shells. However, DWCNTs produced by encapsulating fullerenes into SWCNTs and annealing show PL only from the outer shells. Isolated SWCNTs lying on the substrate show extremely weak PL which has been detected in few studies only. Detachment of the tubes from the substrate drastically increases PL.

- Position of the (S_{22}, S_{11}) PL peaks depends slightly (within 2%) on the nanotube environment (air, dispersant, etc.). However, the shift depends on the (n, m) index, and thus the whole PL map not only shifts, but also warps upon changing the CNT medium.

Applications

- Photoluminescence is used for characterization purposes to measure the quantities of semiconducting nanotube species in a sample. Nanotubes are isolated (dispersed) using an appropriate chemical agent ("dispersant") to reduce the intertube quenching. Then PL is measured, scanning both the excitation and emission energies and thereby producing a PL map. The ovals in the map define (S_{22}, S_{11}) pairs, which unique identify (n, m) index of a tube. The data of Weisman and Bachilo are conventionally used for the identification.

- Nanotube fluorescence has been investigated for the purposes of imaging and sensing in biomedical applications.

Sensitization

Optical properties, including the PL efficiency, can be modified by encapsulating organic dyes (carotene, lycopene, etc.) inside the tubes. Efficient energy transfer occurs between the encapsulated dye and nanotube — light is efficiently absorbed by the dye and without significant loss is transferred to the SWCNT. Thus potentially, optical properties of a carbon nanotube can be controlled by encapsulating certain molecule inside it. Besides, encapsulation allows isolation and characterization of organic molecules which are unstable under ambient conditions. For example, Raman spectra are extremely difficult to measure from dyes because of their strong PL (efficiency close to 100%). However, encapsulation of dye molecules inside SWCNTs completely quenches dye PL, thus allowing measurement and analysis of their Raman spectra.

Cathodoluminescence

Cathodoluminescence (CL) — light emission excited by electron beam — is a process commonly observed in TV screens. An electron beam can be finely focused and scanned across the studied material. This technique is widely used to study defects in semiconductors and nanostructures with nanometer-scale spatial resolution. It would be beneficial to apply this technique to carbon nanotubes. However, no reliable CL, i.e. sharp peaks assignable to certain (n, m) indices, has been detected from carbon nanotubes yet.

Electroluminescence

If appropriate electrical contacts are attached to a nanotube, electron-hole pairs (excitons) can be generated by injecting electrons and holes from the contacts. Subsequent exciton recombination results in electroluminescence (EL). Electroluminescent devices have been produced from single nanotubes and their macroscopic assemblies. Recombination appears to proceed via triplet-triplet annihilation giving distinct peaks corresponding to E_{11} and E_{22} transitions.

Raman Scattering

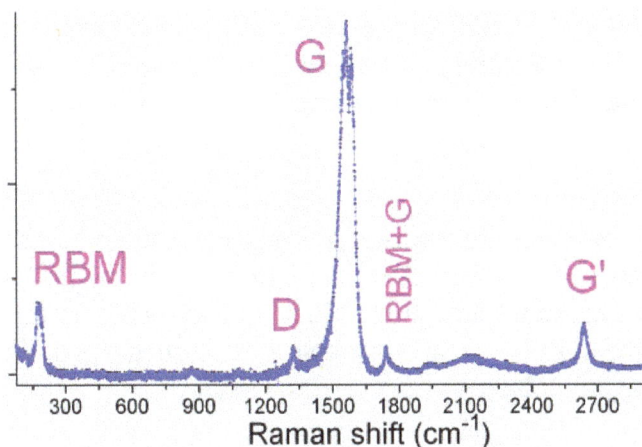

Raman spectrum of single-wall carbon nanotubes

Raman spectroscopy has good spatial resolution (~0.5 micrometers) and sensitivity (single nanotubes); it requires only minimal sample preparation and is rather informative. Consequently, Raman spectroscopy is probably the most popular technique of carbon nanotube characterization. Raman scattering in SWCNTs is resonant, i.e., only those tubes are probed which have one of the bandgaps equal to the exciting laser energy. Several scattering modes dominate the SWCNT spectrum, as discussed below.

Similar to photoluminescence mapping, the energy of the excitation light can be scanned in Raman measurements, thus producing Raman maps. Those maps also contain oval-shaped features uniquely identifying (n, m) indices. Contrary to PL, Raman mapping detects not only semiconducting but also metallic tubes, and it is less sensitive to nanotube bundling than PL. However, requirement of a tunable laser and a dedicated spectrometer is a strong technical impediment.

Radial Breathing Mode

Radial breathing mode (RBM) corresponds to radial expansion-contraction of the nanotube. Therefore, its frequency ν_{RBM} (in cm^{-1}) depends on the nanotube diameter d as, $\nu_{RBM} = A/d + B$ (where A and B are constants dependent on the environment in which the nanotube is present. For example, B=0 for individual nanotubes.) (in nanometers) and can be estimated as $\nu_{RBM} = 234/d + 10$ for SWNT or $\nu_{RBM} = 248/d$ for DWNT, which is very useful in deducing the CNT diameter from the RBM position. Typical RBM range is 100–350 cm^{-1}. If RBM intensity is particularly strong, its weak second overtone can be observed at double frequency.

Bundling Mode

The bundling mode is a special form of RBM supposedly originating from collective vibration in a bundle of SWCNTs.

G Mode

Another very important mode is the G mode (G from graphite). This mode corresponds to planar vibrations of carbon atoms and is present in most graphite-like materials. G band in SWCNT is shifted to lower frequencies relative to graphite (1580 cm^{-1}) and is split into several peaks. The splitting pattern and intensity depend on the tube structure and excitation energy; they can be used, though with much lower accuracy compared to RBM mode, to estimate the tube diameter and whether the tube is metallic or semiconducting.

D Mode

D mode is present in all graphite-like carbons and originates from structural defects. Therefore, the ratio of the G/D modes is conventionally used to quantify the structural quality of carbon nanotubes. High-quality nanotubes have this ratio significantly higher than 100. At a lower functionalisation of the nanotube, the G/D ratio remains almost unchanged. This ratio gives an idea of the functionalisation of a nanotube.

G' Mode

The name of this mode is misleading: it is given because in graphite, this mode is usually the second strongest after the G mode. However, it is actually the second overtone of the defect-induced D mode (and thus should logically be named D'). Its intensity is stronger than that of the D mode due to different selection rules. In particular, D mode is forbidden in the ideal nanotube and requires a structural defect, providing a phonon of certain angular momentum, to be induced. In contrast, G' mode involves a "self-annihilating" pair of phonons and thus does not require defects. The spectral position of G' mode depends on diameter, so it can be used roughly to estimate the SWCNT diameter. In particular, G' mode is a doublet in double-wall carbon nanotubes, but the doublet is often unresolved due to line broadening.

Other overtones, such as a combination of RBM+G mode at ~1750 cm^{-1}, are frequently seen in CNT Raman spectra. However, they are less important and are not considered here.

Anti-Stokes Scattering

All the above Raman modes can be observed both as Stokes and anti-Stokes scattering. As mentioned above, Raman scattering from CNTs is resonant in nature, i.e. only tubes whose band gap energy is similar to the laser energy are excited. The difference between those two energies, and thus the band gap of individual tubes, can be estimated from the intensity ratio of the Stokes/anti-Stokes lines. This estimate however relies on the temperature factor (Boltzmann factor), which is often miscalculated – a focused laser beam is used in the measurement, which can locally heat the nanotubes without changing the overall temperature of the studied sample.

Rayleigh Scattering

Carbon nanotubes have very large aspect ratio, i.e., their length is much larger than their diameter. Consequently, as expected from the classical electromagnetic theory, elastic light scattering (or Rayleigh scattering) by straight CNTs has anisotropic angular dependence, and from its spectrum, the band gaps of individual nanotubes can be deduced.

Another manifestation of Rayleigh scattering is the "antenna effect", an array of nanotubes standing on a substrate has specific angular and spectral distributions of reflected light, and both those distributions depend on the nanotube length.

Synthesis of Carbon Nanotubes

powder of carbon nanotubes

Techniques have been developed to produce carbon nanotubes in sizable quantities, including arc discharge, laser ablation, high-pressure carbon monoxide disproportionation, and chemical vapor deposition (CVD). Most of these processes take place in a vacuum or with process gases. CVD growth of CNTs can occur in vacuum or at atmospheric pressure. Large quantities of nanotubes can be synthesized by these methods; advances in catalysis and continuous growth are making CNTs more commercially viable.

Types

Arc Discharge

Nanotubes were observed in 1991 in the carbon soot of graphite electrodes during an arc discharge, by using a current of 100 amps, that was intended to produce fullerenes. However the first macroscopic production of carbon nanotubes was made in 1992 by two researchers at NEC's Fundamental Research Laboratory. The method used was the same as in 1991. During this process, the carbon contained in the negative electrode sublimates because of the high-discharge temperatures.

The yield for this method is up to 30% by weight and it produces both single- and multi-walled nanotubes with lengths of up to 50 micrometers with few structural defects. Arc-discharge technique uses higher temperatures (above 1,700 °C) for CNT synthesis which typically causes the expansion of CNTs with fewer structural defects in comparison with other methods.

Laser Ablation

In laser ablation, a pulsed laser vaporizes a graphite target in a high-temperature reactor while an inert gas is bled into the chamber. Nanotubes develop on the cooler surfaces of the reactor as the vaporized carbon condenses. A water-cooled surface may be included in the system to collect the nanotubes.

This process was developed by Dr. Richard Smalley and co-workers at Rice University, who at the time of the discovery of carbon nanotubes, were blasting metals with a laser to produce various metal molecules. When they heard of the existence of nanotubes they replaced the metals with graphite to create multi-walled carbon nanotubes. Later that year the team used a composite of graphite and metal catalyst particles (the best yield was from a cobalt and nickel mixture) to synthesize single-walled carbon nanotubes.

The laser ablation method yields around 70% and produces primarily single-walled carbon nanotubes with a controllable diameter determined by the reaction temperature. However, it is more expensive than either arc discharge or chemical vapor deposition.

The effective equation for few cycle optical pulse dynamics was obtained by virtue of the Boltzmann collision-less equation solution for conduction band electrons of semiconductor carbon nanotubes in the case when medium with carbon nanotubes has spatially-modulated refractive index.

Plasma Torch

Single-walled carbon nanotubes can also be synthesized by a thermal plasma method, first invented in 2000 at INRS (Institut national de la recherche scientifique) in Varennes, Canada, by Olivier Smiljanic. In this method, the aim is to reproduce the conditions prevailing in the arc discharge and laser ablation approaches, but a carbon-containing gas is used instead of graphite vapors to supply the necessary carbon. Doing so, the growth of SWNT is more efficient (decomposing the gas can be 10 times less energy-consuming than graphite vaporization). The process is also continuous and low cost. A gaseous mixture of argon, ethylene and ferrocene is introduced into a microwave plasma torch, where it is atomized by the atmospheric pressure plasma, which has the form of an intense 'flame'. The fumes created by the flame contain SWNT, metallic and carbon nanoparticles and amorphous carbon.

Another way to produce single-walled carbon nanotubes with a plasma torch is to use the induction thermal plasma method, implemented in 2005 by groups from the University of Sherbrooke and the National Research Council of Canada. The method is similar to arc discharge in that both use ionized gas to reach the high temperature necessary to vaporize carbon-containing substances and the metal catalysts necessary for the ensuing nanotube growth. The thermal plasma is induced by high-frequency oscillating currents in a coil, and is maintained in flowing inert gas. Typically, a feedstock of carbon black and metal catalyst particles is fed into the plasma, and then cooled down to form single-walled carbon nanotubes. Different single-wall carbon nanotube diameter distributions can be synthesized.

The induction thermal plasma method can produce up to 2 grams of nanotube material per minute, which is higher than the arc discharge or the laser ablation methods.

Chemical Vapor Deposition (CVD)

nanotubes being grown by plasma enhanced chemical vapor deposition

The catalytic vapor phase deposition of carbon was reported in 1952 and 1959, but it was not until 1993 that carbon nanotubes were formed by this process. In 2007, researchers at the University of Cincinnati (UC) developed a process to grow aligned carbon nanotube arrays of length 18 mm on a FirstNano ET3000 carbon nanotube growth system.

During CVD, a substrate is prepared with a layer of metal catalyst particles, most commonly nickel, cobalt, iron, or a combination. The metal nanoparticles can also be produced by other ways, including reduction of oxides or oxides solid solutions. The diameters of the nanotubes that are to be grown are related to the size of the metal particles. This can be controlled by patterned (or masked) deposition of the metal, annealing, or by plasma etching of a metal layer. The substrate is heated to approximately 700 °C. To initiate the growth of nanotubes, two gases are bled into the reactor: a process gas (such as ammonia, nitrogen or hydrogen) and a carbon-containing gas (such as acetylene, ethylene, ethanol or methane). Nanotubes grow at the sites of the metal catalyst; the carbon-containing gas is broken apart at the surface of the catalyst particle, and the carbon is transported to the edges of the particle, where it forms the nanotubes. This mechanism is still being studied. The catalyst particles can stay at the tips of the growing nanotube during growth, or remain at the nanotube base, depending on the adhesion between the catalyst particle and the substrate. Thermal catalytic decomposition of hydrocarbon has become an active area of research and can be a promising route for the bulk production of CNTs. Fluidised bed reactor is the most widely used reactor for CNT preparation. Scale-up of the reactor is the major challenge.

CVD is the most widely used method for the production of carbon nanotubes. For this purpose, the metal nanoparticles are mixed with a catalyst support such as MgO or Al_2O_3 to increase the surface area for higher yield of the catalytic reaction of the carbon feedstock with the metal particles. One issue in this synthesis route is the removal of the catalyst support via an acid treatment, which sometimes could destroy the original structure of the carbon nanotubes. However, alternative catalyst supports that are soluble in water have proven effective for nanotube growth.

If a plasma is generated by the application of a strong electric field during growth (plasma-enhanced chemical vapor deposition), then the nanotube growth will follow the direction of the electric field. By adjusting the geometry of the reactor it is possible to synthesize vertically aligned carbon nanotubes (i.e., perpendicular to the substrate), a morphology that has been of interest to researchers interested in electron emission from nanotubes. Without the plasma, the resulting nanotubes are often randomly oriented. Under certain reaction conditions, even in the absence of a plasma, closely spaced nanotubes will maintain a vertical growth direction resulting in a dense array of tubes resembling a carpet or forest.

Of the various means for nanotube synthesis, CVD shows the most promise for industrial-scale deposition, because of its price/unit ratio, and because CVD is capable of growing nanotubes directly on a desired substrate, whereas the nanotubes must be collected in the other growth techniques. The growth sites are controllable by careful deposition of the catalyst. In 2007, a team from Meijo University demonstrated a high-efficiency CVD technique for growing carbon nanotubes from camphor. Researchers at Rice University, until recently led by the late Richard Smalley, have concentrated upon finding methods to produce large, pure amounts of particular types of nanotubes. Their approach grows long fibers from many small seeds cut from a single nanotube; all of the resulting fibers were found to be of the same diameter as the original nanotube and are expected to be of the same type as the original nanotube.

Super-growth CVD

Super-growth CVD (water-assisted chemical vapor deposition) was developed by Kenji Hata, Sumio Iijima and co-workers at AIST, Japan. In this process, the activity and lifetime of the catalyst are enhanced by addition of water into the CVD reactor. Dense millimeter-tall nanotube "forests", aligned normal to the substrate, were produced. The forests height could be expressed, as

$$H(t) = \beta\tau_o(1 - e^{-t/\tau_o}).$$

In this equation, β is the initial growth rate and τ_o is the characteristic catalyst lifetime.

Their specific surface exceeds 1,000 m²/g (capped) or 2,200 m²/g (uncapped), surpassing the value of 400–1,000 m²/g for HiPco samples. The synthesis efficiency is about 100 times higher than for the laser ablation method. The time required to make SWNT forests of the height of 2.5 mm by this method was 10 minutes in 2004. Those SWNT forests can be easily separated from the catalyst, yielding clean SWNT material (purity >99.98%) without further purification. For comparison, the as-grown HiPco CNTs contain about 5–35% of metal impurities; it is therefore purified through dispersion and centrifugation that damages the nanotubes. Super-growth avoids this problem. Patterned highly organized single-walled nanotube structures were successfully fabricated using the super-growth technique.

The mass density of super-growth CNTs is about 0.037 g/cm^3. It is much lower than that of conventional CNT powders (~1.34 g/cm^3), probably because the latter contain metals and amorphous carbon.

The super-growth method is basically a variation of CVD. Therefore, it is possible to grow material containing SWNT, DWNTs and MWNTs, and to alter their ratios by tuning the growth conditions. Their ratios change by the thinness of the catalyst. Many MWNTs are included so that the diameter of the tube is wide.

The vertically aligned nanotube forests originate from a "zipping effect" when they are immersed in a solvent and dried. The zipping effect is caused by the surface tension of the solvent and the van der Waals forces between the carbon nanotubes. It aligns the nanotubes into a dense material, which can be formed in various shapes, such as sheets and bars, by applying weak compression during the process. Densification increases the Vickers hardness by about 70 times and density is 0.55 g/cm^3. The packed carbon nanotubes are more than 1 mm long and have a carbon purity of 99.9% or higher; they also retain the desirable alignment properties of the nanotubes forest.

Natural, Incidental, and Controlled Flame Environments

Fullerenes and carbon nanotubes are not necessarily products of high-tech laboratories; they are commonly formed in such mundane places as ordinary flames, produced by burning methane, ethylene, and benzene, and they have been found in soot from both indoor and outdoor air. However, these naturally occurring varieties can be highly irregular in size and quality because the environment in which they are produced is often highly uncontrolled. Thus, although they can be used in some applications, they can lack in the high degree of uniformity necessary to satisfy the many needs of both research and industry. Recent efforts have focused on producing more uniform carbon nanotubes in controlled flame environments. Such methods have promise for large-scale, low-cost nanotube synthesis based on theoretical models, though they must compete with rapidly developing large scale CVD production.

Purification

Removal of Catalysts

Nanoscale metal catalysts are important ingredients for fixed- and fluidized-bed CVD synthesis of CNTs. They allow increasing the growth efficiency of CNTs and may give control over their structure and chirality. During synthesis, catalysts can convert carbon precursors into tubular carbon structures but can also form encapsulating carbon overcoats. Together with metal oxide supports they may therefore attach to or become incorporated into the CNT product. The presence of metal impurities can be problematic for many applications. Especially catalyst metals like nickel, cobalt or yttrium may be of toxicological concern. While unencapsulated catalyst metals may be readily removable by acid washing, encapsulated ones require oxidative treatment for opening their carbon shell. The effective removal of catalysts, especially of encapsulated ones, while preserving the CNT structure is a challenge and has been addressed in many studies. A new approach to break carbonaceous catalyst encapsulations is based on rapid thermal annealing.

Application-related Issues

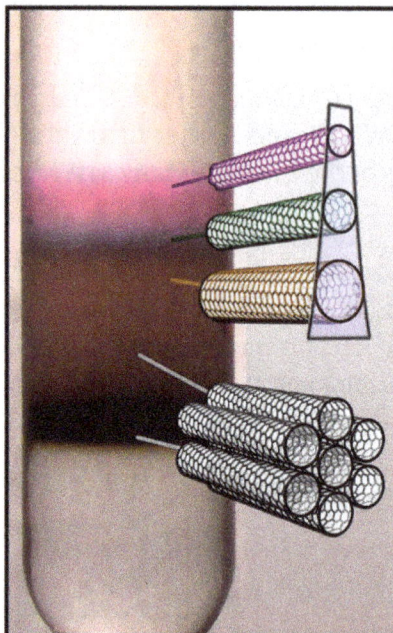

Centrifuge tube with a solution of carbon nanotubes, which were sorted by diameter using density-gradient ultracentrifugation.

Many electronic applications of carbon nanotubes crucially rely on techniques of selectively producing either semiconducting or metallic CNTs, preferably of a certain chirality. Several methods of separating semiconducting and metallic CNTs are known, but most of them are not yet suitable for large-scale technological processes. The most efficient method relies on density-gradient ultracentrifugation, which separates surfactant-wrapped nanotubes by the minute difference in their density. This density difference often translates into difference in the nanotube diameter and (semi)conducting properties. Another method of separation uses a sequence of freezing, thawing, and compression of SWNTs embedded in agarose gel. This process results in a solution containing 70% metallic SWNTs and leaves a gel containing 95% semiconducting SWNTs. The diluted solutions separated by this method show various colors. The separated carbon nanotubes using this method have been applied to electrodes, e.g. electric double-layer capacitor. Moreover, SWNTs can be separated by the column chromatography method. Yield is 95% in semiconductor type SWNT and 90% in metallic type SWNT.

In addition to separation of semiconducting and metallic SWNTs, it is possible to sort SWNTs by length, diameter, and chirality. The highest resolution length sorting, with length variation of <10%, has thus far been achieved by size exclusion chromatography (SEC) of DNA-dispersed carbon nanotubes (DNA-SWNT). SWNT diameter separation has been achieved by density-gradient ultracentrifugation (DGU) using surfactant-dispersed SWNTs and by ion-exchange chromatography (IEC) for DNA-SWNT. Purification of individual chiralities has also been demonstrated with IEC of DNA-SWNT: specific short DNA oligomers can be used to isolate individual SWNT chiralities. Thus far, 12 chiralities have been isolated at purities ranging from 70% for (8,3) and (9,5) SWNTs to 90% for (6,5), (7,5) and (10,5)SWNTs. There have been successful efforts to integrate these purified nanotubes into devices, e. g. FETs.

An alternative to separation is development of a selective growth of semiconducting or metallic CNTs. Recently, a new CVD recipe that involves a combination of ethanol and methanol gases and quartz substrates resulting in horizontally aligned arrays of 95–98% semiconducting nanotubes was announced.

Nanotubes are usually grown on nanoparticles of magnetic metal (Fe, Co), which facilitates production of electronic (spintronic) devices. In particular, control of current through a field-effect transistor by magnetic field has been demonstrated in such a single-tube nanostructure.

Carbon Nanotube Chemistry

Carbon nanotube chemistry involves chemical reactions, which are used to modify the properties of carbon nanotubes (CNTs). CNTs can be functionalized to attain desired properties that can be used in a wide variety of applications. The two main methods of CNT functionalization are covalent and non-covalent modifications.

Because of their hydrophobic nature, CNTs tend to agglomerate hindering their dispersion in solvents or viscous polymer melts. The resulting nanotube bundles or aggregates reduce the mechanical performance of the final composite. The surface of CNTs can be modified to reduce the hydrophobicity and improve interfacial adhesion to a bulk polymer through chemical attachment.

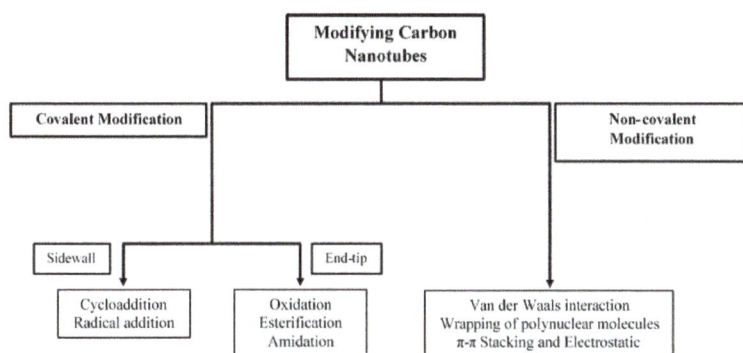

Chart summarizing options for the chemical modification of carbon nanotubes.

Covalent Modification

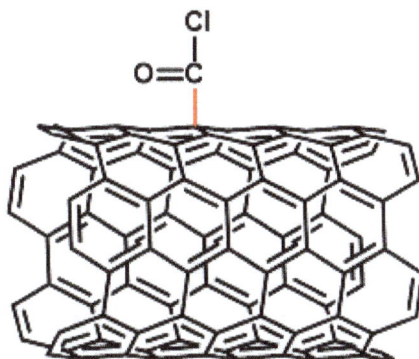

Covalent modification of carbon nanotubes.

Covalent modification attaches a functional group onto the carbon nanotube. The functional groups can be attached onto the side wall or ends of the carbon nanotube. The end caps of the carbon nanotubes have the highest reactivity due to its higher pyrimidization angle and the walls of the carbon nanotubes have lower pyrimidization angles which has lower reactivity. Although covalent modifications are very stable, the bonding process disrupts the sp^2 hybridization of the carbon atoms because a σ-bond is formed. The disruption of the extended sp^2 hybridization typically decreases the conductance of the carbon nanotubes.

Oxidation

The purification and oxidation of carbon nanotubes (CNTs) has been well represented in literature. These processes were essential for low yield production of carbon nanotubes where carbon particles, amorphous carbon particles and coatings comprised a significant percentage of the overall material and are still important for the introduction of surface functional groups. During acid oxidation, the carbon-carbon bonded network of the graphitic layers is broken allowing the introduction of oxygen units in the form of carboxyl, phenolic and lactone groups, which have been extensively exploited for further chemical functionalisation.

First studies on oxidation of carbon nanotubes involved a gas-phase reactions with nitric acid vapor in air, which indiscriminately functionalized the carbon nanotubes with carboxylic, carbonyl or hydroxyl groups. In liquid-phase reactions, carbon nanotubes were treated with oxidizing solutions of nitric acid or a combination of nitric and sulfuric acid to the same effect. However, overoxidation may occur causing the carbon nanotube to break up into fragments, which are known as carbonaceous fragments. Xing et al. revealed sonication assisted oxidation, with sulfuric and nitric acid, of carbon nanotubes and produced carbonyl and carboxyl groups. After the oxidation reaction in acidic solution, treatment with hydrogen peroxide limited the damage on the carbon nanotube network. Single-walled carbon nanotubes can be shortened in a scalable manner using oleum (100% H_2SO_4 with 3% SO_3) and nitric acid. The nitric acid cuts carbon nanotubes while the oleum creates a channel.

In one type of chemical modification, aniline is oxidized to a diazonium intermediate. After expulsion of nitrogen, it forms a covalent bond as an aryl radical:

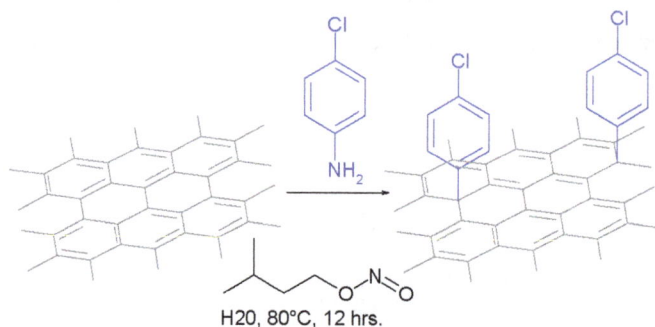

H20, 80°C, 12 hrs.

Esterification/Amidation

Carboxylic groups are used as the precursor for most esterification and amidation reactions. The carboxylic group is converted into an acyl chloride with the use of thionyl or oxalyl chloride which is then reacted with the desired amide, amine, or alcohol. Carbon nanotubes have been deposited on with silver nanoparticles with the aid of amination reactions. Amide functionalized carbon nanotubes have been shown to chelate silver nanoparticles. Carbon nanotubes modified with acyl chloride react readily with highly branched molecules such as poly(amindoamine), which acts as a template for silver ion and later being reduced by formaldehyde. Amino-modified carbon nanotubes can be prepared by reacting ethylenediamine with an acyl chloride functionalized carbon nanotubes.

Halogenation Reactions

Carbon nanotubes can be treated with peroxytrifluroacetic acid to give mainly carboxylic acid and trifluroacetic functional groups. The fluorinated carbon nanotubes, through substitution, can be further functionalized with urea, guanidine, thiourea and aminosilane. Using the Hunsdiecker reaction, carbon nanotubes treated with nitric acid can react with iodosobenzenediacetate to iodate carbon nanotubes.

Cycloaddition

Also known are protocols for cycloadditions such as Diels-Alder reactions, 1,3-dipolar cycloadditions of azomethine ylides and azide–alkyne cycloaddition reactions. One example is a DA reaction assisted by chromium hexacarbonyl and high pressure. The I_D/I_G ratio for reaction with Danishefsky's diene is 2.6.

The most well-known 1,3 cycloaddition reaction involves azomethine ylides reacting with carbon nanotubes, which are of great interest. The addition of a pyrrolidine ring can lead to a variety of functional groups such as second-generation poly(amidoamine) dendrimers, phthalocyanineaddends, perfluoroalkylsilane groups, and amino ethyleneglycol groups. The Diels-cycloaddition reaction can occur, especially on fluorinated carbon nanotubes. They are known to undergo Diels–Alder reaction s with dienes such as 2,3-dimethyl-1,3-butadiene, anthracene, and 2-trimethylsiloxyl-1,3-butadiene.

Radical Addition

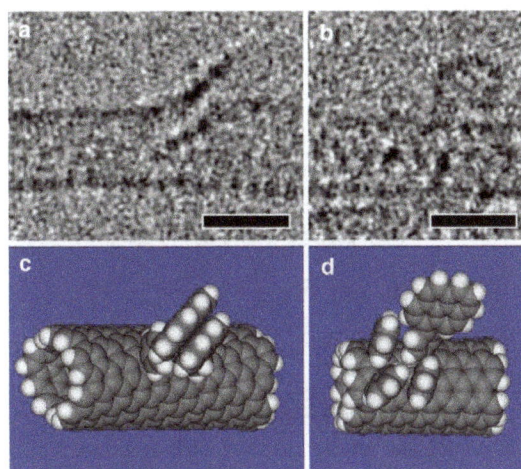

Top: electron micrographs showing interaction of CNTs with 4-(1-pyrenyl)phenyl radical (a) and its boronic ester (b). Bottom: corresponding models.

The modification of carbon nanotubes with aryl diazonium salts was studied first by Tour et al. Due to the harsh conditions needed for the *in situ* generated diazonium compound, other methods have been explored. Stephenson et al. reported using aniline derivatives with sodium nitrite in 96% sulfuric acid and ammonium persulfate. Price et al. demonstrated that stirring carbon nanotubes in water and treating with anilines and oxidizing agents proved to be a milder reaction. The diazonium chemistry functionalized carbon nanotubes which was used as a precursor to further modifications. Suzuki and Heck coupling reactions were performed on iodophenyl-functionalized carbon nanotubes. Wong et al. demonstrated mild photochemical reactions to silylate the carbon nanotubes with trimethoxysilane and hexaphenyldisilane.

Nucleophilic Addition

Hirsch et al. conducted nucleophilic additions with organolithium and organomagnesium compounds onto carbon nanotubes. With further oxidation in air, they were able to create alkyl-modified carbon nanotubes. Hirsch was also able to show the nucleophilic addition of amines by generating lithium amides, leading to amino-modified carbon nanotubes.

Electrophilic Addition

Nanotubes can also be alkylated with alkyl halides using lithium or sodium metal and liquid ammonia (Birch reduction conditions). The initial nanotube salt can function as an polymerization initiator and can react with peroxides to form alkoxy functionalized nanotubes

The alkyl and hydroxyl modification of carbon nanotubes was demonstrated with the electrophilic addition of alkylhalides by microwave irradiation. Tessonnier et al. modified carbon nanotubes with amino groups by deprotonating with butyl lithium and reacting with amino substitution. Balaban et al. applied Friedel-Crafts acylation to carbon nanotubes with nitrobenzene at 180 °C along with aluminum chloride.

Non-covalent Modifications

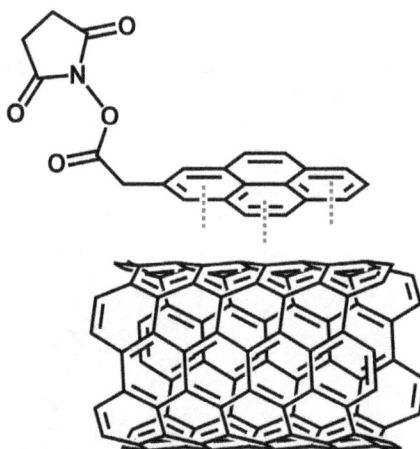

Non-covalent modification of carbon nanotubes.

Non-covalent modifications utilize van der Waals forces and π-π interactions by adsorption of polynuclear aromatic compounds, surfactants, polymers or biomolecules. Non-covalent modifications do not disrupt the natural configuration of carbon nanotubes with the cost of chemical stability, and is prone to phase separation, dissociation in between two phases, in the solid state.

Polynuclear Aromatic Compounds

Some common polynuclear aromatic compounds that are functionalized with hydrophilic or hydrophobic moieties are used to solubilize carbon nanotubes into organic or aqueous solvents. Some of these amphiphiles are phenyl, naphthalene, phenanthrene, pyrene and porphyrin systems. The greater π-π stacking of aromatic amphiphiles such as pyrene amphiphiles had the best solubility compared to phenyl amphiphiles with the worse π-π stacking, lead to more solubility in water. These aromatic systems can be modified with amino and carboxylic acid groups prior to functionalizing the carbon nanotubes.

Biomolecules

The interaction between carbon nanotubes and biomolecules has been widely studied because of their potential to be used in biological applications. The modification of the carbon nanotubes with proteins, carbohydrates, and nucleic acids are built with the bottom-up technique. Proteins have high affinity to carbon nanotubes due to their diversity of amino acids being hydrophobic or hydrophilic. Polysaccharides have been successfully been used to modify carbon nanotubes forming stable hybrids. To make carbon nanotubes soluble in water, phospholipids such as lysoglycerophospholipids have been used. The single phospholipid tail wraps around the carbon nanotube, but the double tailed phospholipids did not have the same ability.

π-π Stacking and Electrostatic Interactions

Molecules that have bifunctionality are used to modify the carbon nanotube. One end of the molecule are polyaromatic compounds that interact with the carbon nanotube through π-π stacking. The other end of the same molecule has a functional group such as amino, carboxyl, or thiol. For example, pyrene derivatives and aryl thiols were used as the linkers for various metal nanobeads such as gold, silver and platinum.

Characterization

A useful tool for the analysis of derivatised nanotubes is Raman spectroscopy which shows a G-band (G for graphite) for the native nanotubes at 1580 cm^{-1} and a D-band (D for defect) at 1350 cm^{-1} when the graphite lattice is disrupted with conversion of sp^2 to sp^3 hybridized carbon. The ratio of both peaks I_D/I_G is taken as a measure of functionalization. Other tools are UV spectroscopy where pristine nanotubes show distinct Van Hove singularities where functionalized tubes do not, and simple TGA analysis.

Selective Chemistry of Single-walled Nanotubes

Selective chemistry of single-walled nanotubes is a field in Carbon nanotube chemistry devoted specifically to the study of functionalization of single-walled carbon nanotubes.

Structure and Reactivity

Reactivity of fullerene molecules with respect to addition chemistries is strongly dependent on the curvature of the carbon framework. Their outer surface (exohedral) reactivity increases with increase in curvature. In comparison with fullerene molecules single-walled nanotubes (SWNTs) are moderately curved. Consequently, nanotubes are expected to be less reactive than most fullerene molecules due to their smaller curvature, but more reactive than a graphene sheet due to pyramidalization and misalignment of pi-orbitals. The strain of a carbon framework is also reflected in the pyramidalization angle (Θ_p) of the carbon constituents. Trigonal carbon atoms (sp² hybridized) prefer a planar orientation with $\Theta_p = 0°$ (i.e. graphene) and fullerene molecules have $\Theta_p = 11.6°$. The (5,5) SWNT has $\Theta_p \sim 6°$ for the sidewall. Values for other (n,n) nanotubes show a trend of increasing Θ_p (sidewall) with decrease in n. Therefore, generally the chemical reactivity of SWNT increases with decrease in diameter (or n, diameter increases with n). Apart from the curvature SWNT reactivity is also highly sensitive to chiral wrapping (n,m) which determine its electronic structure. Nanotubes with $n - m = 3i$ (i is an integer) are all metals and rest are all semiconducting (SC).

Sidewall Functionalization

Carbon nanotubes are metallic or semiconducting, based upon delocalized electrons occupying a 1-D density of states. However, any covalent bond on SWNT sidewall causes localization of these electrons. In the vicinity of localized electrons, the SWNT can no longer be described using a band model that assumes delocalized electrons moving in a periodic potential.

Two important addition reactions of SWNT sidewall are: (1) Fluorination, and (2) Aryl diazonium salt addition. These functional groups on SWNT improve solubility and processibility. Moreover, these reactions allow for combining unique properties of SWNTs with those of other compounds. Above all, the selective diazonium chemistry can be used to separate the semiconducting and metallic nanotubes.

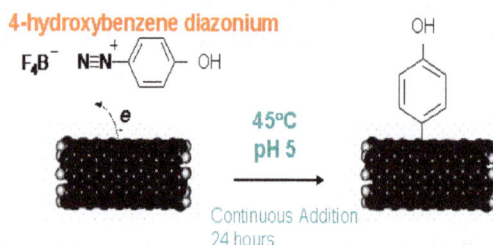

Selective Diazonium Reaction
pH is reduced to prevent electrostatic repulsion.
Temperature is increased to increase the reaction rate.
Diazonium salt is added at a very small rate

Fluorination

The first extensive SWNT sidewall reaction was fluorination in 1998 by Mickleson et al. These fluorine moieties can be removed from the nanotube by treatment in hydrazine and the spectroscopic properties of the SWNT can be restored completely.

Diazonium Chemistry

One of the most important SWNT sidewall reaction is that with diazonium reagent which if done under controlled conditions can be used to do selective covalent chemistry.

Water-soluble diazonium salts react with carbon nanotubes via charge transfer in which they extract electrons from SWNT and form a stable covalent aryl bond. This covalent aryl bond forms with extremely high affinity for electrons with energies near the Fermi level, E_f of the nanotube. Metallic SWNT have a greater electron density near E_f resulting in their higher reactivity over semiconducting nanotubes. The reactant forms a charge-transfer complex at the nanotube surface, where electron donation from the latter stabilizes the transition state and accelerates the forward rate. Once the bond symmetry of the nanotube is disrupted by the formation of this defect, adjacent carbons increase in reactivity and initial selectivity for metallic SWNT is amplified. Under carefully controlled conditions this behavior can be exploited to obtain highly selective functionalization of metallic nanotubes to the near exclusion of the semiconductors.

Selective Reaction Conditions

Primary condition is addition of reactant molecules at a very small rate to SWNT solution for a sufficient long time. This ensures reaction with only metallic SWNTs and with no semiconducting SWNTs as all the reactant molecules are taken up by the metallic SWNTs. Long time injection ensures that all metallic tubes are reacted. For example, one highly selective condition is: addition of 500 µL of 4-hydroxybenzene diazonium tetrafluoroborate solution in water (0.245 mM) at an injection rate of 20.83 µL/h into 5 mL of SWNT solution (1 wt % sodium dodecyl sulfate (SDS)) over 24 hrs. However, if the entire diazonium solution is added all instantaneously then semiconducting SWNTs will also react due to presence of excess reactant.

Spectroscopy and Functionalization

SWNTs have unique optical and spectroscopic properties largely due to one-dimensional confinement of electronic and phonon states, resulting in so-called van Hove singularities in the nanotube density of states (DOS).

Probing Selective Chemistry Via Optical Absorption

Optical absorption monitors the valence (v) to conduction (c) electronic transitions denoted E_{nn} where n is the band index. The E_{11} transitions for the metallic nanotubes occur from ~440 to 645 nm. The E_{11} and E_{22} transitions for the semiconducting nanotubes are found from 830 to 1600 nm and 600 to 800 nm, respectively. These separated absorption features allow for the monitoring of valence electrons in each distinct nanotube. Reaction at the surface result in localization of valence electrons makes them no longer free to participate in photoabsorption which results in decay of the spectrum features.

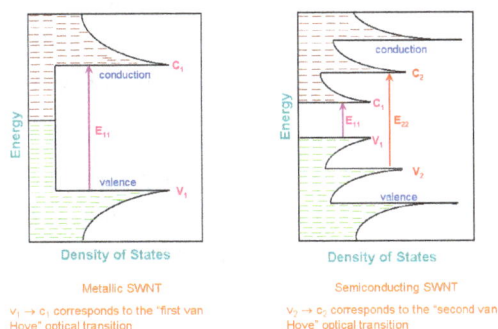

Metallic SWNT

$v_1 \rightarrow c_1$ corresponds to the "first van Hove" optical transition

Semiconducting SWNT

$v_2 \rightarrow c_2$ corresponds to the "second van Hove" optical transition

Selective diazonium chemistry abruptly decreases the peak intensities that represent the first Van Hove transition of metallic species (E_{11}, metal), while the peak intensities representing the second (E_{22}, semiconducting) and first (E_{11}, semiconducting) Van Hove transition of the semiconducting species show little or no change. A relative decrease in metallic SWNT absorption features over semiconducting features represents a highly preferential functionalization of the metallic nanotubes.

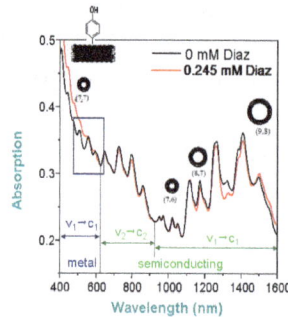

Ultraviolet– visible–near-infrared (UV-vis-nIR) Absorption Spectrum Allows Probing of Valence Electrons
Absorption spectra of aqueous suspended nanotubes after successive additions of 4-hydroxybenzenediazonium tetrafluoroborate under controlled conditions. Transition for metallic SWNT decay almost completely due to covalent reaction. Transitions for SC nanotubes remain almost unchanges.

Raman Spectroscopy

Raman spectroscopy is a powerful technique with wide ranging applications in carbon nanotube studies. Some important Raman features are radial breathing mode (RBM), tangential mode (G-band), and disorder-related mode.

RBM features correspond to the coherent vibration of the C atoms in the radial direction of the nanotube. These features are unique in carbon nanotubes and occur with the frequencies ω_{RBM} between 120 and 350 cm^{-1} for SWNT in the diameter range (0.7 nm-2 nm). They can be used to probe the SWNT diameter, electronic structure through their frequency and intensity (I_{RBM}) respectively and hence perform an (n,m) assignment to their peaks. The addition of the moiety to the sidewall of the nanotube disrupts the oscillator strength that gives rise to RBM feature and hence causes decay of these features. These features are distinct for species of a particular nanotube (n,m) and hence enables to probe which SWNTs are functionalized and to what extent.

Two main components of the tangential mode include G$^+$ at 1590 cm^{-1} and G$^-$ at 1570 cm^{-1}. G$^+$ feature is associated with carbon atom vibrations along the nanotube axis. The G$^-$ feature is associated with vibrations of carbon atoms along the circumferential direction. The G-band frequency can be used (1) to distinguish between metallic and semiconducting SWNTs, and (2) to probe charge transfer arising from doping a SWNT. Frequency of G$^+$ is sensitive to charge transfer. It upshifts for acceptors and downshifts for donors. Lineshape of G$^-$ is highly sensitive to whether SWNT is metallic (Breit-Wigner-Fano lineshape) or semiconducting (Lorentzian lineshape).

The disorder-related mode (D peak) is a phonon mode at 1300 cm^{-1} and involves the resonantly enhanced scattering of an electron via phonon emission by a defect that breaks the basic symmetry of the graphene plane. This mode corresponds to the conversion of a sp^2-hybridized carbon to a sp^3-hybridized on the surface. Intensity of D peak measures covalent bond made with the nanotube surface. This feature does not increase as a result of surfactant or hydronium ion adsorption on the nanotube surface.

Selective Reaction and Raman Features

Selective functionalization increases the intensity of the D peak due to formation of aryl-nanotube bond and decreases the tangential mode due loss of electronic resonance. These two effects are generally summed together as increase in their peaks ratio (D/G). RBM peaks of metallic nanotubes decay and the peaks corresponding to those of semiconducting nanotubes remain almost unchanged.

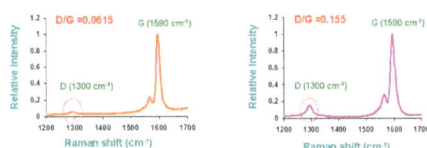

Raman Spectrum and Covalent Reaction
Disorder mode grows with increase in functionalization.
(Raman spectra taken at 785-nm excitation)

Reaction Mechanism

Diazonium reagent and SWNT reaction has a two step mechanism. First, the diazonium reagent adsorbs noncovalently to an empty site on the nanotube surface, forming a charge-transfer complex. This is a fast, selective noncovalent adsorption and diazonium group in this complex partially dopes the nanotube, diminishing the tangential mode in the Raman spectrum. Desorption of A from nanotube is negligible ($k_{-1} \sim 0$). In second step complex B then decomposes to form a covalent bond with the nanotube surface. This is a slower step that need not be selective and is represented by the restoration of the G peak and increase in the D band.

Reversibility of Diazonium Chemistry

Nanotubes reacted with the diazonium reagent can be converted back into pristine nanotubes when thermally treated at 300 °C in an atmosphere of inert gas. This cleaves the aryl hydroxyl moieties from the nanotube sidewall and restores the spectroscopic feature (Raman and absorption spectra) of pristine nanotube.

Chemical Separation of Metallic and Semiconducting SWNTs

Metallic and semiconducting carbon nanotubes generally coexist in as-grown materials. To get only semiconducting or only metallic nanotubes selective functionalization of metallic SWNTs via 4-hydroxybenzene diazonium can be used. Separation can be done in solution by deprotonation of the p-hydroxybenzene group on the reacted nanotubes (metallic) in alkaline solution followed by electrophoretic separation of these charged species from the neutral species (semiconducting nanotubes). This followed by annealing would give separated pristine semiconducting and metallic SWNT.

References

- "Molecular interactions on single-walled carbon nanotubes revealed by high-resolution transmission microscopy". Nature Communications. 6: 7732. 2015. doi:10.1038/ncomms8732.

- K. Takeuchi, T. Hayashi, Y. A. Kim, K. Fujisawa, M. Endo "The state-of-the-art science and applications of carbon nanotubes", February 2014, Volume 5, Issue 1, pp 15

- D. Janas; et al. (2014). "Direct evidence of delayed electroluminescence from carbon nanotubes on the macroscale". Applied Physics Letters. 104 (26): 261107. Bibcode:2014ApPhL.104z1107J. doi:10.1063/1.4886800.

- K. Takeuchi, T. Hayashi, Y. A. Kim, K. Fujisawa, M. Endo "The state-of-the-art science and applications of carbon nanotubes", February 2014, Volume 5, Issue 1, pp 15

- M.B. Belonenko; et al. (2014). "Few cycle pulses in the bragg medium containing carbon nanotubes" (PDF). NANOSYSTEMS: PHYSICS, CHEMISTRY, MATHEMATICS. 14 (5): 644.

- D. Janas; et al. (2013). "Electroluminescence from carbon nanotube films resistively heated in air". Applied Physics Letters. 102 (18): 181104. Bibcode:2013ApPhL.102r1104J. doi:10.1063/1.4804296.

- Meyer-Plath A, Orts-Gil G, Petrov S, et al. (2012). "Plasma-thermal purification and annealing of carbon nanotubes". Carbon. 50 (10): 3934–3942. doi:10.1016/j.carbon.2012.04.049.

- "Pirahna USV built using nano-enhanced carbon prepreg". ReinforcedPlastics.com. 19 February 2009. Archived from the original on 3 March 2012.

Understanding Metallurgy

Metallurgy studies the physical and chemical behavior of metallic elements. It is the technology of metals. The process of working on these metals to create structures and assemblies is known as metalworking. The processes of metalworking that have been explained in this chapter are casting, forging, rolling, cladding, metal fabrication etc. This section elucidates the basic concepts of metallurgy.

Metallurgy

Metallurgy is a domain of materials science and engineering that studies the physical and chemical behaviour of metallic elements, their intermetallic compounds, and their mixtures, which are called alloys. Metallurgy is also the technology of metals: the way in which science is applied to the production of metals, and the engineering of metal components for usage in products for consumers and manufacturers. The production of metals involves the processing of ores to extract the metal they contain, and the mixture of metals, sometimes with other elements, to produce alloys. Metallurgy is distinguished from the craft of metalworking, although metalworking relies on metallurgy, as medicine relies on medical science, for technical advancement.

Georgius Agricola, author of *De re metallica*, an important early work on metal extraction

Metallurgy is subdivided into *ferrous metallurgy* (sometimes also known as black metallurgy) and *non-ferrous metallurgy* or *colored metallurgy*. Ferrous metallurgy involves processes and alloys based on iron while non-ferrous metallurgy involves processes and alloys based on other metals. The production of ferrous metals accounts for 95 percent of world metal production.

Smelting Gold in Nicaragua in the La Luz Gold Mine in Siuna and Bonanza about 1959.
Smelting is a basic step in obtaining usable quantities of most metals.

Pouring Smelted Gold into an ingot at the La Luz Gold Mine in Siuna, Nicaragua about 1959.

History

The earliest recorded metal employed by humans appears to be gold, which can be found free or "native". Small amounts of natural gold have been found in Spanish caves used during the late Paleolithic period, c. 40,000 BC. Silver, copper, tin and meteoric iron can also be found in native form, allowing a limited amount of metalworking in early cultures. Egyptian weapons made from meteoric iron in about 3000 BC were highly prized as "daggers from heaven".

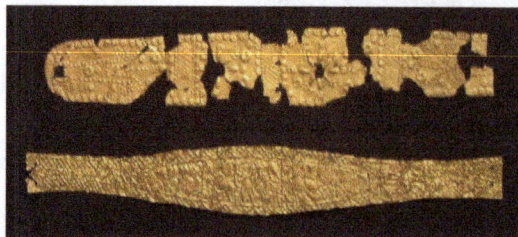

Gold headband from Thebes 750–700 BC

Certain metals, notably tin, lead and (at a higher temperature) copper, can be recovered from their ores by simply heating the rocks in a fire or blast furnace, a process known as smelting. The first evidence of this extractive metallurgy dates from the 5th and 6th millennia BC and was found in the archaeological sites of Majdanpek, Yarmovac and Plocnik, all three in Serbia. To date, the earliest evidence of copper smelting is found at the Belovode site, including a copper axe from 5500 BC belonging to the Vinča culture. Other signs of early metals are found from the third millennium BC in places like Palmela (Portugal), Los Millares (Spain), and Stonehenge (United Kingdom). However, the ultimate beginnings cannot be clearly ascertained and new discoveries are both continuous and ongoing.

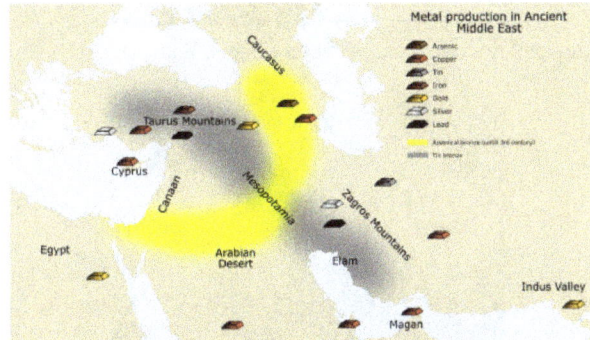

Mining areas of the ancient Middle East. Boxes colors: arsenic is in brown, copper in red, tin in grey, iron in reddish brown, gold in yellow, silver in white and lead in black. Yellow area stands for arsenic bronze, while grey area stands for tin bronze.

These first metals were single ones or as found. About 3500 BC, it was discovered that by combining copper and tin, a superior metal could be made, an alloy called bronze, representing a major technological shift known as the Bronze Age.

The extraction of iron from its ore into a workable metal is much more difficult than for copper or tin. The process appears to have been invented by the Hittites in about 1200 BC, beginning the Iron Age. The secret of extracting and working iron was a key factor in the success of the Philistines.

Historical developments in ferrous metallurgy can be found in a wide variety of past cultures and civilizations. This includes the ancient and medieval kingdoms and empires of the Middle East and Near East, ancient Iran, ancient Egypt, ancient Nubia, and Anatolia (Turkey), Ancient Nok, Carthage, the Greeks and Romans of ancient Europe, medieval Europe, ancient and medieval China, ancient and medieval India, ancient and medieval Japan, amongst others. Many applications, practices, and devices associated or involved in metallurgy were established in ancient China, such as the innovation of the blast furnace, cast iron, hydraulic-powered trip hammers, and double acting piston bellows.

A 16th century book by Georg Agricola called *De re metallica* describes the highly developed and complex processes of mining metal ores, metal extraction and metallurgy of the time. Agricola has been described as the "father of metallurgy".

Extraction

Extractive metallurgy is the practice of removing valuable metals from an ore and refining the extracted raw metals into a purer form. In order to convert a metal oxide or sulfide to a purer metal, the ore must be reduced physically, chemically, or electrolytically.

Furnace bellows operated by waterwheels, Yuan Dynasty, China.

Aluminium plant in Žiar nad Hronom (Central Slovakia)

Extractive metallurgists are interested in three primary streams: feed, concentrate (valuable metal oxide/sulfide), and tailings (waste). After mining, large pieces of the ore feed are broken through crushing and/or grinding in order to obtain particles small enough where each particle is either mostly valuable or mostly waste. Concentrating the particles of value in a form supporting separation enables the desired metal to be removed from waste products.

Mining may not be necessary if the ore body and physical environment are conducive to leaching. Leaching dissolves minerals in an ore body and results in an enriched solution. The solution is collected and processed to extract valuable metals.

Ore bodies often contain more than one valuable metal. Tailings of a previous process may be used as a feed in another process to extract a secondary product from the original ore. Additionally, a concentrate may contain more than one valuable metal. That concentrate would then be processed to separate the valuable metals into individual constituents.

Alloys

Casting bronze

Common engineering metals include aluminium, chromium, copper, iron, magnesium, nickel, titanium and zinc. These are most often used as alloys. Much effort has been placed on understanding the iron-carbon alloy system, which includes steels and cast irons. Plain carbon steels (those that contain essentially only carbon as an alloying element) are used in low-cost, high-strength applications where weight and corrosion are not a problem. Cast irons, including ductile iron, are also part of the iron-carbon system.

Stainless steel or galvanized steel are used where resistance to corrosion is important. Aluminium alloys and magnesium alloys are used for applications where strength and lightness are required.

Copper-nickel alloys (such as Monel) are used in highly corrosive environments and for non-magnetic applications. Nickel-based superalloys like Inconel are used in high-temperature applications such as gas turbines, turbochargers, pressure vessels, and heat exchangers. For extremely high temperatures, single crystal alloys are used to minimize creep.

Production

In production engineering, metallurgy is concerned with the production of metallic components for use in consumer or engineering products. This involves the production of alloys, the shaping, the heat treatment and the surface treatment of the product. The task of the metallurgist is to achieve balance between material properties such as cost, weight, strength, toughness, hardness, corrosion, fatigue resistance, and performance in temperature extremes. To achieve this goal, the operating environment must be carefully considered. In a saltwater environment, ferrous metals and some aluminium alloys corrode quickly. Metals exposed to cold or cryogenic conditions may endure a ductile to brittle transition and lose their toughness, becoming more brittle and prone to cracking. Metals under continual cyclic loading can suffer from metal fatigue. Metals under constant stress at elevated temperatures can creep.

Metalworking Processes

Metals are shaped by processes such as:

- casting – molten metal is poured into a shaped mold.

- forging – a red-hot billet is hammered into shape.

- rolling – a billet is passed through successively narrower rollers to create a sheet.

- laser cladding – metallic powder is blown through a movable laser beam (e.g. mounted on a NC 5-axis machine). The resulting melted metal reaches a substrate to form a melt pool. By moving the laser head, it is possible to stack the tracks and build up a three-dimensional piece.

- extrusion – a hot and malleable metal is forced under pressure through a die, which shapes it before it cools.

- sintering – a powdered metal is heated in a non-oxidizing environment after being compressed into a die.

- machining – lathes, milling machines, and drills cut the cold metal to shape.

- fabrication – sheets of metal are cut with guillotines or gas cutters and bent and welded into structural shape.

- 3D printing – Sintering or melting powder metal in a very small point on a moving 'print head' moving in 3D space to make any object to shape.

Cold-working processes, in which the product's shape is altered by rolling, fabrication or other processes while the product is cold, can increase the strength of the product by a process called work hardening. Work hardening creates microscopic defects in the metal, which resist further changes of shape.

Various forms of casting exist in industry and academia. These include sand casting, investment casting (also called the "lost wax process"), die casting, and continuous casting.

Heat Treatment

Metals can be heat-treated to alter the properties of strength, ductility, toughness, hardness and/or resistance to corrosion. Common heat treatment processes include annealing, precipitation strengthening, quenching, and tempering. The annealing process softens the metal by heating it and then allowing it to cool very slowly, which gets rid of stresses in the metal and makes the grain structure large and soft-edged so that when the metal is hit or stressed it dents or perhaps bends, rather than breaking; it is also easier to sand, grind, or cut annealed metal. Quenching is the process of cooling a high-carbon steel very quickly after heating, thus "freezing" the steel's molecules in the very hard martensite form, which makes the metal harder. There is a balance between hardness and toughness in any steel; the harder the steel, the less tough or impact-resistant it is, and the more impact-resistant it is, the less hard it is. Tempering relieves stresses in the metal that were caused by the hardening process; tempering makes the metal less hard while making it better able to sustain impacts without breaking.

Often, mechanical and thermal treatments are combined in what are known as thermo-mechanical treatments for better properties and more efficient processing of materials. These processes are common to high-alloy special steels, super alloys and titanium alloys.

Plating

Electroplating is a chemical surface-treatment technique. It involves bonding a thin layer of another metal such as gold, silver, chromium or zinc to the surface of the product. It is used to reduce corrosion as well as to improve the product's aesthetic appearance.

Thermal Spraying

Thermal spraying techniques are another popular finishing option, and often have better high temperature properties than electroplated coatings.

Microstructure

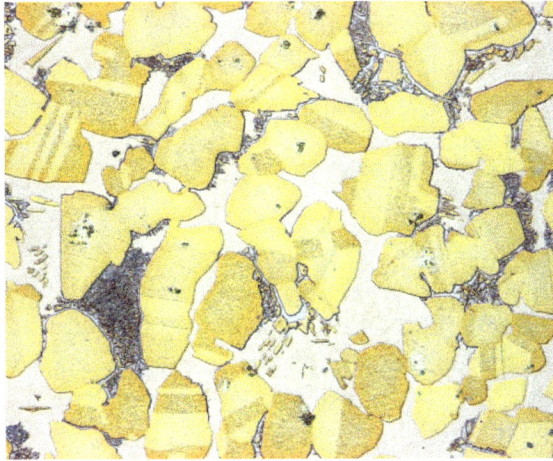

Metallography allows the metallurgist to study the microstructure of metals.

Metallurgists study the microscopic and macroscopic properties using metallography, a technique invented by Henry Clifton Sorby. In metallography, an alloy of interest is ground flat and polished to a mirror finish. The sample can then be etched to reveal the microstructure and macrostructure of the metal. The sample is then examined in an optical or electron microscope, and the image contrast provides details on the composition, mechanical properties, and processing history.

Crystallography, often using diffraction of x-rays or electrons, is another valuable tool available to the modern metallurgist. Crystallography allows identification of unknown materials and reveals the crystal structure of the sample. Quantitative crystallography can be used to calculate the amount of phases present as well as the degree of strain to which a sample has been subjected.

Conferences

EMC, the European Metallurgical Conference has developed to the most important networking business event dedicated to the non-ferrous metals industry in Europe. From the start of the conference sequence in 2001 at Friedrichshafen it was host of the most relevant metallurgists from all countries of the world. The European Metallurgical Conference is organized by GDMB Society of Metallurgists and Miners.

Extractive Metallurgy

Extractive metallurgy is a branch of metallurgical engineering wherein process and methods of extraction of metals from their natural mineral deposits are studied. The field is a materials science, covering all aspects of the types of ore, washing, concentration, separation, chemical processes and extraction of pure metal and their alloying to suit various applications, sometimes for direct use as a finished product, but more often in a form that requires further working to achieve the given properties to suit the applications.

The field of ferrous and non-ferrous extractive metallurgy have specialties that are generically grouped into the categories of mineral processing, hydrometallurgy, pyrometallurgy, and electrometallurgy based on the process adopted to extract the metal. Several processes are used for extraction of same metal depending on occurrence and chemical requirements.

Mineral Processing

Mineral processing begins with beneficiation, consisting of initially breaking down the ore to required sizes depending on the concentration process to be followed, by crushing, grinding, sieving etc. Thereafter, the ore is physically separated from any unwanted impurity, depending on the form of occurrence and/or further process involved. Separation processes take advantage of physical properties of the materials. These physical properties can include density, particle size and shape, electrical and magnetic properties, and surface properties. Major physical and chemical methods include magnetic separation, froth floatation, leaching etc., whereby the impurities and unwanted materials are removed from the ore and the base ore of the metal is concentrated, meaning the percentage of metal in the ore is increased. This concentrate is then either processed to remove moisture or else used as is for extraction of the metal or made into shapes and forms that can undergo further processing, with ease of handling.

Ore bodies often contain more than one valuable metal. Tailings of a previous process may be used as a feed in another process to extract a secondary product from the original ore. Additionally, a concentrate may contain more than one valuable metal. That concentrate would then be processed to separate the valuable metals into individual constituents.

Hydrometallurgy

Hydrometallurgy is concerned with processes involving aqueous solutions to extract metals from ores. The first step in the hydrometallurgical process is leaching, which involves dissolution of the valuable metals into the aqueous solution and /or a suitable solvent. After the solution is separated from the ore solids, the extract is often subjected to various processes of purification and concentration before the valuable metal is recovered either in its metallic state or as a chemical compound. This may include precipitation, distillation, adsorption, and solvent extraction. The final recovery step may involve precipitation, cementation, or an electrometallurgical process. Sometimes, hydrometallurgical processes may be carried out directly on the ore material without any pretreatment steps. More often, the ore must be pretreated by various mineral processing steps, and sometimes by pyrometallurgical processes.

Pyrometallurgy

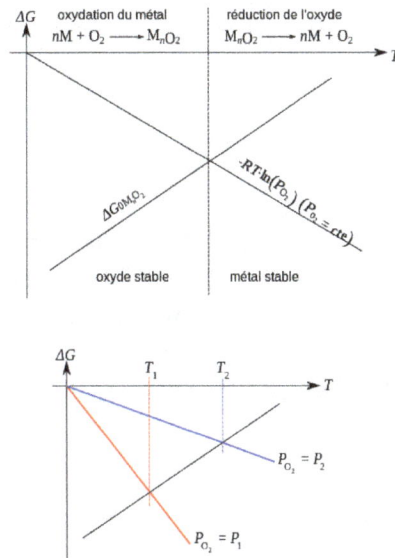

Ellingham diagram for high temperature oxidation

Pyrometallurgy involves high temperature processes where chemical reactions take place among gases, solids, and molten materials. Solids containing valuable metals are treated to form intermediate compounds for further processing or converted into their elemental or metallic state. Pyrometallurgical processes that involve gases and solids are typified by calcining and roasting operations. Processes that produce molten products are collectively referred to as smelting operations. The energy required to sustain the high temperature pyrometallurgical processes may derive from the exothermic nature of the chemical reactions taking place. Typically, these reactions are oxidation, e.g. of sulfide to sulfur dioxide . Often, however, energy must be added to the process by combustion of fuel or, in the case of some smelting processes, by the direct application of electrical energy.

Ellingham Diagrams are a useful way of analysing the possible reactions, and so predicting their outcome.

Electrometallurgy

Electrometallurgy involves metallurgical processes that take place in some form of electrolytic cell. The most common types of electrometallurgical processes are electrowinning and electro-refining. Electrowinning is an electrolysis process used to recover metals in aqueous solution, usually as the result of an ore having undergone one or more hydrometallurgical processes. The metal of interest is plated onto the cathode, while the anode is an inert electrical conductor. Electro-refining is used to dissolve an impure metallic anode (typically from a smelting process) and produce a high purity cathode. Fused salt electrolysis is another electrometallurgical process whereby the valuable metal has been dissolved into a molten salt which acts as the electrolyte, and the valuable metal collects on the cathode of the cell. The fused salt electrolysis process is conducted at temperatures sufficient to keep both the electrolyte and the metal being produced in the molten state. The scope of electrometallurgy has significant overlap with the areas of hydrometallurgy and (in the case of fused salt electrolysis) pyrometallurgy. Additionally, electrochemical phenomena play a considerable role in many mineral processing and hydrometallurgical processes.

Hydrometallurgy

Hydrometallurgy is a method for obtaining metals from their ores. It is a technique within the field of extractive metallurgy involving the use of aqueous chemistry for the recovery of metals from ores, concentrates, and recycled or residual materials. Metal chemical processing techniques that complement hydrometallurgy are pyrometallurgy, vapour metallurgy and molten salt electrometallurgy. Hydrometallurgy is typically divided into three general areas:

- Leaching

- Solution concentration and purification

- Metal or metal compound recovery

Leaching

Leaching involves the use of aqueous solutions to extract metal from metal bearing materials which is brought into contact with a material containing a valuable metal. The lixiviant solution conditions vary in terms of pH, oxidation-reduction potential, presence of chelating agents and temperature, to optimize the rate, extent and selectivity of dissolution of the desired metal component into the aqueous phase. Through the use of chelating agents, one can selectively extract certain metals. Such chelating agents are typically amines of schiff bases.

The five basic leaching reactor configurations are in-situ, heap, vat, tank and autoclave.

In-situ Leaching

In-situ leaching is also called "solution mining." The process initially involves drilling of holes into the ore deposit. Explosives or hydraulic fracturing are used to create open pathways within the deposit for solution to penetrate into. Leaching solution is pumped into the deposit where it makes contact with the ore. The solution is then collected and processed. The Beverley uranium deposit is an example of in-situ leaching.

Heap Leaching

In heap leaching processes, crushed (and sometimes agglomerated) ore is piled in a heap which is lined with an impervious layer. Leach solution is sprayed over the top of the heap, and allowed to percolate downward through the heap. The heap design usually incorporates collection sumps, which allow the "pregnant" leach solution (i.e. solution with dissolved valuable metals) to be pumped for further processing. An example is gold cyanidation, where pulverized ores are extracted with a solution of sodium cyanide, which, in the presence of air, dissolves the gold, leaving behind the nonprecious residue.

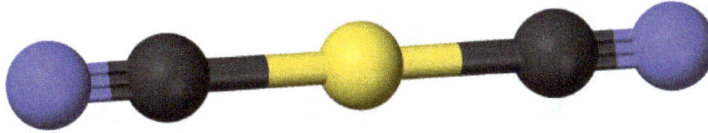

Ball-and-stick model of the aurocyanide or dicyanoaurate(I) complex anion, [Au(CN)$_2$]$^-$.

Vat Leaching

Vat leaching, also called agitation leaching, involves contacting material, which has usually undergone size reduction and classification, with leach solution in large vats.

Tank Leaching

Stirred tank, also called agitation leaching, involves contacting material, which has usually undergone size reduction and classification, with leach solution in agitated tanks. The agitation can enhance reaction kinetics by enhancing mass transfer. Tanks are often configured as reactors in series.

Autoclave Leaching

Autoclave reactors are used for reactions at higher temperatures, which can enhance the rate of the reaction. Similarly, autoclaved enable the use gaseous reagents in the system.

Solution Concentration and Purification

After leaching, the leach liquor must normally undergo concentration of the metal ions that are to be recovered. Additionally, undesirable metal ions sometimes require removal.

- Precipitation is the selective removal of a compound of the targeted metal or removal of a major impurity by precipitation of one of its compounds. Copper is precipitated as its sulfide as a means to purify nickel leachates.

- Cementation is the conversion of the metal ion to the metal by a redox reaction. A typical application involves addition of scrap iron to a solution of copper ions. Iron dissolves and copper metal is deposited.

- Solvent Extraction

- Ion Exchange

- Gas reduction. Treating a solution of nickel and ammonia with hydrogen affords nickel metal as its powder.

- Electrowinning is a particularly selective if expensive electrolysis process applied to the isolation of precious metals. Gold can be electroplated from its solutions.

Solvent Extraction

In the solvent extraction is a mixture of an extractant in a diluent is used to extract a metal from one phase to another. In solvent extraction this mixture is often referred to as the "organic" because the main constituent (diluent) is some type of oil.

The PLS (pregnant leach solution) is mixed to emulsification with the stripped organic and allowed to separate. The metal will be exchanged from the PLS to the organic they are modified. The resulting streams will be a loaded organic and a raffinate. When dealing with electrowinning, the loaded organic is then mixed to emulsification with a lean electrolyte and allowed to separate. The metal will be exchanged from the organic to the electrolyte. The resulting streams will be a stripped organic and a rich electrolyte. The organic stream is recycled through the solvent extraction process while the aqueous streams cycle through leaching and electrowinning processes respectively.

Ion Exchange

Chelating agents, natural zeolite, activated carbon, resins, and liquid organics impregnated with chelating agents are all used to exchange cations or anions with the solution. Selectivity and recovery are a function of the reagents used and the contaminants present.

Metal Recovery

Metal recovery is the final step in a hydrometallurgical process. Metals suitable for sale as raw materials are often directly produced in the metal recovery step. Sometimes, however, further refining is required if ultra-high purity metals are to be produced. The primary types of metal recovery processes are electrolysis, gaseous reduction, and precipitation. For example, a major target of hydrometallurgy is copper, which is conveniently obtained by electrolysis. Cu^{2+} ions reduce at mild potentials, leaving behind other contaminating metals such as Fe^{2+} and Zn^{2+}.

Electrolysis

Electrowinning and electrorefining respectively involve the recovery and purification of metals using electrodeposition of metals at the cathode, and either metal dissolution or a competing oxidation reaction at the anode.

Precipitation

Precipitation in hydrometallurgy involves the chemical precipitation of either metals and their compounds or of the contaminants from aqueous solutions. Precipitation will proceed when, through reagent addition, evaporation, pH change or temperature manipulation, any given species exceeds its limit of solubility.

Metalworking

Metalworking is the process of working with metals to create individual parts, assemblies, or large-scale structures. The term covers a wide range of work from large ships and bridges to precise engine parts and delicate jewelry. It therefore includes a correspondingly wide range of skills, processes, and tools.

Turning a bar of metal on a lathe

Metalworking is a science, art, hobby, industry and trade. Its historical roots span cultures, civilizations, and millennia. Metalworking has evolved from the discovery of smelting various ores, producing malleable and ductile metal useful for tools and adornments. Modern metalworking processes, though diverse and specialized, can be categorized as forming, cutting, or joining processes. Today's machine shop includes a number of machine tools capable of creating a precise, useful workpiece.

Prehistory

The oldest archaeological evidence of copper mining and working was the discovery of a copper pendant in northern Iraq from 8,700 BCE. The earliest substantiated and dated evidence of metalworking in the Americas was the processing of copper in Wisconsin, near Lake Michigan. Copper was hammered until brittle then heated so it could be worked some more. This technology is dated to about 4000-5000 BCE. The oldest gold artifacts in the world come from the Bulgarian Varna Necropolis and date from 4450 BCE.

Not all metal required fire to obtain it or work it. Isaac Asimov speculated that gold was the "first metal." His reasoning is that by its chemistry it is found in nature as nuggets of pure gold. In other words, gold, as rare as it is, is sometimes found in nature as the metal that it is. There are a few other metals that sometimes occur natively, and as a result of meteors. Almost all other metals are found in ores, a mineral-bearing rock, that require heat or some other process to liberate the metal. Another feature of gold is that it is workable as it is found, meaning that no technology beyond a stone hammer and anvil to work the metal is needed. This is a result of gold's properties of malleability and ductility. The earliest tools were stone, bone, wood, and sinew, all of which sufficed to work gold.

At some unknown point the connection between heat and the liberation of metals from rock became clear, rocks rich in copper, tin, and lead came into demand. These ores were mined wherever they were recognized. Remnants of such ancient mines have been found all over Southwestern Asia. Metalworking was being carried out by the South Asian inhabitants of Mehrgarh between 7000–3300 BCE. The end of the beginning of metalworking occurs sometime around 6000 BCE when copper smelting became common in Southwestern Asia.

Ancient civilisations knew of seven metals. Here they are arranged in order of their oxidation potential (in volts):

- Iron +0.44 V,

- Tin +0.14 V

- Lead +0.13 V

- Copper −0.34 V

- Mercury −0.79 V

- Silver −0.80 V

- Gold −1.50 V.

The oxidation potential is important because it is one indicator of how tightly bound to the ore the metal is likely to be. As can be seen, iron is significantly higher than the other six metals while gold is dramatically lower than the six above it. Gold's low oxidation is one of the main reasons that gold is found in nuggets. These nuggets are relatively pure gold and are workable as they are found.

Copper ore, being relatively abundant, and tin ore became the next important players in the story of metalworking. Using heat to smelt copper from ore, a great deal of copper was produced. It was used for both jewelry and simple tools. However, copper by itself was too soft for tools requiring edges and stiffness. At some point tin was added into the molten copper and bronze was born. Bronze is an alloy of copper and tin. Bronze was an important advance because it had the edge-durability and stiffness that pure copper lacked. Until the advent of iron, bronze was the most advanced metal for tools and weapons in common use.

Outside Southwestern Asia, these same advances and materials were being discovered and used around the world. China and Great Britain jumped into the use of bronze with little time being devoted to copper. Japan began the use of bronze and iron almost simultaneously. In the Americas things were different. Although the peoples of the Americas knew of metals, it wasn't until the European colonisation that metalworking for tools and weapons became common. Jewelry and art were the principal uses of metals in the Americas prior to European influence.

Around 2700 BCE, production of bronze was common in locales where the necessary materials could be assembled for smelting, heating, and working the metal. Iron was beginning to be smelted and began its emergence as an important metal for tools and weapons. The Iron Age was dawning.

History

By the historical periods of the Pharaohs in Egypt, the Vedic Kings in India, the Tribes of Israel, and the Maya civilization in North America, among other ancient populations, precious metals began to have value attached to them. In some cases rules for ownership, distribution, and trade were created, enforced, and agreed upon by the respective peoples. By the above periods metalworkers were very skilled at creating objects of adornment, religious artifacts, and trade instruments of precious metals (non-ferrous), as well as weaponry usually of ferrous metals and/or alloys. These skills were finely honed and well executed. The techniques were practiced by artisans, blacksmiths, atharvavedic practitioners, alchemists, and other categories of metalworkers around the globe. For example, the ancient technique of granulation is found around the world in numerous ancient cultures before the historic record shows people traveled to far regions to share this process that is still being used by metalsmiths today.

A turret lathe operator machining parts for transport planes at the Consolidated Aircraft Corporation plant, Fort Worth, Texas, USA in the 1940s

As time progressed metal objects became more common, and ever more complex. The need to further acquire and work metals grew in importance. Skills related to extracting metal ores from the earth began to evolve, and metalsmiths became more knowledgeable. Metalsmiths became important members of society. Fates and economies of entire civilizations were greatly affected by the availability of metals and metalsmiths. The metalworker depends on the extraction of precious metals to make jewelry, build more efficient electronics, and for industrial and technological applications from construction to shipping containers to rail, and air transport. Without metals, goods and services would cease to move around the globe on the scale we know today.

General Metalworking Processes

A combination square used for transferring designs.

A caliper is used to precisely measure a short length.

Metalworking generally is divided into the following categories, *forming*, *cutting*, and, *joining*. Each of these categories contain various processes.

Prior to most operations, the metal must be marked out and/or measured, depending on the desired finished product.

Marking out (also known as layout) is the process of transferring a design or pattern to a workpiece and is the first step in the handcraft of metalworking. It is performed in many industries or hobbies, although in industry, the repetition eliminates the need to mark out every individual piece. In the metal trades area, marking out consists of transferring the engineer's plan to the workpiece in preparation for the next step, machining or manufacture.

Calipers are hand tools designed to precisely measure the distance between two points. Most calipers have two sets of flat, parallel edges used for inner or outer diameter measurements. These calipers can be accurate to within one-thousandth of an inch (25.4 μm). Different types of calipers have different mechanisms for displaying the distance measured. Where larger objects need to be measured with less precision, a tape measure is often used.

Compatibility chart of materials versus processes											
	Material										
Process	Iron	Steel	Aluminium	Copper	Magnesium	Nickel	Refractory metals	Titanium	Zinc	Brass	Bronze
Sand casting	X	X	X	X	X	X			0	0	X
Permanent mold casting	X	0	X	0	X	0			0	0	X
Die casting			X	0	X				X		
Investment casting		X	X	X	0	0				0	X
Ablation casting		X	X	X	0	0					
Closed-die forging		X	0	0	0	0	0	0			
Extrusion		0	X	X	X	0	0	0			
Cold heading		X	X	X		0					
Stamping & deep drawing		X	X	X	0	X			0	0	
Screw machine	0	X	X	X	0	X	0		0	X	X
Powder metallurgy	X	X	0	X		0	X	0			
Key: **X** = Routinely performed, **0** = Performed with difficulty, caution, or some sacrifice, **blank** = Not recommended											

Casting

A sand casting mold

Casting achieves a specific form by pouring molten metal into a mold and allowing it to cool, with no mechanical force. Forms of casting include:

- Investment casting (called lost wax casting in art)

- Centrifugal casting

- Die casting

- Sand casting

- Shell casting

- Spin casting

Forming Processes

These *forming* processes modify metal or workpiece by deforming the object, that is, without removing any material. Forming is done with a system of mechanical forces and, especially for bulk metal forming, with heat.

Bulk Forming Processes

A red-hot metal workpiece is inserted into a forging press.

Plastic deformation involves using heat or pressure to make a workpiece more conductive to mechanical force. Historically, this and casting were done by blacksmiths, though today the process has been industrialized. In bulk metal forming, the workpiece is generally heated up.

- Cold sizing

- Extrusion

- Drawing

- Forging

- Powder metallurgy

- Friction drilling

- Rolling

Sheet (and Tube) Forming Processes

These types of forming process involve the application of mechanical force at room temperature. However, some recent developments involve the heating of dies and/or parts.

- Bending

- Coining

- Decambering

- Deep drawing (DD)

- Flowforming

- Hydroforming (HF)

- Hot metal gas forming

- Hot press hardening

- Incremental forming (IF)

- Spinning, Shear forming or Flowforming

- Raising

- Roll forming

- Roll bending

- Repoussé and chasing

- Rubber pad forming

- Shearing

- Stamping

- Superplastic forming (SPF)

- Wheeling using an English wheel (wheeling machine)

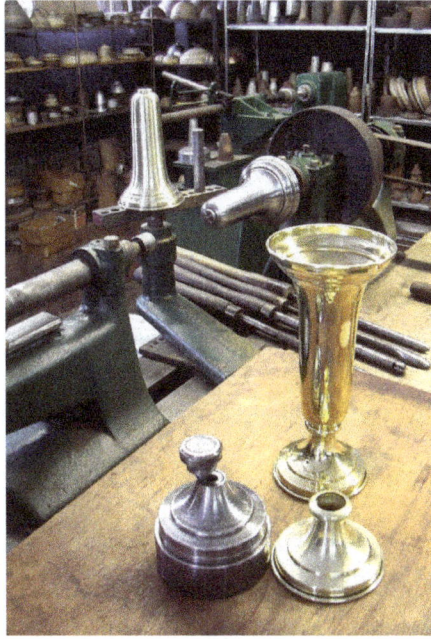
A metal spun brass vase

Cutting Processes

A CNC plasma cutting machining

Cutting is a collection of processes wherein material is brought to a specified geometry by removing excess material using various kinds of tooling to leave a finished part that meets specifications. The net result of cutting is two products, the waste or excess material, and the finished part. In woodworking, the waste would be sawdust and excess wood. In cutting metals the waste is chips or swarf and excess metal.

Cutting processes fall into one of three major categories:

- Chip producing processes most commonly known as machining

- Burning, a set of processes wherein the metal is cut by oxidizing a kerf to separate pieces of metal

- Miscellaneous specialty process, not falling easily into either of the above categories

Drilling a hole in a metal part is the most common example of a chip producing process. Using an oxy-fuel cutting torch to separate a plate of steel into smaller pieces is an example of burning. Chemical milling is an example of a specialty process that removes excess material by the use of etching chemicals and masking chemicals.

There are many technologies available to cut metal, including:

- Manual technologies: saw, chisel, shear or snips

- Machine technologies: turning, milling, drilling, grinding, sawing

- Welding/burning technologies: burning by laser, oxy-fuel burning, and plasma

- Erosion technologies: by water jet, electric discharge, or abrasive flow machining.

- Chemical technologies: Photochemical machining

Cutting fluid or coolant is used where there is significant friction and heat at the cutting interface between a cutter such as a drill or an end mill and the workpiece. Coolant is generally introduced by a spray across the face of the tool and workpiece to decrease friction and temperature at the cutting tool/workpiece interface to prevent excessive tool wear. In practice there are many methods of delivering coolant.

Milling

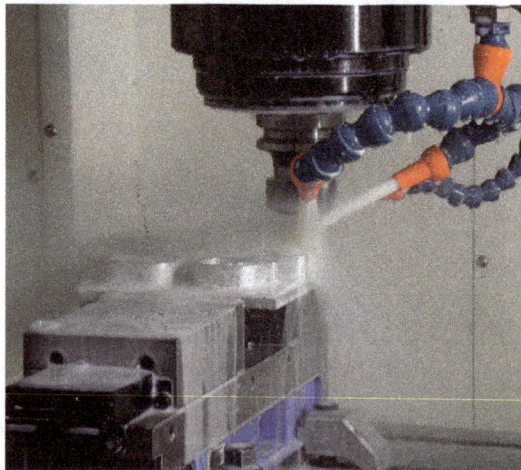

A milling machine in operation, including coolant hoses.

Milling is the complex shaping of metal or other materials by removing material to form the final shape. It is generally done on a milling machine, a power-driven machine that in its basic form consists of a milling cutter that rotates about the spindle axis (like a drill), and a worktable that can move in multiple directions (usually two dimensions [x and y axis] relative to the workpiece). The spindle usually moves in the z axis. It is possible to raise the table (where the workpiece rests). Milling machines may be operated manually or under computer numerical control (CNC), and can perform a vast number of complex operations, such as slot cutting, planing, drilling and threading, rabbeting, routing, etc. Two common types of mills are the horizontal mill and vertical mill.

The pieces produced are usually complex 3D objects that are converted into x, y, and z coordinates that are then fed into the CNC machine and allow it to complete the tasks required. The milling machine can produce most parts in 3D, but some require the objects to be rotated around the x, y, or z coordinate axis (depending on the need). Tolerances are usually in the thousandths of an inch (Unit known as Thou), depending on the specific machine.

In order to keep both the bit and material cool, a high temperature coolant is used. In most cases the coolant is sprayed from a hose directly onto the bit and material. This coolant can either be machine or user controlled, depending on the machine.

Materials that can be milled range from aluminum to stainless steel and almost everything in between. Each material requires a different speed on the milling tool and varies in the amount of material that can be removed in one pass of the tool. Harder materials are usually milled at slower speeds with small amounts of material removed. Softer materials vary, but usually are milled with a high bit speed.

The use of a milling machine adds costs that are factored into the manufacturing process. Each time the machine is used coolant is also used, which must be periodically added in order to prevent breaking bits. A milling bit must also be changed as needed in order to prevent damage to the material. Time is the biggest factor for costs. Complex parts can require hours to complete, while very simple parts take only minutes. This in turn varies the production time as well, as each part will require different amounts of time.

Safety is key with these machines. The bits are traveling at high speeds and removing pieces of usually scalding hot metal. The advantage of having a CNC milling machine is that it protects the machine operator.

Turning

A lathe cutting material from a workpiece.

Turning is a metal cutting process for producing a cylindrical surface with a single point tool. The workpiece is rotated on a spindle and the cutting tool is fed into it radially, axially or both. Producing surfaces perpendicular to the workpiece axis is called facing. Producing surfaces using both radial and axial feeds is called profiling.

A *lathe* is a machine tool which spins a block or cylinder of material so that when abrasive, cutting, or deformation tools are applied to the workpiece, it can be shaped to produce an object which has rotational symmetry about an axis of rotation. Examples of objects that can be produced on a lathe include candlestick holders, crankshafts, camshafts, and bearing mounts.

Lathes have four main components: the bed, the headstock, the carriage, and the tailstock. The bed is a precise & very strong base which all of the other components rest upon for alignment. The headstock's spindle secures the workpiece with a chuck, whose jaws (usually three or four) are tightened around the piece. The spindle rotates at high speed, providing the energy to cut the material. While historically lathes were powered by belts from a line shaft, modern examples uses electric motors. The workpiece extends out of the spindle along the axis of rotation above the flat bed. The carriage is a platform that can be moved, precisely and independently parallel and perpendicular to the axis of rotation. A hardened cutting tool is held at the desired height (usually the middle of the workpiece) by the toolpost. The carriage is then moved around the rotating workpiece, and the cutting tool gradually removes material from the workpiece. The tailstock can be slid along the axis of rotation and then locked in place as necessary. It may hold centers to further secure the workpiece, or cutting tools driven into the end of the workpiece.

Other operations that can be performed with a single point tool on a lathe are:

Chamfering: Cutting an angle on the corner of a cylinder.Parting: The tool is fed radially into the workpiece to cut off the end of a part.Threading: A tool is fed along and across the outside or inside surface of rotating parts to produce external or internal threads.Boring: A single-point tool is fed linearly and parallel to the axis of rotation to create a round hole.Drilling: Feeding the drill into the workpiece axially. Knurling: Uses a tool to produce a rough surface texture on the work piece. Frequently used to allow grip by hand on a metal part.

Modern computer numerical control (CNC) lathes and (CNC) machining centres can do secondary operations like milling by using driven tools. When driven tools are used the work piece stops rotating and the driven tool executes the machining operation with a rotating cutting tool. The CNC machines use x, y, and z coordinates in order to control the turning tools and produce the product. Most modern day CNC lathes are able to produce most turned objects in 3D.

Nearly all types of metal can be turned, although more time & specialist cutting tools are needed for harder workpieces.

Threading

There are many threading processes including: cutting threads with a tap or die, thread milling, single-point thread cutting, thread rolling and forming, and thread grinding. A *tap* is used to cut a female thread on the inside surface of a pre-drilled hole, while a *die* cuts a male thread on a preformed cylindrical rod.

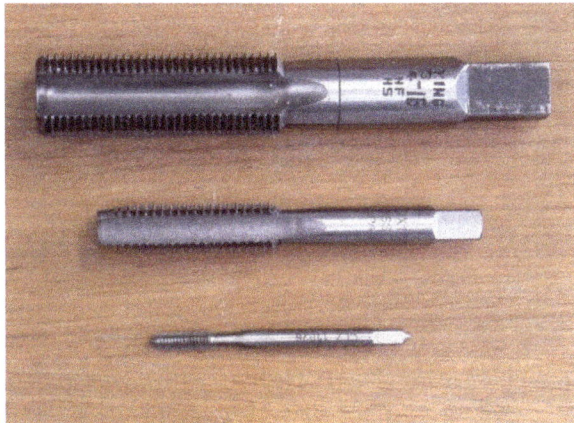

Three different types and sizes of taps.

Grinding

A surface grinder

Grinding uses an abrasive process to remove material from the workpiece. A grinding machine is a machine tool used for producing very fine finishes, making very light cuts, or high precision forms using an abrasive wheel as the cutting device. This wheel can be made up of various sizes and types of stones, diamonds or inorganic materials.

The simplest grinder is a bench grinder or a hand-held angle grinder, for deburring parts or cutting metal with a zip-disc.

Grinders have increased in size and complexity with advances in time and technology. From the old days of a manual toolroom grinder sharpening endmills for a production shop, to today's 30000 RPM CNC auto-loading manufacturing cell producing jet turbines, grinding processes vary greatly.

Grinders need to be very rigid machines to produce the required finish. Some grinders are even used to produce glass scales for positioning CNC machine axis. The common rule is the machines used to produce scales be 10 times more accurate than the machines the parts are produced for.

In the past grinders were used for finishing operations only because of limitations of tooling. Modern grinding wheel materials and the use of industrial diamonds or other man-made coatings (cubic boron nitride) on wheel forms have allowed grinders to achieve excellent results in production environments instead of being relegated to the back of the shop.

Modern technology has advanced grinding operations to include CNC controls, high material removal rates with high precision, lending itself well to aerospace applications and high volume production runs of precision components.

Filing

A file is an abrasive surface like this one that allows machinists to remove small, imprecise amounts of metal.

Filing is combination of grinding and saw tooth cutting using a file. Prior to the development of modern machining equipment it provided a relatively accurate means for the production of small parts, especially those with flat surfaces. The skilled use of a file allowed a machinist to work to fine tolerances and was the hallmark of the craft. Today filing is rarely used as a production technique in industry, though it remains as a common method of deburring.

Other

Broaching is a machining operation used to cut keyways into shafts. Electron beam machining (EBM) is a machining process where high-velocity electrons are directed toward a work piece, creating heat and vaporizing the material. Ultrasonic machining uses ultrasonic vibrations to machine very hard or brittle materials.

Joining Processes

Mig welding

Welding

Welding is a fabrication process that joins materials, usually metals or thermoplastics, by causing coalescence. This is often done by melting the workpieces and adding a filler material to form a pool of molten material that cools to become a strong joint, but sometimes pressure is used in conjunction with heat, or by itself, to produce the weld.

Many different energy sources can be used for welding, including a gas flame, an electric arc, a laser, an electron beam, friction, and ultrasound. While often an industrial process, welding can be done in many different environments, including open air, underwater and in space. Regardless of location, however, welding remains dangerous, and precautions must be taken to avoid burns, electric shock, poisonous fumes, and overexposure to ultraviolet light.

Brazing

Brazing is a joining process in which a filler metal is melted and drawn into a capillary formed by the assembly of two or more work pieces. The filler metal reacts metallurgically with the workpiece(s) and solidifies in the capillary, forming a strong joint. Unlike welding, the work piece is not melted. Brazing is similar to soldering, but occurs at temperatures in excess of 450 °C (842 °F). Brazing has the advantage of producing less thermal stresses than welding, and brazed assemblies tend to be more ductile than weldments because alloying elements can not segregate and precipitate.

Brazing techniques include, flame brazing, resistance brazing, furnace brazing, diffusion brazing, inductive brazing and vacuum brazing.

Soldering

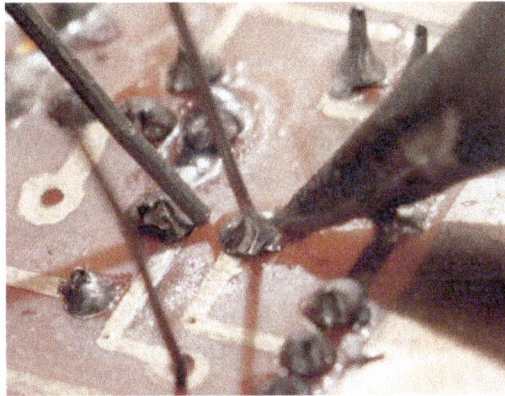

Soldering a printed circuit board.

Soldering is a joining process that occurs at temperatures below 450 °C (842 °F). It is similar to brazing in the way that a filler is melted and drawn into a capillary to form a join, although at a lower temperature. Because of this lower temperature and different alloys used as fillers, the metallurgical reaction between filler and work piece is minimal, resulting in a weaker joint.

Riveting

Riveting is one of the most ancient metalwork joining processes. Its use has declined markedly during the second half of the 20th century, but it still retains important uses in industry and construction, and in artisan crafts such as jewellery, medieval armouring and Metal Couture into the 21st century. The earlier use of rivets is being superseded by improvements in welding and component fabrication techniques.

A rivet is essentially a two-headed and unthreaded bolt which holds two other pieces of metal together. Holes are drilled or punched through the two pieces of metal to be joined. The holes being aligned, a rivet is passed through the holes and permanent heads are formed onto the ends of the rivet utilizing hammers and forming dies (by either coldworking or hotworking). Rivets are commonly purchased with one head already formed.

When it is necessary to remove rivets, one of the rivet's heads is sheared off with a cold chisel. The rivet is then driven out with a hammer and punch.

Associated Processes

While these processes are not primary metalworking processes, they are often performed before or after metalworking processes.

Heat Treatment

Metals can be heat treated to alter the properties of strength, ductility, toughness, hardness or resistance to corrosion. Common heat treatment processes include annealing, precipitation hardening, quenching, and tempering. The annealing process softens the metal by allowing recovery of cold work and grain growth. Quenching can be used to harden alloy steels, or in precipitation hardenable alloys, to trap dissolved solute atoms in solution. Tempering will cause the dissolved alloying elements to precipitate, or in the case of quenched steels, improve impact strength and ductile properties.

Often, mechanical and thermal treatments are combined in what is known as thermo-mechanical treatments for better properties and more efficient processing of materials. These processes are common to high alloy special steels, super alloys and titanium alloys.

Plating

Electroplating is a common surface-treatment technique. It involves bonding a thin layer of another metal such as gold, silver, chromium or zinc to the surface of the product. It is used to reduce corrosion as well as to improve the product's aesthetic appearance.

Thermal Spraying

Thermal spraying techniques are another popular finishing option, and often have better high temperature properties than electroplated coatings.

Process of Metalworking

Casting (Metalworking)

In metalworking, casting means a process, in which liquid metal is poured into a mold, that contains a hollow cavity of the desired shape, and then allowed to cool and solidify. The solidified part is also known as a casting, which is ejected or broken out of the mold to complete the process. Casting is most often used for making complex shapes that would be difficult or uneconomical to make by other methods.

Molten metal before casting

Casting iron in a sand mold

Casting processes have been known for thousands of years, and widely used for sculpture, especially in bronze, jewellery in precious metals, and weapons and tools. Traditional techniques include lost-wax casting, plaster mold casting and sand casting.

The modern casting process is subdivided into two main categories: expendable and non-expendable casting. It is further broken down by the mold material, such as sand or metal, and pouring method, such as gravity, vacuum, or low pressure.

Expendable Mold Casting

Expendable mold casting is a generic classification that includes sand, plastic, shell, plaster, and investment (lost-wax technique) moldings. This method of mold casting involves the use of temporary, non-reusable molds.

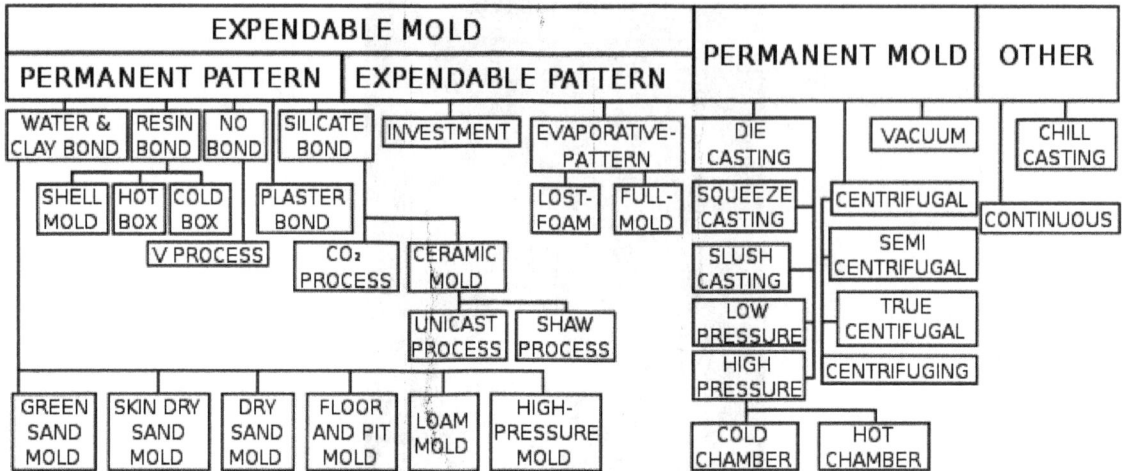

EXPENDABLE MOLD		PERMANENT MOLD	OTHER
PERMANENT PATTERN	EXPENDABLE PATTERN		

- PERMANENT PATTERN
  - WATER & CLAY BOND
    - SHELL MOLD
    - GREEN SAND MOLD
    - SKIN DRY SAND MOLD
    - DRY SAND MOLD
    - FLOOR AND PIT MOLD
    - LOAM MOLD
    - HIGH-PRESSURE MOLD
  - RESIN BOND
    - HOT BOX
    - COLD BOX
    - V PROCESS
  - NO BOND
  - SILICATE BOND
    - PLASTER BOND
    - CO₂ PROCESS
    - CERAMIC MOLD
      - UNICAST PROCESS
      - SHAW PROCESS
- EXPENDABLE PATTERN
  - INVESTMENT
  - EVAPORATIVE-PATTERN
    - LOST-FOAM
    - FULL-MOLD
- PERMANENT MOLD
  - DIE CASTING
    - SQUEEZE CASTING
    - SLUSH CASTING
    - LOW PRESSURE
    - HIGH PRESSURE
      - COLD CHAMBER
      - HOT CHAMBER
  - VACUUM
    - CENTRIFUGAL
      - SEMI CENTRIFUGAL
      - TRUE CENTIFUGAL
      - CENTRIFUGING
- OTHER
  - CHILL CASTING
    - CONTINUOUS

Sand Casting

Sand casting is one of the most popular and simplest types of casting, and has been used for centuries. Sand casting allows for smaller batches than permanent mold casting and at a very reasonable cost. Not only does this method allow manufacturers to create products at a low cost, but there are other benefits to sand casting, such as very small-size operations. From castings that fit in the palm of your hand to train beds (one casting can create the entire bed for one rail car), it can all be done with sand casting. Sand casting also allows most metals to be cast depending on the type of sand used for the molds.

Sand casting requires a lead time of days, or even weeks sometimes, for production at high output rates (1–20 pieces/hr-mold) and is unsurpassed for large-part production. Green (moist) sand has almost no part weight limit, whereas dry sand has a practical part mass limit of 2,300–2,700 kg (5,100–6,000 lb). Minimum part weight ranges from 0.075–0.1 kg (0.17–0.22 lb). The sand is bonded together using clays, chemical binders, or polymerized oils (such as motor oil). Sand can be recycled many times in most operations and requires little maintenance.

Plaster Mold Casting

Plaster casting is similar to sand casting except that plaster of paris is substituted for sand as a mold material. Generally, the form takes less than a week to prepare, after which a production rate of 1–10 units/hr·mold is achieved, with items as massive as 45 kg (99 lb) and as small as 30 g (1 oz) with very good surface finish and close tolerances. Plaster casting is an inexpensive alternative to other molding processes for complex parts due to the low cost of the plaster and its ability to produce near net shape castings. The biggest disadvantage is that it can only be used with low melting point non-ferrous materials, such as aluminium, copper, magnesium, and zinc.

Shell Molding

Shell molding is similar to sand casting, but the molding cavity is formed by a hardened "shell" of sand instead of a flask filled with sand. The sand used is finer than sand casting sand and is mixed with a resin so that it can be heated by the pattern and hardened into a shell around the pattern. Because of the resin and finer sand, it gives a much finer surface finish. The process is easily automated and more precise than sand casting. Common metals that are cast include cast iron, aluminium, magnesium, and copper alloys. This process is ideal for complex items that are small to medium-sized.

Investment Casting

An investment-cast valve cover

Investment casting (known as lost-wax casting in art) is a process that has been practiced for thousands of years, with the lost-wax process being one of the oldest known metal forming techniques. From 5000 years ago, when beeswax formed the pattern, to today's high technology waxes, refractory materials and specialist alloys, the castings ensure high-quality components are produced with the key benefits of accuracy, repeatability, versatility and integrity.

Investment casting derives its name from the fact that the pattern is invested, or surrounded, with a refractory material. The wax patterns require extreme care for they are not strong enough to withstand forces encountered during the mold making. One advantage of investment casting is that the wax can be reused.

The process is suitable for repeatable production of net shape components from a variety of different metals and high performance alloys. Although generally used for small castings, this process has been used to produce complete aircraft door frames, with steel castings of up to 300 kg and aluminium castings of up to 30 kg. Compared to other casting processes such as die casting or sand casting, it can be an expensive process. However, the components that can be produced using investment casting can incorporate intricate contours, and in most cases the components are cast near net shape, so require little or no rework once cast.

Waste Molding of Plaster

A durable plaster intermediate is often used as a stage toward the production of a bronze sculpture or as a pointing guide for the creation of a carved stone. With the completion of a plaster, the work is more durable (if stored indoors) than a clay original which must be kept moist to avoid cracking. With the low cost plaster at hand, the expensive work of bronze casting or stone carving may be deferred until a patron is found, and as such work is considered to be a technical, rather than artistic process, it may even be deferred beyond the lifetime of the artist.

In waste molding a simple and thin plaster mold, reinforced by sisal or burlap, is cast over the original clay mixture. When cured, it is then removed from the damp clay, incidentally destroying the fine details in undercuts present in the clay, but which are now captured in the mold. The mold may then at any later time (but only once) be used to cast a plaster positive image, identical to the original clay. The surface of this plaster may be further refined and may be painted and waxed to resemble a finished bronze casting.

Evaporative-pattern Casting

This is a class of casting processes that use pattern materials that evaporate during the pour, which means there is no need to remove the pattern material from the mold before casting. The two main processes are lost-foam casting and full-mold casting.

Lost-foam Casting

Lost-foam casting is a type of evaporative-pattern casting process that is similar to investment casting except foam is used for the pattern instead of wax. This process takes advantage of the low boiling point of foam to simplify the investment casting process by removing the need to melt the wax out of the mold.

Full-mold Casting

Full-mold casting is an evaporative-pattern casting process which is a combination of sand casting and lost-foam casting. It uses an expanded polystyrene foam pattern which is then surrounded by sand, much like sand casting. The metal is then poured directly into the mold, which vaporizes the foam upon contact.

Non-expendable mold casting differs from expendable processes in that the mold need not be reformed after each production cycle. This technique includes at least four different methods: permanent, die, centrifugal, and continuous casting. This form of casting also results in improved repeatability in parts produced and delivers Near Net Shape results.

Permanent Mold Casting

Permanent mold casting is a metal casting process that employs reusable molds ("permanent molds"), usually made from metal. The most common process uses gravity to fill the mold. However, gas pressure or a vacuum are also used. A variation on the typical gravity casting process, called slush casting, produces hollow castings. Common casting metals are aluminum, magnesium, and copper alloys. Other materials include tin, zinc, and lead alloys and iron and steel are also cast in graphite molds. Permanent molds, while lasting more than one casting still have a limited life before wearing out.

Die casting

The die casting process forces molten metal under high pressure into mold cavities (which are machined into dies). Most die castings are made from nonferrous metals, specifically zinc, copper, and aluminium-based alloys, but ferrous metal die castings are possible. The die casting method is especially suited for applications where many small to medium-sized parts are needed with good detail, a fine surface quality and dimensional consistency.

Semi-solid Metal Casting

Semi-solid metal (SSM) casting is a modified die casting process that reduces or eliminates the residual porosity present in most die castings. Rather than using liquid metal as the feed material, SSM casting uses a higher viscosity feed material that is partially solid and partially liquid. A modified die casting machine is used to inject the semi-solid slurry into re-usable hardened steel dies. The high viscosity of the semi-solid metal, along with the use of controlled die filling conditions, ensures that the semi-solid metal fills the die in a non-turbulent manner so that harmful porosity can be essentially eliminated.

Used commercially mainly for aluminium and magnesium alloys, SSM castings can be heat treated to the T4, T5 or T6 tempers. The combination of heat treatment, fast cooling rates (from using un-coated steel dies) and minimal porosity provides excellent combinations of strength and ductility. Other advantages of SSM casting include the ability to produce complex shaped parts net shape, pressure tightness, tight dimensional tolerances and the ability to cast thin walls.

Centrifugal Casting

In this process molten metal is poured in the mold and allowed to solidify while the mold is rotating. Metal is poured into the center of the mold at its axis of rotation. Due to centrifugal force the liquid metal is thrown out towards the periphery.

Centrifugal casting is both gravity- and pressure-independent since it creates its own force feed using a temporary sand mold held in a spinning chamber at up to 900 N. Lead time varies with the application. Semi- and true-centrifugal processing permit 30–50 pieces/hr-mold to be produced, with a practical limit for batch processing of approximately 9000 kg total mass with a typical per-item limit of 2.3–4.5 kg.

Industrially, the centrifugal casting of railway wheels was an early application of the method developed by the German industrial company Krupp and this capability enabled the rapid growth of the enterprise.

Small art pieces such as jewelry are often cast by this method using the lost wax process, as the forces enable the rather viscous liquid metals to flow through very small passages and into fine details such as leaves and petals. This effect is similar to the benefits from vacuum casting, also applied to jewelry casting.

Continuous Casting

Continuous casting is a refinement of the casting process for the continuous, high-volume production of metal sections with a constant cross-section. Molten metal is poured into an open-ended, water-cooled mold, which allows a 'skin' of solid metal to form over the still-liquid centre, gradually solidifying the metal from the outside in. After solidification, the strand, as it is sometimes called, is continuously withdrawn from the mold. Predetermined lengths of the strand can be cut off by either mechanical shears or traveling oxyacetylene torches and transferred to further forming processes, or to a stockpile. Cast sizes can range from strip (a few millimeters thick by about five meters wide) to billets (90 to 160 mm square) to slabs (1.25 m wide by 230 mm thick). Sometimes, the strand may undergo an initial hot rolling process before being cut.

Continuous casting is used due to the lower costs associated with continuous production of a standard product, and also increased quality of the final product. Metals such as steel, copper, aluminum and lead are continuously cast, with steel being the metal with the greatest tonnages cast using this method.

Terminology

Metal casting processes uses the following terminology:

- Pattern: An approximate duplicate of the final casting used to form the mold cavity.

- Molding material: The material that is packed around the pattern and then the pattern is removed to leave the cavity where the casting material will be poured.

- Flask: The rigid wood or metal frame that holds the molding material.

  o Cope: The top half of the pattern, flask, mold, or core.

  o Drag: The bottom half of the pattern, flask, mold, or core.

- Core: An insert in the mold that produces internal features in the casting, such as holes.

  o Core print: The region added to the pattern, core, or mold used to locate and support the core.

- Mold cavity: The combined open area of the molding material and core, where the metal is poured to produce the casting.

- Riser: An extra void in the mold that fills with molten material to compensate for shrinkage during solidification.

- Gating system: The network of connected channels that deliver the molten material to the mold cavities.

  o Pouring cup or pouring basin: The part of the gating system that receives the molten material from the pouring vessel.

  o Sprue: The pouring cup attaches to the sprue, which is the vertical part of the gating system. The other end of the sprue attaches to the runners.

  o Runners: The horizontal portion of the gating system that connects the sprues to the gates.

- o   Gates: The controlled entrances from the runners into the mold cavities.

- Vents: Additional channels that provide an escape for gases generated during the pour.

- Parting line or parting surface: The interface between the cope and drag halves of the mold, flask, or pattern.

- Draft: The taper on the casting or pattern that allow it to be withdrawn from the mold

- Core box: The mold or die used to produce the cores.

Some specialized processes, such as die casting, use additional terminology.

Theory

Casting is a solidification process, which means the solidification phenomenon controls most of the properties of the casting. Moreover, most of the casting defects occur during solidification, such as *gas porosity* and *solidification shrinkage.*

Solidification occurs in two steps: *nucleation* and *crystal growth.* In the nucleation stage solid particles form within the liquid. When these particles form their internal energy is lower than the surrounded liquid, which creates an energy interface between the two. The formation of the surface at this interface requires energy, so as nucleation occurs the material actually undercools, that is it cools below its freezing temperature, because of the extra energy required to form the interface surfaces. It then recalescences, or heats back up to its freezing temperature, for the crystal growth stage. Note that nucleation occurs on a pre-existing solid surface, because not as much energy is required for a partial interface surface, as is for a complete spherical interface surface. This can be advantageous because fine-grained castings possess better properties than coarse-grained castings. A fine grain structure can be induced by *grain refinement* or *inoculation,* which is the process of adding impurities to induce nucleation.

All of the nucleations represent a crystal, which grows as the heat of fusion is extracted from the liquid until there is no liquid left. The direction, rate, and type of growth can be controlled to maximize the properties of the casting. Directional solidification is when the material solidifies at one end and proceeds to solidify to the other end; this is the most ideal type of grain growth because it allows liquid material to compensate for shrinkage.

Cooling Curves

Intermediate cooling rates from melt result in a dendritic microstructure. Primary and secondary dendrites can be seen in this image.

Cooling curves are important in controlling the quality of a casting. The most important part of the cooling curve is the *cooling rate* which affects the microstructure and properties. Generally speaking, an area of the casting which is cooled quickly will have a fine grain structure and an area which cools slowly will have a coarse grain structure. Below is an example cooling curve of a pure metal or eutectic alloy, with defining terminology.

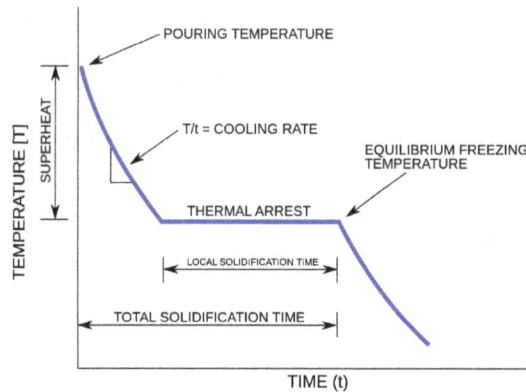

Note that before the thermal arrest the material is a liquid and after it the material is a solid; during the thermal arrest the material is converting from a liquid to a solid. Also, note that the greater the superheat the more time there is for the liquid material to flow into intricate details.

The above cooling curve depicts a basic situation with a pure alloy, however, most castings are of alloys, which have a cooling curve shaped as shown below.

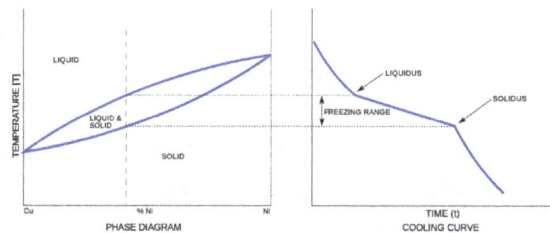

Note that there is no longer a thermal arrest, instead there is a freezing range. The freezing range corresponds directly to the liquidus and solidus found on the phase diagram for the specific alloy.

Chvorinov's Rule

The local solidification time can be calculated using Chvorinov's rule, which is:

$$t = B\left(\frac{V}{A}\right)^n$$

Where t is the solidification time, V is the volume of the casting, A is the surface area of the casting that contacts the mold, n is a constant, and B is the mold constant. It is most useful in determining if a riser will solidify before the casting, because if the riser does solidify first then it is worthless.

The Gating System

The gating system serves many purposes, the most important being conveying the liquid material to the mold, but also controlling shrinkage, the speed of the liquid, turbulence, and trapping dross. The gates are usually attached to the thickest part of the casting to assist in controlling shrinkage. In especially large castings multiple gates or runners may be required to introduce metal to more than one point in the mold cavity. The speed of the material is important because if the material is traveling too slowly it can cool before completely filling, leading to misruns and cold shuts. If the material is moving too fast then the liquid material can erode the mold and contaminate the final casting. The shape and length of the gating system can also control how quickly the material cools; short round or square channels minimize heat loss.

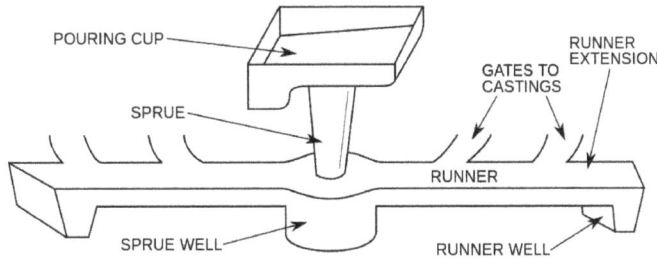

A simple gating system for a horizontal parting mold.

The gating system may be designed to minimize turbulence, depending on the material being cast. For example, steel, cast iron, and most copper alloys are turbulent insensitive, but aluminium and magnesium alloys are turbulent sensitive. The turbulent insensitive materials usually have a short and open gating system to fill the mold as quickly as possible. However, for turbulent sensitive materials short sprues are used to minimize the distance the material must fall when entering the mold. Rectangular pouring cups and tapered sprues are used to prevent the formation of a vortex as the material flows into the mold; these vortices tend to suck gas and oxides into the mold. A large sprue well is used to dissipate the kinetic energy of the liquid material as it falls down the sprue, decreasing turbulence. The *choke*, which is the smallest cross-sectional area in the gating system used to control flow, can be placed near the sprue well to slow down and smooth out the flow. Note that on some molds the choke is still placed on the gates to make separation of the part easier, but induces extreme turbulence. The gates are usually attached to the bottom of the casting to minimize turbulence and splashing.

The gating system may also be designed to trap dross. One method is to take advantage of the fact that some dross has a lower density than the base material so it floats to the top of the gating system. Therefore, long flat runners with gates that exit from the bottom of the runners can trap dross in the runners; note that long flat runners will cool the material more rapidly than round or square runners. For materials where the dross is a similar density to the base material, such as aluminium, *runner extensions* and *runner wells* can be advantageous. These take advantage of the fact that the dross is usually located at the beginning of the pour, therefore the runner is extended past the last gate(s) and the contaminates are contained in the wells. Screens or filters may also be used to trap contaminates.

It is important to keep the size of the gating system small, because it all must be cut from the casting and remelted to be reused. The efficiency, or *yield*, of a casting system can be calculated by dividing the weight of the casting by the weight of the metal poured. Therefore, the higher the number the more efficient the gating system/risers.

Shrinkage

There are three types of shrinkage: *shrinkage of the liquid, solidification shrinkage* and *patternmaker's shrinkage*. The shrinkage of the liquid is rarely a problem because more material is flowing into the mold behind it. Solidification shrinkage occurs because metals are less dense as a liquid than a solid, so during solidification the metal density dramatically increases. Patternmaker's shrinkage refers to the shrinkage that occurs when the material is cooled from the solidification temperature to room temperature, which occurs due to thermal contraction.

Solidification Shrinkage

Solidification shrinkage of various metals	
Metal	**Percentage**
Aluminium	6.6
Copper	4.9
Magnesium	4.0 or 4.2
Zinc	3.7 or 6.5
Low carbon steel	2.5–3.0
High carbon steel	4.0
White cast iron	4.0–5.5
Gray cast iron	−2.5–1.6
Ductile cast iron	−4.5–2.7

Most materials shrink as they solidify, but, as the adjacent table shows, a few materials do not, such as gray cast iron. For the materials that do shrink upon solidification the type of shrinkage depends on how wide the freezing range is for the material. For materials with a narrow freezing range, less than 50 °C (122 °F), a cavity, known as a *pipe*, forms in the center of the casting, because the outer shell freezes first and progressively solidifies to the center. Pure and eutectic metals usually have narrow solidification ranges. These materials tend to form a *skin* in open air molds, therefore they are known as *skin forming alloys*. For materials with a wide freezing range, greater than 110 °C (230 °F), much more of the casting occupies the *mushy* or *slushy* zone (the temperature range between the solidus and the liquidus), which leads to small pockets of liquid trapped throughout and ultimately porosity. These castings tend to have poor ductility, toughness, and fatigue resistance. Moreover, for these types of materials to be fluid-tight a secondary operation is required to impregnate the casting with a lower melting point metal or resin.

For the materials that have narrow solidification ranges pipes can be overcome by designing the casting to promote directional solidification, which means the casting freezes first at the point farthest from the gate, then progressively solidifies towards the gate. This allows a continuous feed of liquid material to be present at the point of solidification to compensate for the shrinkage. Note that there is still a shrinkage void where the final material solidifies, but if designed properly this will be in the gating system or riser.

Risers and Riser Aids

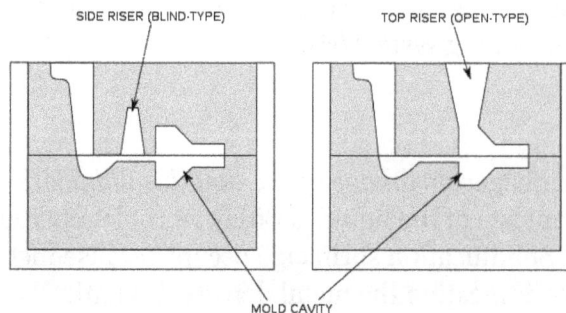

Different types of risers

Risers, also known as *feeders*, are the most common way of providing directional solidification. It supplies liquid metal to the solidifying casting to compensate for solidification shrinkage. For a riser to work properly the riser must solidify after the casting, otherwise it cannot supply liquid metal to shrinkage within the casting. Risers add cost to the casting because it lowers the *yield* of each casting; i.e. more metal is lost as scrap for each casting. Another way to promote directional solidification is by adding chills to the mold. A chill is any material which will conduct heat away from the casting more rapidly that the material used for molding.

Risers are classified by three criteria. The first is if the riser is open to the atmosphere, if it is then it is called an *open* riser, otherwise it is known as a *blind* type. The second criterion is where the riser is located; if it is located on the casting then it is known as a *top riser* and if it is located next to the casting it is known as a *side riser*. Finally, if riser is located on the gating system so that it fills after the molding cavity, it is known as a *live riser* or *hot riser*, but if the riser fills with materials that's already flowed through the molding cavity it is known as a *dead riser* or *cold riser*.

Riser aids are items used to assist risers in creating directional solidification or reducing the number of risers required. One of these items are *chills* which accelerate cooling in a certain part of the mold. There are two types: external and internal chills. External chills are masses of high-heat-capacity and high-thermal-conductivity material that are placed on an edge of the molding cavity. Internal chills are pieces of the same metal that is being poured, which are placed inside the mold cavity and become part of the casting. Insulating sleeves and toppings may also be installed around the riser cavity to slow the solidification of the riser. Heater coils may also be installed around or above the riser cavity to slow solidification.

Patternmaker's Shrink

Typical patternmaker's shrinkage of various metals		
Metal	**Percentage**	**in/ft**
Aluminium	1.0–1.3	$\frac{1}{8}$–$\frac{5}{32}$
Brass	1.5	$\frac{3}{16}$
Magnesium	1.0–1.3	$\frac{1}{8}$–$\frac{5}{32}$
Cast iron	0.8–1.0	$\frac{1}{10}$–$\frac{1}{8}$
Steel	1.5–2.0	$\frac{3}{16}$–$\frac{1}{4}$

Shrinkage after solidification can be dealt with by using an oversized pattern designed specifically for the alloy used. *Contraction rules*, or *shrink rules*, are used to make the patterns oversized to compensate for this type of shrinkage. These rulers are up to 2.5% oversize, depending on the material being cast. These rulers are mainly referred to by their percentage change. A pattern made to match an existing part would be made as follows: First, the existing part would be measured using a standard ruler, then when constructing the pattern, the pattern maker would use a contraction rule, ensuring that the casting would contract to the correct size.

Note that patternmaker's shrinkage does not take phase change transformations into account. For example, eutectic reactions, martensitic reactions, and graphitization can cause expansions or contractions.

Mold Cavity

The mold cavity of a casting does not reflect the exact dimensions of the finished part due to a number of reasons. These modifications to the mold cavity are known as *allowances* and account for pattern-maker's shrinkage, draft, machining, and distortion. In non-expendable processes, these allowances are imparted directly into the permanent mold, but in expendable mold processes they are imparted into the patterns, which later form the mold cavity. Note that for non-expendable molds an allowance is required for the dimensional change of the mold due to heating to operating temperatures.

For surfaces of the casting that are perpendicular to the parting line of the mold a draft must be included. This is so that the casting can be released in non-expendable processes or the pattern can be released from the mold without destroying the mold in expendable processes. The required draft angle depends on the size and shape of the feature, the depth of the mold cavity, how the part or pattern is being removed from the mold, the pattern or part material, the mold material, and the process type. Usually the draft is not less than 1%.

The machining allowance varies drastically from one process to another. Sand castings generally have a rough surface finish, therefore need a greater machining allowance, whereas die casting has a very fine surface finish, which may not need any machining tolerance. Also, the draft may provide enough of a machining allowance to begin with.

The distortion allowance is only necessary for certain geometries. For instance, U-shaped castings will tend to distort with the legs splaying outward, because the base of the shape can contract while the legs are constrained by the mold. This can be overcome by designing the mold cavity to slope the leg inward to begin with. Also, long horizontal sections tend to sag in the middle if ribs are not incorporated, so a distortion allowance may be required.

Cores may be used in expendable mold processes to produce internal features. The core can be of metal but it is usually done in sand.

Filling

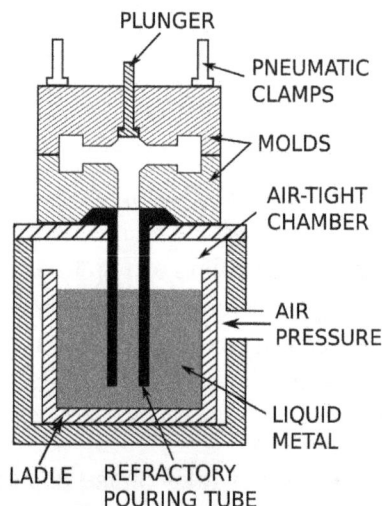

Schematic of the low-pressure permanent mold casting process

There are a few common methods for filling the mold cavity: *gravity*, *low-pressure*, *high-pressure*, and *vacuum*.

Vacuum filling, also known as *counter-gravity* filling, is more metal efficient than gravity pouring because less material solidifies in the gating system. Gravity pouring only has a 15 to 50% metal yield as compared to 60 to 95% for vacuum pouring. There is also less turbulence, so the gating system can be simplified since it does not have to control turbulence. Plus, because the metal is drawn from below the top of the pool the metal is free from dross and slag, as these are lower density (lighter) and float to the top of the pool. The pressure differential helps the metal flow into every intricacy of the mold. Finally, lower temperatures can be used, which improves the grain structure. The first patented vacuum casting machine and process dates to 1879.

Low-pressure filling uses 5 to 15 psig (35 to 100 kPag) of air pressure to force liquid metal up a feed tube into the mold cavity. This eliminates turbulence found in gravity casting and increases density, repeatability, tolerances, and grain uniformity. After the casting has solidified the pressure is released and any remaining liquid returns to the crucible, which increases yield.

Tilt Filling

Tilt filling, also known as *tilt casting*, is an uncommon filling technique where the crucible is attached to the gating system and both are slowly rotated so that the metal enters the mold cavity with little turbulence. The goal is to reduce porosity and inclusions by limiting turbulence. For most uses tilt filling is not feasible because the following inherent problem: if the system is rotated slow enough to not induce turbulence, the front of the metal stream begins to solidify, which results in mis-runs. If the system is rotated faster then it induces turbulence, which defeats the purpose. Durville of France was the first to try tilt casting, in the 1800s. He tried to use it to reduce surface defects when casting coinage from aluminium bronze.

Macrostructure

The grain macrostructure in ingots and most castings have three distinct regions or zones: the chill zone, columnar zone, and equiaxed zone. The image below depicts these zones.

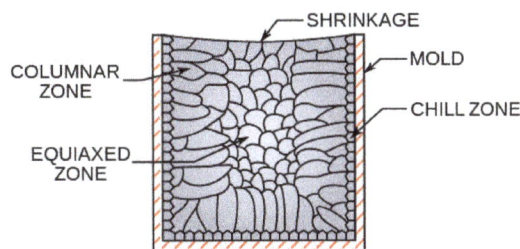

The chill zone is named so because it occurs at the walls of the mold where the wall *chills* the material. Here is where the nucleation phase of the solidification process takes place. As more heat is removed the grains grow towards the center of the casting. These are thin, long *columns* that are perpendicular to the casting surface, which are undesirable because they have anisotropic properties. Finally, in the center the equiaxed zone contains spherical, randomly oriented crystals. These are desirable because they have isotropic properties. The creation of this zone can be promoted by using a low pouring temperature, alloy inclusions, or inoculants.

Inspection

Common inspection methods for steel castings are *magnetic particle testing* and *liquid penetrant testing*. Common inspection methods for aluminum castings are *radiography*, *ultrasonic testing*, and *liquid penetrant testing*.

Defects

There are a number of problems that can be encountered during the casting process. The main types are: *gas porosity*, *shrinkage defects*, *mold material defects*, *pouring metal defects*, and *metallurgical defects*.

Casting Process Simulation

Casting process simulation uses numerical methods to calculate cast component quality considering mold filling, solidification and cooling, and provides a quantitative prediction of casting mechanical properties, thermal stresses and distortion. Simulation accurately describes a cast component's quality up-front before production starts. The casting rigging can be designed with respect to the required component properties. This has benefits beyond a reduction in pre-production sampling, as the precise layout of the complete casting system also leads to energy, material, and tooling savings.

The software supports the user in component design, the determination of melting practice and casting methoding through to pattern and mold making, heat treatment, and finishing. This saves costs along the entire casting manufacturing route.

Casting process simulation was initially developed at universities starting from the early '70s, mainly in Europe and in the U.S., and is regarded as the most important innovation in casting technology over the last 50 years. Since the late '80s, commercial programs are available which make it possible for foundries to gain new insight into what is happening inside the mold or die during the casting process.

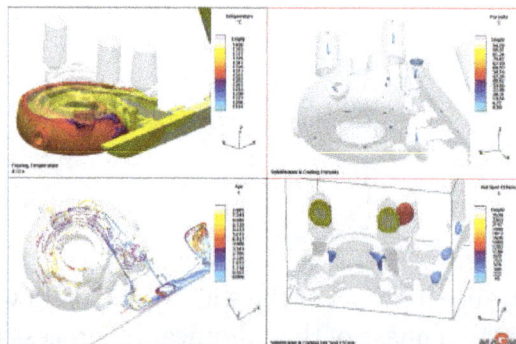

A high-performance software for the simulation of casting processes provides opportunities for an interactive or automated evaluation of results (here, for example, of mold filling and solidification, porosity and flow characteristics). Picture: Componenta B.V., The Netherlands)

Forging

Hot metal ingot being loaded into a hammer forge

Forging is a manufacturing process involving the shaping of metal using localized compressive forces. The blows are delivered with a hammer (often a power hammer) or a die. Forging is often classified according to the temperature at which it is performed: cold forging (a type of cold working), warm forging, or hot forging (a type of hot working). For the latter two, the metal is heated, usually in a forge. Forged parts can range in weight from less than a kilogram to hundreds of metric tons. Forging has been done by smiths for millennia; the traditional products were kitchenware, hardware, hand tools, edged weapons, and jewellery. Since the Industrial Revolution, forged parts are widely used in mechanisms and machines wherever a component requires high strength; such forgings usually require further processing (such as machining) to achieve a finished part. Today, forging is a major worldwide industry.

History

Forging is one of the oldest known metalworking processes. Traditionally, forging was performed by a smith using hammer and anvil, though introducing water power to the production and working of iron in the 12th century allowed the use of large trip hammers or power hammers that exponentially increased the amount and size of iron that could be produced and forged easily. The smithy or forge has evolved over centuries to become a facility with engineered processes, production equipment, tooling, raw materials and products to meet the demands of modern industry.

In modern times, industrial forging is done either with presses or with hammers powered by compressed air, electricity, hydraulics or steam. These hammers may have reciprocating weights in the thousands of pounds. Smaller power hammers, 500 lb (230 kg) or less reciprocating weight, and hydraulic presses are common in art smithies as well. Some steam hammers remain in use, but they became obsolete with the availability of the other, more convenient, power sources.

Advantages and Disadvantages

Forging can produce a piece that is stronger than an equivalent cast or machined part. As the metal is shaped during the forging process, its internal grain deforms to follow the general shape of the part. As a result, the grain is continuous throughout the part, giving rise to a piece with improved strength characteristics. Additionally, forgings can target a lower total cost when compared to a casting or fabrication. When you consider all the costs that are involved in a product's lifecycle from procurement to lead time to rework, then factor in the costs of scrap, downtime and further quality issues, the long-term benefits of forgings can outweigh the short-term cost-savings that castings or fabrications might offer.

Some metals may be forged cold, but iron and steel are almost always hot forged. Hot forging prevents the work hardening that would result from cold forging, which would increase the difficulty of performing secondary machining operations on the piece. Also, while work hardening may be desirable in some circumstances, other methods of hardening the piece, such as heat treating, are generally more economical and more controllable. Alloys that are amenable to precipitation hardening, such as most aluminium alloys and titanium, can be hot forged, followed by hardening.

Production forging involves significant capital expenditure for machinery, tooling, facilities and personnel. In the case of hot forging, a high-temperature furnace (sometimes referred to as the forge) is required to heat ingots or billets. Owing to the massiveness of large forging hammers and presses and the parts they can produce, as well as the dangers inherent in working with hot metal, a special building is frequently required to house the operation. In the case of drop forging operations, provisions must be made to absorb the shock and vibration generated by the hammer. Most forging operations use metal-forming dies, which must be precisely machined and carefully heat-treated to correctly shape the workpiece, as well as to withstand the tremendous forces involved.

Processes

There are many different kinds of forging processes available, however they can be grouped into three main classes:

- Drawn out: length increases, cross-section decreases

- Upset: length decreases, cross-section increases

- Squeezed in closed compression dies: produces multidirectional flow

Common forging processes include: roll forging, swaging, cogging, open-die forging, impression-die forging, press forging, automatic hot forging and upsetting.

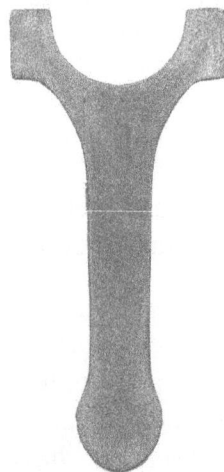

A cross-section of a forged connecting rod that has been etched to show the grain flow

Temperature

All of the following forging processes can be performed at various temperatures, however they are generally classified by whether the metal temperature is above or below the recrystallization temperature. If the temperature is above the material's recrystallization temperature it is deemed *hot forging*; if the temperature is below the material's recrystallization temperature but above 30% of the recrystallization temperature (on an absolute scale) it is deemed *warm forging*; if below 30% of the recrystallization temperature (usually room temperature) then it is deemed *cold forging*. The main advantage of hot forging is that it can be done faster and more precise, and as the metal is deformed work hardening effects are negated by the recrystallization process. Cold forging typically results in work hardening of the piece.

Drop Forging

Boat nail production in Hainan, China

Drop forging is a forging process where a hammer is raised and then "dropped" onto the workpiece to deform it according to the shape of the die. There are two types of drop forging: open-die drop forging and closed-die drop forging. As the names imply, the difference is in the shape of the die, with the former not fully enclosing the workpiece, while the latter does.

Open-die Drop Forging

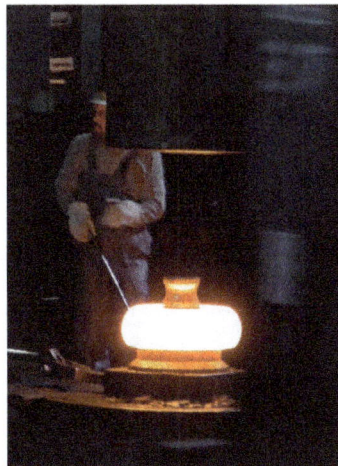

Open-die drop forging (with two dies) of an ingot to be further processed into a wheel

Open-die forging is also known as *smith forging*. In open-die forging, a hammer strikes and deforms the workpiece, which is placed on a stationary anvil. Open-die forging gets its name from the fact that the dies (the surfaces that are in contact with the workpiece) do not enclose the workpiece, allowing it to flow except where contacted by the dies. The operator therefore needs to orient and position the workpiece to get the desired shape. The dies are usually flat in shape, but some have a specially shaped surface for specialized operations. For example, a die may have a round, concave, or convex surface or be a tool to form holes or be a cut-off tool. Open-die forgings can be worked into shapes which include discs, hubs, blocks, shafts (including step shafts or with flanges), sleeves, cylinders, flats, hexes, rounds, plate, and some custom shapes. Open-die forging lends itself to short runs and is appropriate for art smithing and custom work. In some cases, open-die forging may be employed to rough-shape ingots to prepare them for subsequent operations. Open-die forging may also orient the grain to increase strength in the required direction.

Advantages of Open-die Forging

- Reduced chance of voids

- Better fatigue resistance

- Improved microstructure

- Continuous grain flow

- Finer grain size

- Greater strength

"Cogging" is the successive deformation of a bar along its length using an open-die drop forge. It is commonly used to work a piece of raw material to the proper thickness. Once the proper thickness is achieved the proper width is achieved via "edging". "Edging" is the process of concentrating material using a concave shaped open-die. The process is called "edging" because it is usually carried out on the ends of the workpiece. "Fullering" is a similar process that thins out sections of the forging using a convex shaped die. These processes prepare the workpieces for further forging processes.

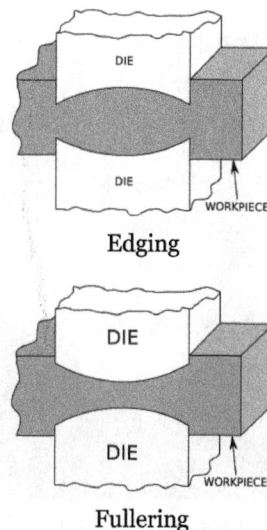

Edging

Fullering

Impression-die Forging

Impression-die forging is also called "closed-die forging". In impression-die forging, the metal is placed in a die resembling a mold, which is attached to an anvil. Usually, the hammer die is shaped as well. The hammer is then dropped on the workpiece, causing the metal to flow and fill the die cavities. The hammer is generally in contact with the workpiece on the scale of milliseconds. Depending on the size and complexity of the part, the hammer may be dropped multiple times in quick succession. Excess metal is squeezed out of the die cavities, forming what is referred to as "flash". The flash cools more rapidly than the rest of the material; this cool metal is stronger than the metal in the die, so it helps prevent more flash from forming. This also forces the metal to completely fill the die cavity. After forging, the flash is removed. In commercial impression-die forging, the workpiece is usually moved through a series of cavities in a die to get from an ingot to the final form. The first impression is used to distribute the metal into the rough shape in accordance to the needs of later cavities; this impression is called an "edging", "fullering", or "bending" impression. The following cavities are called "blocking" cavities, in which the piece is working into a shape that more closely resembles the final product. These stages usually impart the workpiece with generous bends and large fillets. The final shape is forged in a "final" or "finisher" impression cavity. If there is only a short run of parts to be done, then it may be more economical for the die to lack a final impression cavity and instead machine the final features. Impression-die forging has been improved in recent years through increased automation which includes induction heating, mechanical feeding, positioning and manipulation, and the direct heat treatment of parts after forging. One variation of impression-die forging is called "flashless forging", or "true closed-die forging". In this type of forging, the die cavities are completely closed, which keeps the workpiece from forming flash. The major advantage to this process is that less metal is lost to flash. Flash can account for 20 to 45% of the starting material. The disadvantages of this process include additional cost due to a more complex die design and the need for better lubrication and workpiece placement. There are other variations of part formation that integrate impression-die forging. One method incorporates casting a forging preform from liquid metal. The casting is removed after it has solidified, but while still hot. It is then finished in a single cavity die. The flash is trimmed, then the part is quench hardened. Another variation follows the same process as outlined above, except the preform is produced by the spraying deposition of metal droplets into shaped collectors (similar to the Osprey process). Closed-die forging has a high initial cost due to the creation of dies and required design work to make working die cavities. However, it has low recurring costs for each part, thus forgings become more economical with more volume. This is one of the major reasons closed-die forgings are often used in the automotive and tool industries. Another reason forgings are common in these industrial sectors is that forgings generally have about a 20 percent higher strength-to-weight ratio compared to cast or machined parts of the same material.

Design of Impression-die Forgings and Tooling

Forging dies are usually made of high-alloy or tool steel. Dies must be impact resistant, wear resistant, maintain strength at high temperatures, and have the ability to withstand cycles of rapid heating and cooling. In order to produce a better, more economical die the following standards are maintained:

- The dies part along a single, flat plane whenever possible. If not, the parting plane follows the contour of the part.

- The parting surface is a plane through the center of the forging and not near an upper or lower edge.

- Adequate draft is provided; usually at least 3° for aluminium and 5° to 7° for steel.

- Generous fillets and radii are used.

- Ribs are low and wide.

- The various sections are balanced to avoid extreme difference in metal flow.

- Full advantage is taken of fiber flow lines.

- Dimensional tolerances are not closer than necessary.

The dimensional tolerances of a steel part produced using the impression-die forging method are outlined in the table below. The dimensions across the parting plane are affected by the closure of the dies, and are therefore dependent on die wear and the thickness of the final flash. Dimensions that are completely contained within a single die segment or half can be maintained at a significantly greater level of accuracy.

Dimensional tolerances for impression-die forgings		
Mass [kg (lb)]	**Minus tolerance [mm (in)]**	**Plus tolerance [mm (in)]**
0.45 (1)	0.15 (0.006)	0.46 (0.018)
0.91 (2)	0.20 (0.008)	0.61 (0.024)
2.27 (5)	0.25 (0.010)	0.76 (0.030)
4.54 (10)	0.28 (0.011)	0.84 (0.033)
9.07 (20)	0.33 (0.013)	0.99 (0.039)
22.68 (50)	0.48 (0.019)	1.45 (0.057)
45.36 (100)	0.74 (0.029)	2.21 (0.087)

A lubricant is used when forging to reduce friction and wear. It is also used as a thermal barrier to restrict heat transfer from the workpiece to the die. Finally, the lubricant acts as a parting compound to prevent the part from sticking in the dies.

Press Forging

Press forging works by slowly applying a continuous pressure or force, which differs from the near-instantaneous impact of drop-hammer forging. The amount of time the dies are in contact with the workpiece is measured in seconds (as compared to the milliseconds of drop-hammer forges). The press forging operation can be done either cold or hot.

The main advantage of press forging, as compared to drop-hammer forging, is its ability to deform the complete workpiece. Drop-hammer forging usually only deforms the surfaces of the work piece in contact with the hammer and anvil; the interior of the workpiece will stay relatively undeformed. Another advantage to the process includes the knowledge of the new part's strain rate. We specifically know what kind of strain can be put on the part, because the compression rate of the press forging operation is controlled.

There are a few disadvantages to this process, most stemming from the workpiece being in contact with the dies for such an extended period of time. The operation is a time-consuming process due to the amount and length of steps. The workpiece will cool faster because the dies are in contact with workpiece; the dies facilitate drastically more heat transfer than the surrounding atmosphere. As the workpiece cools it becomes stronger and less ductile, which may induce cracking if deformation continues. Therefore, heated dies are usually used to reduce heat loss, promote surface flow, and enable the production of finer details and closer tolerances. The workpiece may also need to be reheated.

When done in high productivity, press forging is more economical than hammer forging. The operation also creates closer tolerances. In hammer forging a lot of the work is absorbed by the machinery, when in press forging, the greater percentage of work is used in the work piece. Another advantage is that the operation can be used to create any size part because there is no limit to the size of the press forging machine. New press forging techniques have been able to create a higher degree of mechanical and orientation integrity. By the constraint of oxidation to the outer layers of the part, reduced levels of microcracking occur in the finished part.

Press forging can be used to perform all types of forging, including open-die and impression-die forging. Impression-die press forging usually requires less draft than drop forging and has better dimensional accuracy. Also, press forgings can often be done in one closing of the dies, allowing for easy automation.

Upset forging

Upset forging increases the diameter of the workpiece by compressing its length. Based on number of pieces produced, this is the most widely used forging process. A few examples of common parts produced using the upset forging process are engine valves, couplings, bolts, screws, and other fasteners.

Upset forging is usually done in special high-speed machines called *crank presses*. The machines are usually set up to work in the horizontal plane, to facilitate the quick exchange of workpieces from one station to the next, but upsetting can also be done in a vertical crank press or a hydraulic press. The initial workpiece is usually wire or rod, but some machines can accept bars up to 25 cm (9.8 in) in diameter and a capacity of over 1000 tons. The standard upsetting machine employs split dies that contain multiple cavities. The dies open enough to allow the workpiece to move from one cavity to the next; the dies then close and the heading tool, or ram, then moves longitudinally against the bar, upsetting it into the cavity. If all of the cavities are utilized on every cycle, then a finished part will be produced with every cycle, which makes this process advantageous for mass production.

These rules must be followed when designing parts to be upset forged:

- The length of unsupported metal that can be upset in one blow without injurious buckling should be limited to three times the diameter of the bar.

- Lengths of stock greater than three times the diameter may be upset successfully, provided that the diameter of the upset is not more than 1.5 times the diameter of the stock.

- In an upset requiring stock length greater than three times the diameter of the stock, and where the diameter of the cavity is not more than 1.5 times the diameter of the stock, the length of unsupported metal beyond the face of the die must not exceed the diameter of the bar.

Automatic Hot Forging

The automatic hot forging process involves feeding mill-length steel bars (typically 7 m (23 ft) long) into one end of the machine at room temperature and hot forged products emerge from the other end. This all occurs rapidly; small parts can be made at a rate of 180 parts per minute (ppm) and larger can be made at a rate of 90 ppm. The parts can be solid or hollow, round or symmetrical, up to 6 kg (13 lb), and up to 18 cm (7.1 in) in diameter. The main advantages to this process are its high output rate and ability to accept low-cost materials. Little labor is required to operate the machinery.

There is no flash produced so material savings are between 20 and 30% over conventional forging. The final product is a consistent 1,050 °C (1,920 °F) so air cooling will result in a part that is still easily machinable (the advantage being the lack of annealing required after forging). Tolerances are usually ±0.3 mm (0.012 in), surfaces are clean, and draft angles are 0.5 to 1°. Tool life is nearly double that of conventional forging because contact times are on the order of 0.06-second. The downside is that this process is only feasible on smaller symmetric parts and cost; the initial investment can be over $10 million, so large quantities are required to justify this process.

The process starts by heating the bar to 1,200 to 1,300 °C (2,190 to 2,370 °F) in less than 60 seconds using high-power induction coils. It is then descaled with rollers, sheared into blanks, and transferred through several successive forming stages, during which it is upset, preformed, final forged, and pierced (if necessary). This process can also be coupled with high-speed cold-forming operations. Generally, the cold forming operation will do the finishing stage so that the advantages of cold-working can be obtained, while maintaining the high speed of automatic hot forging.

Examples of parts made by this process are: wheel hub unit bearings, transmission gears, tapered roller bearing races, stainless steel coupling flanges, and neck rings for LP gas cylinders. Manual transmission gears are an example of automatic hot forging used in conjunction with cold working.

Roll Forging

Roll forging is a process where round or flat bar stock is reduced in thickness and increased in length. Roll forging is performed using two cylindrical or semi-cylindrical rolls, each containing one or more shaped grooves. A heated bar is inserted into the rolls and when it hits a spot the rolls rotate and the bar is progressively shaped as it is rolled through the machine. The piece is then transferred to the next set of grooves or turned around and reinserted into the same grooves. This continues until the desired shape and size is achieved. The advantage of this process is there is no flash and it imparts a favorable grain structure into the workpiece.

Examples of products produced using this method include axles, tapered levers and leaf springs.

Net-shape and Near-net-shape Forging

This process is also known as *precision forging*. It was developed to minimize cost and waste associated with post-forging operations. Therefore, the final product from a precision forging needs little or no final machining. Cost savings are gained from the use of less material, and thus less scrap, the overall decrease in energy used, and the reduction or elimination of machining. Precision forging also requires less of a draft, 1° to 0°. The downside of this process is its cost, therefore it is only implemented if significant cost reduction can be achieved.

Cold Forging

Near net shape forging is most common when parts are forged without heating the slug, bar or billet. Aluminum is a common material that can be cold forged depending on final shape. Lubrication of the parts being formed is critical to increase the life of the mating dies.

Cost Implications

To achieve a low-cost net shape forging for demanding applications that are subject to a high degree of scrutiny, i.e. non-destructive testing by way of a dye-penetrant inspection technique, it is crucial that basic forging process disciplines be implemented. If the basic disciplines are not met, subsequent material removal operations will likely be necessary to remove material defects found at non-destructive testing inspection. Hence low-cost parts will not be achievable.

Example disciplines are: die-lubricant management (Use of uncontaminated and homogeneous mixtures, amount and placement of lubricant). Tight control of die temperatures and surface finish / friction.

Induction Forging

Unlike the above processes, induction forging is based on the type of heating style used. Many of the above processes can be used in conjunction with this heating method.

Multidirectional Forging

Multidirectional Forging is forming of a work piece in a single step in several directions. The multidirectional forming takes place through constructive measures of the tool. The vertical movement of the press ram is redirected using wedges which distributes and redirects the force of the forging press in horizontal directions.

Materials and Applications

Forging of Steel

Depending on the forming temperature steel forging can be divided into:

- Hot forging of steel

    o Forging temperatures above the recrystallization temperature between 950 - 1250 °C

    o Good formability

    o Low forming forces

- o   Constant tensile strength of the workpieces

- Warm forging of steel

    - o   Forging temperatures between 750 – 950 °C

    - o   Less or no scaling at the workpiece surface

    - o   Narrower tolerances achievable than in hot forging

    - o   Limited formability and higher forming forces than for hot forging

    - o   Lower forming forces than in cold forming

- Cold forging of steel

    - o   Forging temperatures at room conditions, self-heating up to 150 °C due to the forming energy

    - o   Narrowest tolerances achievable

    - o   No scaling at workpiece surface

    - o   Increase of strength and decrease of ductility due to strain hardening

    - o   Low formability and high forming forces are necessary

For industrial processes steel alloys are primarily forged in hot condition. Brass, bronze, copper, precious metals and their alloys are manufactured by cold forging processes, while each metal requires a different forging temperature.

Forging of Aluminium

- Aluminium forging is performed at a temperature range between 350 and 550 °C

- Forging temperatures above 550 °C are too close to the solidus temperature of the alloys and lead in conjunction with varying effective strains to unfavorable workpiece surfaces and potentially to a partial melting as well as fold formation.

- Forging temperatures below 350 °C reduce formability by increasing the yield stress, which can lead to unfilled dies, cracking at the workpiece surface and increased die forces

Due to the narrow temperature range and high thermal conductivity, aluminium forging can only be realized in a particular process window. To provide good forming conditions a homogeneous temperature distribution in the entire workpiece is necessary. Therefore, the control of the tool temperature has a major influence to the process. For example, by optimizing the preform geometries the local effective strains can be influenced to reduce local overheating for a more homogeneous temperature distribution.

Application of Aluminium Forged Parts

High-strength aluminium alloys have the tensile strength of medium strong steel alloys while providing significant weight advantages. Therefore, aluminium forged parts are mainly used in aerospace, automotive industry and many other fields of engineering especially in those fields, where highest safety standards against failure by abuse, by shock or vibratory stresses are needed. Such parts are for example chassis parts, steering components and brake parts. Commonly used alloys are AlSi1MgMn (EN AW-6082) and AlZnMgCu1,5 (EN AW-7075). About 80% of all aluminium forged parts are made of AlSi1MgMn. The high-strength alloy AlZnMgCu1,5 is mainly used for aerospace applications.

Equipment

Hydraulic drop-hammer

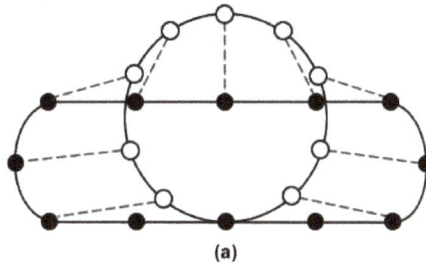

(a)

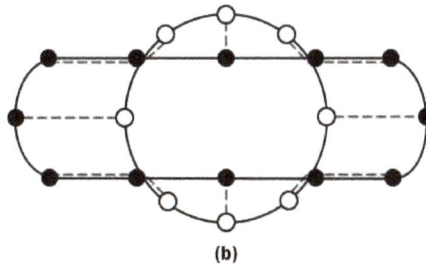

(b)

(a) Material flow of a conventionally forged disc; (b) Material flow of an impactor forged disc

The most common type of forging equipment is the hammer and anvil. Principles behind the hammer and anvil are still used today in *drop-hammer* equipment. The principle behind the machine is simple: raise the hammer and drop it or propel it into the workpiece, which rests on the anvil. The main variations between drop-hammers are in the way the hammer is powered; the most common being air and steam hammers. Drop-hammers usually operate in a vertical position. The main reason for this is excess energy (energy that isn't used to deform the workpiece) that isn't released as heat or sound needs to be transmitted to the foundation. Moreover, a large machine base is needed to absorb the impacts.

To overcome some shortcomings of the drop-hammer, the *counterblow machine* or *impactor* is used. In a counterblow machine both the hammer and anvil move and the workpiece is held between them. Here excess energy becomes recoil. This allows the machine to work horizontally and have a smaller base. Other advantages include less noise, heat and vibration. It also produces a distinctly different flow pattern. Both of these machines can be used for open-die or closed-die forging.

Forging Presses

A *forging press*, often just called a press, is used for press forging. There are two main types: mechanical and hydraulic presses. Mechanical presses function by using cams, cranks and/or toggles to produce a preset (a predetermined force at a certain location in the stroke) and reproducible stroke. Due to the nature of this type of system, different forces are available at different stroke positions. Mechanical presses are faster than their hydraulic counterparts (up to 50 strokes per minute). Their capacities range from 3 to 160 MN (300 to 18,000 short tons-force). Hydraulic presses use fluid pressure and a piston to generate force. The advantages of a hydraulic press over a mechanical press are its flexibility and greater capacity. The disadvantages include a slower, larger, and costlier machine to operate.

The roll forging, upsetting, and automatic hot forging processes all use specialized machinery.

List of large forging presses, by ingot size			
Force (tonnes)	Ingot size (tonnes)	Company	Location
16,500	600	Shanghai Electric Group	Shanghai, China
16,000	600	China National Erzhong Group	Deyang, China
14,000	600	Japan Steel Works	Japan
15,000	580	China First Heavy Industries Group	Heilongjiang, China
13,000		Doosan	South Korea

List of large forging presses, by force				
Force (tonnes)	Force (tons)	Ingot size (tonnes)	Company	Location
80,000	(88,200)	>150	China Erzhong	Deyang, China
75,000	(82,690)		VSMPO-AVISMA	Russia
65,000	(71,660)		Aubert & Duval	Issoire, France
(45,350)	50,000	20	Alcoa, Wyman Gordon	USA
40,000	(44,100)		Aubert & Duval	Pamiers, France
30,000	(33,080)	8	Wyman Gordon	Livingston, Scotland
30,000	(33,070)		Weber Metals, Inc.	California, United States
30,000	(30,108)		Firth Rixson	Georgia, United States

Rolling (Metalworking)

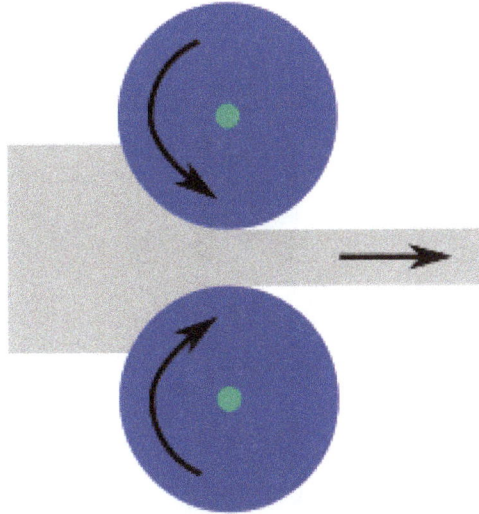

Rolling schematic view

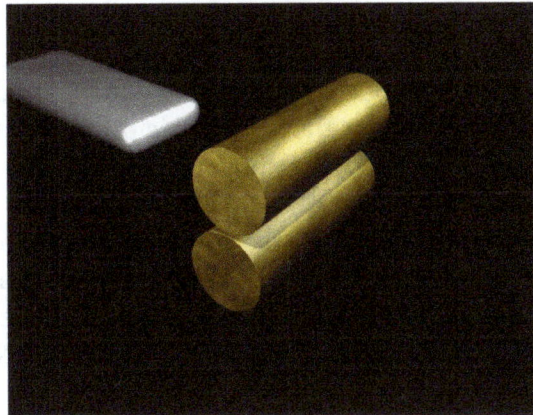

Rolling visualization. (Click on image to view animation.)

In metalworking, rolling is a metal forming process in which metal stock is passed through one or more pairs of rolls to reduce the thickness and to make the thickness uniform. The concept is similar to the rolling of dough. Rolling is classified according to the temperature of the metal rolled. If the temperature of the metal is above its recrystallization temperature, then the process is known as hot rolling. If the temperature of the metal is below its recrystallization temperature, the process is known as cold rolling. In terms of usage, hot rolling processes more tonnage than any other manufacturing process, and cold rolling processes the most tonnage out of all cold working processes. Roll stands holding pairs of rolls are grouped together into rolling mills that can quickly process metal, typically steel, into products such as structural steel (I-beams, angle stock, channel stock, and so on), bar stock, and rails. Most steel mills have rolling mill divisions that convert the semi-finished casting products into finished products.

There are many types of rolling processes, including *ring rolling, roll bending, roll forming, profile rolling,* and *controlled rolling.*

History

Iron and Steel

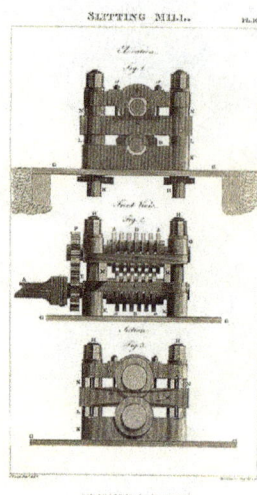

Slitting mill, 1813.

The invention of the rolling mill in Europe may be attributed to Leonardo da Vinci in his drawings. The earliest rolling mills in crude form but the same basic principles were found in Middle East and South Asia as early as 600 BCE. Earliest rolling mills were slitting mills, which were introduced from what is now Belgium to England in 1590. These passed flat bars between rolls to form a plate of iron, which was then passed between grooved rolls (slitters) to produce rods of iron. The first experiments at rolling iron for tinplate took place about 1670. In 1697, Major John Hanbury erected a mill at Pontypool to roll 'Pontypool plates'—blackplate. Later this began to be rerolled and tinned to make tinplate. The earlier production of plate iron in Europe had been in forges, not rolling mills.

The slitting mill was adapted to producing hoops (for barrels) and iron with a half-round or other sections by means that were the subject of two patents of c. 1679.,

Some of the earliest literature on rolling mills can be traced back to Christopher Polhem in 1761 in *Patriotista Testamente*, where he mentions rolling mills for both plate and bar iron. He also explains how rolling mills can save on time and labor because a rolling mill can produce 10 to 20 or more bars at the same time.

A patent was granted to Thomas Blockley of England in 1759 for the polishing and rolling of metals. Another patent was granted in 1766 to Richard Ford of England for the first tandem mill. A tandem mill is one in which the metal is rolled in successive stands; Ford's tandem mill was for hot rolling of wire rods.

Other Metals

Rolling mills for lead seem to have existed by the late 17th century. Copper and brass were also rolled by the late 18th century.

Modern Rolling

Properzi roller, Museo Nazionale della Scienza e della Tecnologia "Leonardo da Vinci", Milan

Modern rolling practice can be attributed to the pioneering efforts of Henry Cort of Funtley Iron Mills, near Fareham, England. In 1783 a patent was issued to Henry Cort for his use of grooved rolls for rolling iron bars. With this new design mills were able to produce 15 times more output per day than with a hammer. Although Cort was not the first to use grooved rolls, he was first to combine the use of many of the best features of various ironmaking and shaping processes known at the time. Thus modern writers have called him "father of modern rolling."

The first rail rolling mill was established by John Birkenshaw in 1820, where he produced fish bellied wrought iron rails in lengths of 15 to 18 feet. With the advancement of technology in rolling mills the size of rolling mills grew rapidly along with the size products being rolled. Example of this was at The Great Exhibition in 1851 a plate 20 feet long, 3 ½ feet wide, and 7/16 of inch thick, weighed 1,125 pounds was exhibited by the Consett Iron Company. Further evolution of the rolling mill came with the introduction of Three-high mills in 1853 used for rolling heavy sections.

Hot and Cold Rolling

Hot Rolling

A coil of hot-rolled steel

Hot rolling is a metalworking process that occurs above the recrystallization temperature of the material. After the grains deform during processing, they recrystallize, which maintains an equi-axed microstructure and prevents the metal from work hardening. The starting material is usually large pieces of metal, like semi-finished casting products, such as slabs, blooms, and billets. If these products came from a continuous casting operation the products are usually fed directly into the rolling mills at the proper temperature. In smaller operations the material starts at room temperature and must be heated. This is done in a gas- or oil-fired soaking pit for larger workpieces and for smaller workpieces induction heating is used. As the material is worked the temperature must be monitored to make sure it remains above the recrystallization temperature. To maintain a safety factor a *finishing temperature* is defined above the recrystallization temperature; this is usually 50 to 100 °C (90 to 180 °F) above the recrystallization temperature. If the temperature does drop below this temperature the material must be re-heated before more hot rolling.

Hot rolled metals generally have little directionality in their mechanical properties and deformation induced residual stresses. However, in certain instances non-metallic inclusions will impart some directionality and workpieces less than 20 mm (0.79 in) thick often have some directional properties. Also, non-uniform cooling will induce a lot of residual stresses, which usually occurs in shapes that have a non-uniform cross-section, such as I-beams. While the finished product is of good quality, the surface is covered in mill scale, which is an oxide that forms at high temperatures. It is usually removed via pickling or the smooth clean surface process, which reveals a smooth surface. Dimensional tolerances are usually 2 to 5% of the overall dimension.

Hot rolled mild steel seems to have a wider tolerance for amount of included carbon than does cold rolled steel, and is therefore more difficult for a blacksmith to use. Also for similar metals, hot rolled products seem to be less costly than cold-rolled ones.

Hot rolling is used mainly to produce sheet metal or simple cross sections, such as rail tracks. Other typical uses for hot rolled metal includes truck frames, automotive wheels, pipe and tubular, water heaters, agriculture equipment, strappings, stampings, compressor shells, railcar components, wheel rims, metal buildings, railroad hopper cars, doors, shelving, discs, guard rails, automotive clutch plates.

Shape Rolling Design

Rolling mills are often divided into roughing, intermediate and finishing rolling cages. During shape rolling, an initial billet (round or square) with edge of diameter typically ranging between 100–140 mm is continuously deformed to produce a certain finished product with smaller cross section dimension and geometry. Different sequences can be adopted to produce a certain final product starting from a given billet. However,since each rolling mill is significantly expensive (up to 2 million of euros), a typical requirement is to contract the number or rolling passes. Different approaches have been achieved including, empirical knowledge, employment of numerical models and Artificial Intelligence techniques. Lambiase et al. validated a finite element model (FE) for predicting the final shape of a rolled bar in round-flat pass. one of the major concern when designing rolling mills is to reduce the number of passes; a possible solution to such requirement is represented by the slit pass also called split pass which divided an incoming bar in two or more subpart thus virtually increasing the cross section reduction ratio per pass as reported by Lambiase. Another solution for reducing the number of passes in the rolling mills is the employment of automated systems for Roll Pass Design as that proposed by Lambiase and Langella. subsequently, Lambiase further developed an Automated System based on Artificial Intelligence and particularly an integrated system including an inferential engine based on Genetic Algorithms a knowledge database based on an Artificial Neural Network trained by a parametric Finite element model and to optimize and automatically design rolling mills.

Cold Rolling

Cold rolling occurs with the metal below its recrystallization temperature (usually at room temperature), which increases the strength via strain hardening up to 20%. It also improves the surface finish and holds tighter tolerances. Commonly cold-rolled products include sheets, strips, bars, and rods; these products are usually smaller than the same products that are hot rolled. Because of the smaller size of the workpieces and their greater strength, as compared to hot rolled stock, four-high or cluster mills are used. Cold rolling cannot reduce the thickness of a workpiece as much as hot rolling in a single pass.

Cold-rolled sheets and strips come in various conditions: *full-hard*, *half-hard*, *quarter-hard*, and *skin-rolled*. Full-hard rolling reduces the thickness by 50%, while the others involve less of a reduction. Skin-rolling, also known as a *skin-pass*, involves the least amount of reduction: 0.5-1%. It is used to produce a smooth surface, a uniform thickness, and reduce the yield point phenomenon (by preventing Lüders bands from forming in later processing). It locks dislocations at the surface and thereby reduces the possibility of formation of Lüders bands. To avoid the formation of Lüders bands it is necessary to create substantial density of unpinned dislocations in ferrite matrix. It is also used to break up the spangles in galvanized steel. Skin-rolled stock is usually used in subsequent cold-working processes where good ductility is required.

Other shapes can be cold-rolled if the cross-section is relatively uniform and the transverse dimension is relatively small. Cold rolling shapes requires a series of shaping operations, usually along the lines of sizing, breakdown, roughing, semi-roughing, semi-finishing, and finishing.

If processed by a blacksmith, the smoother, more consistent, and lower levels of carbon encapsulated in the steel makes it easier to process, but at the cost of being more expensive.

Typical uses for cold-rolled steel include metal furniture, desks, filing cabinets, tables, chairs, motorcycle exhaust pipes, computer cabinets and hardware, home appliances and components, shelving, lighting fixtures, hinges, tubing, steel drums, lawn mowers, electronic cabinetry, water heaters, metal containers, and a variety of construction-related products.

Processes

Roll Bending

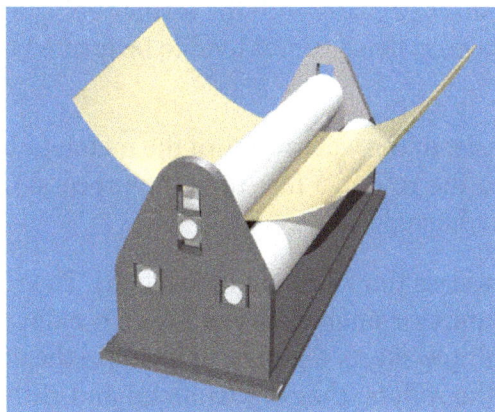

Roll bending

Roll bending produces a cylindrical shaped product from plate or steel metals .

Roll Forming

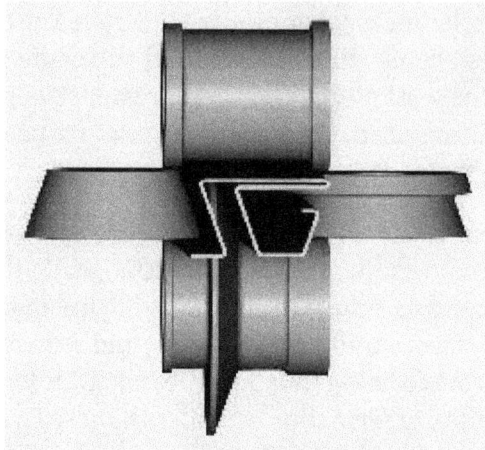

Roll forming

Roll forming, roll bending or plate rolling is a continuous bending operation in which a long strip of metal (typically coiled steel) is passed through consecutive sets of rolls, or stands, each performing only an incremental part of the bend, until the desired cross-section profile is obtained. Roll forming is ideal for producing parts with long lengths or in large quantities. There are 3 main processes: 4 rollers, 3 rollers and 2 rollers, each of which has as different advantages according to the desired specifications of the output plate.

Flat Rolling

Flat rolling is the most basic form of rolling with the starting and ending material having a rectangular cross-section. The material is fed in between two *rollers*, called *working rolls*, that rotate in opposite directions. The gap between the two rolls is less than the thickness of the starting material, which causes it to deform. The decrease in material thickness causes the material to elongate. The friction at the interface between the material and the rolls causes the material to be pushed through. The amount of deformation possible in a single pass is limited by the friction between the rolls; if the change in thickness is too great the rolls just slip over the material and do not draw it in. The final product is either sheet or plate, with the former being less than 6 mm (0.24 in) thick and the latter greater than; however, heavy plates tend to be formed using a press, which is termed *forming*, rather than rolling.

Often the rolls are heated to assist in the workability of the metal. Lubrication is often used to keep the workpiece from sticking to the rolls. To fine-tune the process, the speed of the rolls and the temperature of the rollers are adjusted.

h is sheet metal with a thickness less than 200 μm (0.0079 in). The rolling is done in a *cluster mill* because the small thickness requires a small diameter rolls. To reduce the need for small rolls *pack rolling* is used, which rolls multiple sheets together to increase the effective starting thickness. As the foil sheets come through the rollers, they are trimmed and slitted with circular or razor-like knives. Trimming refers to the edges of the foil, while slitting involves cutting it into several sheets.

Aluminum foil is the most commonly produced product via pack rolling. This is evident from the two different surface finishes; the shiny side is on the roll side and the dull side is against the other sheet of foil.

Ring Rolling

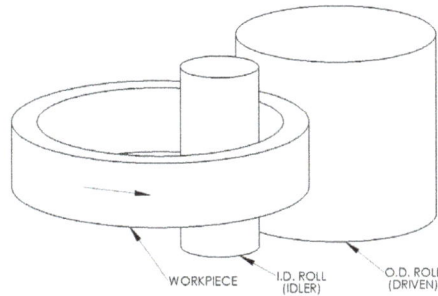

A schematic of ring rolling

Ring rolling is a specialized type of hot rolling that increases the diameter of a ring. The starting material is a thick-walled ring. This workpiece is placed between two rolls, an inner *idler roll* and a *driven roll*, which presses the ring from the outside. As the rolling occurs the wall thickness decreases as the diameter increases. The rolls may be shaped to form various cross-sectional shapes. The resulting grain structure is circumferential, which gives better mechanical properties. Diameters can be as large as 8 m (26 ft) and face heights as tall as 2 m (79 in). Common applications include bearings, gears, rockets, turbines, airplanes, pipes, and pressure vessels.

Structural Shape Rolling

Cross-sections of continuously rolled structural shapes, showing the change induced by each rolling mill.

Controlled Rolling

Controlled rolling is a type of thermomechanical processing which integrates controlled deformation and heat treating. The heat which brings the workpiece above the recrystallization temperature is also used to perform the heat treatments so that any subsequent heat treating is unnecessary. Types of heat treatments include the production of a fine grain structure; controlling the nature, size, and distribution of various transformation products (such as ferrite, austenite, pearlite, bainite, and martensite in steel); inducing precipitation hardening; and, controlling the toughness. In order to achieve this the entire process must be closely monitored and controlled. Common variables in controlled rolling include the starting material composition and structure, deformation levels, temperatures at various stages, and cool-down conditions. The benefits of controlled rolling include better mechanical properties and energy savings.

Forge Rolling

Forge rolling is a longitudinal rolling process to reduce the cross-sectional area of heated bars or billets by leading them between two contrary rotating roll segments. The process is mainly used to provide optimized material distribution for subsequent die forging processes. Owing to this a better material utilization, lower process forces and better surface quality of parts can be achieved in die forging processes.

Basically any forgeable metal can also be forge-rolled. Forge rolling is mainly used to preform long-scaled billets through targeted mass distribution for parts such as crankshafts, connection rods, steering knuckles and vehicle axles. Narrowest manufacturing tolerances can only partially be achieved by forge rolling. This is the main reason why forge rolling is rarely used for finishing, but mainly for preforming.

Characteristics of forge rolling:

- high productivity and high material utilization

- good surface quality of forge-rolled workpieces

- extended tool life-time

- small tools and low tool costs

- improved mechanical properties due to optimized grain flow compared to exclusively die forged workpieces

Mills

A *rolling mill*, also known as a *reduction mill* or *mill*, has a common construction independent of the specific type of rolling being performed:

Rolling mills

Rolling mill for cold rolling metal sheet like this piece of brass sheet

- Work rolls

- Backup rolls - are intended to provide rigid support required by the working rolls to prevent bending under the rolling load

- Rolling balance system - to ensure that the upper work and back up rolls are maintained in proper position relative to lower rolls

- Roll changing devices - use of an overhead crane and a unit designed to attach to the neck of the roll to be removed from or inserted into the mill.

- Mill protection devices - to ensure that forces applied to the backup roll chocks are not of such a magnitude to fracture the roll necks or damage the mill housing

- Roll cooling and lubrication systems

- Pinions - gears to divide power between the two spindles, rotating them at the same speed but in different directions

- Gearing - to establish desired rolling speed

- Drive motors - rolling narrow foil product to thousands of horsepower

- Electrical controls - constant and variable voltages applied to the motors

- Coilers and uncoilers - to unroll and roll up coils of metal

Slabs are the feed material for hot strip mills or plate mills and blooms are rolled to billets in a billet mill or large sections in a structural mill. The output from a strip mill is coiled and, subsequently, used as the feed for a cold rolling mill or used directly by fabricators. Billets, for re-rolling, are subsequently rolled in either a merchant, bar or rod mill. Merchant or bar mills produce a variety of shaped products such as angles, channels, beams, rounds (long or coiled) and hexagons.

Configurations

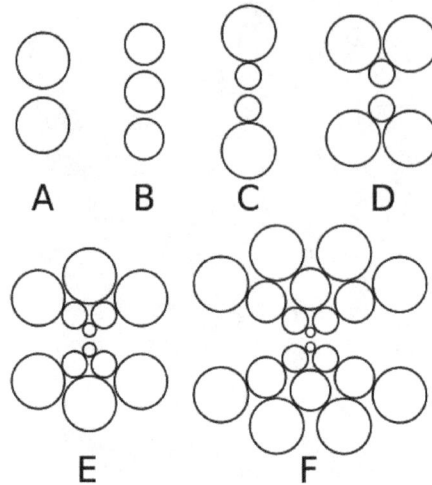

Various rolling configurations. Key: A. 2-high B. 3-high C. 4-high D. 6-high E. 12-high cluster & F. 20-high Sendzimir Mill cluster

Mills are designed in different types of configurations, with the most basic being a *two-high non-reversing*, which means there are two rolls that only turn in one direction. The *two-high reversing* mill has rolls that can rotate in both directions, but the disadvantage is that the rolls must be stopped, reversed, and then brought back up to rolling speed between each pass. To resolve this, the *three-high* mill was invented, which uses three rolls that rotate in one direction; the metal is fed through two of the rolls and then returned through the other pair. The disadvantage to this system is the workpiece must be lifted and lowered using an elevator. All of these mills are usually used for primary rolling and the roll diameters range from 60 to 140 cm (24 to 55 in).

To minimize the roll diameter a *four-high* or *cluster* mill is used. A small roll diameter is advantageous because less roll is in contact with the material, which results in a lower force and power requirement. The problem with a small roll is a reduction of stiffness, which is overcome using *backup rolls*. These backup rolls are larger and contact the back side of the smaller rolls. A four-high mill has four rolls, two small and two large. A cluster mill has more than 4 rolls, usually in three tiers. These types of mills are commonly used to hot roll wide plates, most cold rolling applications, and to roll foils.

Historically mills were classified by the product produced:

- Blooming, cogging and slabbing mills, being the preparatory mills to rolling finished rails, shapes or plates, respectively. If reversing, they are from 34 to 48 inches in diameter, and if three-high, from 28 to 42 inches in diameter.

- Billet mills, three-high, rolls from 24 to 32 inches in diameter, used for the further reduction of blooms down to 1.5x1.5-inch billets, being the preparatory mills for the bar and rod

- Beam mills, three-high, rolls from 28 to 36 inches in diameter, for the production of heavy beams and channels 12 inches and over.

- Rail mills with rolls from 26 to 40 inches in diameter.

- Shape mills with rolls from 20 to 26 inches in diameter, for smaller sizes of beams and channels and other structural shapes.

- Merchant bar mills with rolls from 16 to 20 inches in diameter.

- Small merchant bar mills with finishing rolls from 8 to 16 inches in diameter, generally arranged with a larger size roughing stand.

- Rod and wire mills with finishing rolls from 8 to 12 inches in diameter, always arranged with larger size roughing stands.

- Hoop and cotton tie mills, similar to small merchant bar mills.

- Armour plate mills with rolls from 44 to 50 inches in diameter and 140 to 180-inch body.

- Plate mills with rolls from 28 to 44 inches in diameter.

- Sheet mills with rolls from 20 to 32 inches in diameter.

- Universal mills for the production of square-edged or so-called universal plates and various wide flanged shapes by a system of vertical and horizontal rolls.

Tandem Mill

A tandem mill is a special type of modern rolling mill where rolling is done in one pass. In a traditional rolling mill rolling is done in several passes, but in tandem mill there are several *stands* (>=2 stands) and reductions take place successively. The number of stands ranges from 2 to 18. Tandem mills can be either of hot or cold rolling mill types.

Defects

In hot rolling, if the temperature of the workpiece is not uniform the flow of the material will occur more in the warmer parts and less in the cooler. If the temperature difference is great enough cracking and tearing can occur.

Flatness and Shape

In a flat metal workpiece, the flatness is a descriptive attribute characterizing the extent of the geometric deviation from a reference plane. The deviation from complete flatness is the direct result of the workpiece relaxation after hot or cold rolling, due to the internal stress pattern caused by the non-uniform transversal compressive action of the rolls and the uneven geometrical properties of the entry material. The transverse distribution of differential strain/ elongation-induced stress with respect to the material's average applied stress is commonly referenced to as shape. Due to the strict relationship between shape and flatness, these terms can be used in an interchangeable manner. In the case of metal strips and sheets, the flatness reflects the differential fiber elongation across the width of the workpiece. This property must be subject to an accurate feedback-based control in order to guarantee the machinability of the metal sheets in the final transformation processes. Some technological details about the feedback control of flatness are given in.

Profile

Profile is made up of the measurements of crown and wedge. Crown is the thickness in the center as compared to the average thickness at the edges of the workpiece. Wedge is a measure of the thickness at one edge as opposed to the other edge. Both may be expressed as absolute measurements or as relative measurements. For instance, one could have 2 mil of crown (the center of the workpiece is 2 mil thicker than the edges), or one could have 2% crown (the center of the workpiece is 2% thicker than the edges).

It is typically desirable to have some crown in the workpiece as this will cause the workpiece to tend to pull to the center of the mill, and thus will run with higher stability.

Flatness

Maintaining a uniform gap between the rolls is difficult because the rolls deflect under the load required to deform the workpiece. The deflection causes the workpiece to be thinner on the edges and thicker in the middle. This can be overcome by using a crowned roller (parabolic crown), however the crowned roller will only compensate for one set of conditions, specifically the material, temperature, and amount of deformation.

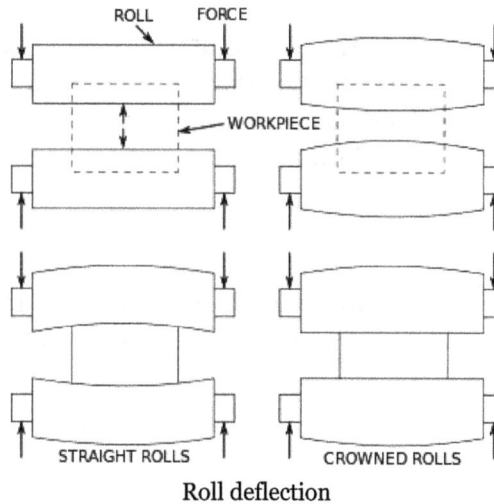

Roll deflection

Other methods of compensating for roll deformation include continual varying crown (CVC), pair cross rolling, and work roll bending. CVC was developed by SMS-Siemag AG and involves grinding a third order polynomial curve into the work rolls and then shifting the work rolls laterally, equally, and opposite to each other. The effect is that the rolls will have a gap between them that is parabolic in shape, and will vary with lateral shift, thus allowing for control of the crown of the rolls dynamically. Pair cross rolling involves using either flat or parabolically crowned rolls, but shifting the ends at an angle so that the gap between the edges of the rolls will increase or decrease, thus allowing for dynamic crown control. Work roll bending involves using hydraulic cylinders at the ends of the rolls to counteract roll deflection.

Another way to overcome deflection issues is by decreasing the load on the rolls, which can be done by applying a longitudinal force; this is essentially drawing. Other method of decreasing roll deflection include increasing the elastic modulus of the roll material and adding back-up supports to the rolls.

The different classifications for flatness defects are:

- Symmetrical edge wave - the edges on both sides of the workpiece are "wavy" due to the material at the edges being longer than the material in the center.

- Asymmetrical edge wave - one edge is "wavy" due to the material at one side being longer than the other side.

- Center buckle - The center of the strip is "wavy" due to the strip in the center being longer than the strip at the edges.

- Quarter buckle - This is a rare defect where the fibers are elongated in the quarter regions (the portion of the strip between the center and the edge). This is normally attributed to using excessive roll bending force since the bending force may not compensate for the roll deflection across the entire length of the roll.

It is important to note that one could have a flatness defect even with the workpiece having the same thickness across the width. Also, one could have fairly high crown or wedge, but still produce material that is flat. In order to produce flat material, the material must be reduced by the same percentage across the width. This is important because mass flow of the material must be preserved, and the more a material is reduced, the more it is elongated. If a material is elongated in the same manner across the width, then the flatness coming into the mill will be preserved at the exit of the mill.

Surface Defect Remediation

Many surface defects can be scarfed off the surface of semi-finished rolled products before further rolling. Methods of scarfing have included hand-chipping with chisels (18th and 19th centuries); powered chipping and grinding with air chisels and grinders; burning with an oxy-fuel torch, whose gas pressure blows away the metal or slag melted by the flame; and laser scarfing.

Cladding (metalworking)

Cladding is the bonding together of dissimilar metals. It is different from fusion welding or gluing as a method to fasten the metals together. Cladding is often achieved by extruding two metals through a die as well as pressing or rolling sheets together under high pressure.

The United States Mint uses cladding to manufacture coins from different metals. This allows a cheaper metal to be used as a filler.

Roll Bonding

In roll bonding, two or more layers of different metals are thoroughly cleaned and passed through a pair of rollers under sufficient pressure to bond the layers. The pressure is high enough to deform the metals and reduce the combined thickness of the clad material. Heat may be applied, especially when metal are not ductile enough. As an example of application, bonding of the sheets can be controlled by painting a pattern on one sheet; only the bare metal surfaces bond, and the un-bonded portion can be inflated if the sheet is heated and the coating vaporizes. This is used to make heat exchangers for refrigeration equipment.

Explosive Welding

In explosive welding, the pressure to bond the two layers is provided by detonation of a sheet of chemical explosive. No heat-affected zone is produced in the bond between metals. The explosion propagates across the sheet, which tends to expel impurities and oxides from between the sheets. Pieces up to 4 x 16 metres can be manufactured. The process is useful for cladding metal sheets with a corrosion-resistant layer.

Laser Cladding

Laser cladding is a method of depositing material by which a powdered or wire feedstock material is melted and consolidated by use of a laser in order to coat part of a substrate or fabricate a near-net shape part (additive manufacturing technology) .

It is often used to improve mechanical properties or increase corrosion resistance, repair worn out parts, and fabricate metal matrix composites.

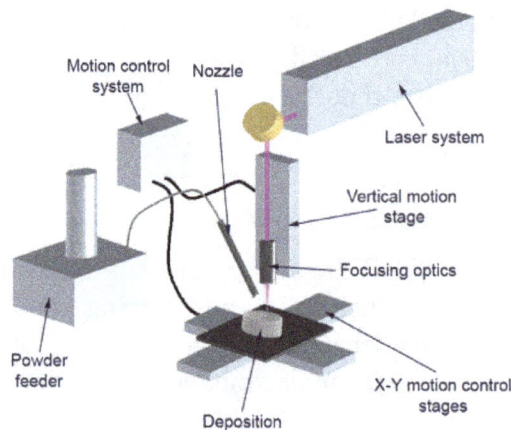

A schematic of the equipment

Process

The powder used in laser cladding is normally of a metallic nature, and is injected into the system by either coaxial or lateral nozzles. The interaction of the metallic powder stream and the laser causes melting to occur, and is known as the melt pool. This is deposited onto a substrate; moving the substrate allows the melt pool to solidify and thus produces a track of solid metal. This is the most common technique, however some processes involve moving the laser/nozzle assembly over a stationary substrate to produce solidified tracks. The motion of the substrate is guided by a CAD system which interpolates solid objects into a set of tracks, thus producing the desired part at the end of the trajectory.

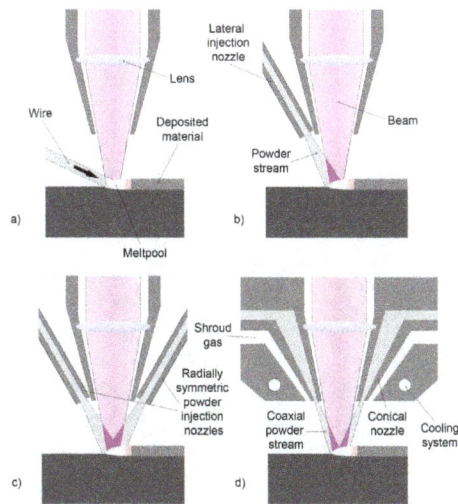

The different feeding systems available

A great deal of research is now being concentrated on developing automatic laser cladding machines. Many of the process parameters must be manually set, such as laser power, laser focal point, substrate velocity, powder injection rate, etc., and thus require the attention of a specialized technician to ensure proper results. However, many groups are focusing their attention on developing sensors to measure the process online. Such sensors monitor the clad's geometry (height and width of deposited track), metallurgical properties (such as the rate of solidification, and hence the final microstructure), and temperature information of both the immediate melt pool and its surrounding areas. With such sensors, control strategies are being designed such that constant observation from a technician is no longer required to produce a final product. Further research has been directed to forward processing where system parameters are developed around specific metallurgical properties for user defined applications (such as microstructure, internal stresses, dilution zone gradients, and clad contact angle).

Advantages

- Best technique for coating any shape => increase life-time of wearing parts.

- Particular dispositions for repairing parts (ideal if the mould of the part no longer exist or too long time needed for a new fabrication).

- Most suited technique for graded material application.

- Well adapted for near-net-shape manufacturing.

- Low dilution between track and substrate (unlike other welding processes and strong metallurgical bond.

- Low deformation of the substrate and small heat affected zone (HAZ).

- High cooling rate => fine microstructure.

- A lot of material flexibility (metal, ceramic, even polymer).

- Built part is free of crack and porosity.

- Compact technology.

Metal Fabrication

A set of six-axis welding robots

Metal fabrication is the building of metal structures by cutting, bending, and assembling processes. It is a value added process that involves the construction of machines and structures from various raw materials. A fab shop will bid on a job, usually based on the engineering drawings, and if awarded the contract will build the product. Large fab shops will employ a multitude of value added processes in one plant or facility including welding, cutting, forming and machining. These large fab shops offer additional value to their customers by limiting the need for purchasing personnel to locate multiple vendors for different services. Metal fabrication jobs usually start with shop drawings including precise measurements then move to the fabrication stage and finally to the installation of the final project. Fabrication shops are employed by contractors, OEMs and VARs. Typical projects include loose parts, structural frames for buildings and heavy equipment, and stairs and hand railings for buildings.

Processes

- *Cutting* is done by sawing, shearing, or chiseling (all with manual and powered variants); torching with hand-held torches (such as oxy-fuel torches or plasma torches); and via numerical control (CNC) cutters (using a laser, mill bits, torch, or water jet).

- *Bending* is done by hammering (manual or powered) or via press brakes and similar tools. Modern metal fabricators use press brakes to either coin or air-bend metal sheet into form. CNC-controlled backgauges use hard stops to position cut parts in order to place bend lines in the correct position. Off-line programing software now makes programing the CNC-controlled press brakes seamless and very efficient.

- *Assembling* (joining of the pieces) is done by welding, binding with adhesives, riveting, threaded fasteners, or even yet more bending in the form of a crimped seam. Structural steel and sheet metal are the usual starting materials for fabrication, along with the welding wire, flux, and fasteners that will join the cut pieces. As with other manufacturing processes, both human labor and automation are commonly used. The product resulting from fabrication may be called a fabrication. Shops that specialize in this type of metal work are called *fab shops*. The end products of other common types of metalworking, such as machining, metal stamping, forging, and casting, may be similar in shape and function, but those processes are not classified as fabrication.

Overlap

Fabrication comprises or overlaps with various metalworking specialties:

- Fabrication shops and machine shops have overlapping capabilities, but fabrication shops generally concentrate on metal preparation and assembly as described above. By comparison, machine shops also cut metal, but they are more concerned with the machining of parts on machine tools. Firms that encompass both fab work and machining are also common.

- Blacksmithing has always involved fabrication, although it was not always called by that name.

- The products produced by welders, which are often referred to as weldments, are an example of fabrication.

- Boilermakers originally specialized in boilers, leading to their trade's name, but the term as used today has a broader meaning.

- Similarly, millwrights originally specialized in setting up grain mills and saw mills, but today they may be called upon for a broad range of fabrication work.

- Ironworkers, also known as steel erectors, also engage in fabrication. Often the fabrications for structural work begin as prefabricated segments in a fab shop, then are moved to the site by truck, rail, or barge, and finally are installed by erectors.

Raw Materials

Standard raw materials used by metal fabricators are;

- plate metal
- formed and expanded metal
    - tube stock,
- welding wire/welding rod
- casting

Cutting and Burning

The raw material has to be cut to size. This is done with a variety of tools.

The most common way to cut material is by Shearing (metalworking);

Special band saws designed for cutting metal have hardened blades and a feed mechanism for even cutting. Abrasive cut-off saws, also known as chop saws, are similar to miter saws but with a steel cutting abrasive disk. Cutting torches can cut very large sections of steel *with little effort.*

Burn tables are CNC cutting torches, usually natural gas powered. Plasma and laser cutting tables, and Water jet cutters, are also common. Plate steel is loaded on a table and the parts are cut out as programmed. The support table is made of a grid of bars that can be replaced. Some very expensive burn tables also include CNC punch capability, with a carousel of different punches and taps. Fabrication of structural steel by plasma and laser cutting introduces robots to move the cutting head in three dimensions around the material to be cut.

Forming

Forming is a process of material deformation. Forming is typically applied to metals. To define the process, a raw material piece is formed by applying force to an object. The force must be great enough to change the shape of the object from its initial shape. The process of forming can be controlled with the use of tools such as punches or dies. Machinery can also be used to regulate force magnitude and direction. An example of machine based forming can also combine forming and welding to produce lengths of fabricated sheeting, most commonly seen in the form of linear grating (used principally for water drainage).

Proper design and use of tools with machinery creates a repeatable form which can be used to create products for many industries, including jewelry, aerospace, automotive, construction, civil and architectural, etc.

Machining

Machining is the process of removing unwanted material from the block of metal to get the desire shape. Machining is a trade, in and of itself, although Fab shops will generally entail a limited machining capability including; metal lathes, mills, magnetic based drills, along with other portable metal working tools.

Welding

Welding is the main focus of steel fabrication. The formed and machined parts will be assembled and tack welded into place then re-checked for accuracy. A fixture may be used to locate parts for welding if multiple weldments have been ordered.

The welder then completes welding as per the engineering drawings if welding is detailed, or as per his/her own judgement if no welding details are provided.

Special precautions may be needed to prevent warping of the weldment due to heat. These may include re-designing the weldment to use less weld, welding in a staggered fashion, using a stout fixture, covering the weldment in sand during cooling, and straightening operations after welding.

Straightening of warped steel weldments is done with an Oxy-acetylene torch and is somewhat of an art. Heat is selectively applied to the steel in a slow, linear sweep. The steel will have a net contraction, upon cooling, in the direction of the sweep. A highly skilled welder can remove significant warpage using this technique.

Steel weldments are occasionally annealed in a low temperature oven to relieve residual stresses. Such weldments, particularly those employed for engine blocks, may be line-bored after heat treatment.

Final Assembly

After the weldment has cooled it is generally sand blasted, primed and painted. Any additional manufacturing specified by the customer is then completed. The finished product is then inspected and shipped.

Specialities

Many fabrication shops have speciality processes which they develop or invest in, based on their customers needs and their expertise:

- casting

- chipping

- powder coating

- powder metallurgy

- welding

Work Hardening

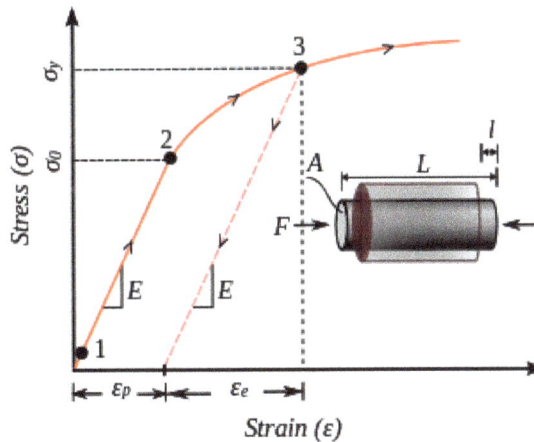

A phenomenological uniaxial stress-strain curve showing typical work hardening plastic behavior of materials in uniaxial compression. For work hardening materials the yield stress increases with increasing plastic deformation.

Work hardening, also known as strain hardening or cold working, is the strengthening of a metal by plastic deformation. This strengthening occurs because of dislocation movements and dislocation generation within the crystal structure of the material. Many non-brittle metals with a reasonably high melting point as well as several polymers can be strengthened in this fashion. Alloys not amenable to heat treatment, including low-carbon steel, are often work-hardened. Some materials cannot be work-hardened at low temperatures, such as indium, however others can only be strengthened via work hardening, such as pure copper and aluminum.

Work hardening may be desirable or undesirable depending on the context. An example of undesirable work hardening is during machining when early passes of a cutter inadvertently work-harden the workpiece surface, causing damage to the cutter during the later passes. Certain alloys are more prone to this than others; superalloys such as Inconel require machining strategies that take it into account. An example of desirable work hardening is that which occurs in metalworking processes that intentionally induce plastic deformation to exact a shape change. These processes are known as cold working or cold forming processes. They are characterized by shaping the workpiece at a temperature below its recrystallization temperature, usually at ambient temperature. Cold forming techniques are usually classified into four major groups: squeezing, bending, drawing, and shearing. Applications include the heading of bolts and cap screws and the finishing of cold rolled steel. In cold forming, metal is formed at high speed and high pressure using tool steel or carbide dies. The cold working of the metal increasing the hardness, yield strength, and tensile strength.

History

Copper was the first metal in common use for tools and containers since it is one of the few metals available in non-oxidized form, not requiring the smelting of an ore. Copper is easily softened by heating and then cooling (it does not harden by quenching, as in cool water). In this annealed state it may then be hammered, stretched and otherwise formed, progressing toward the desired final shape, but becoming harder and less ductile as work progresses. If work continues beyond a certain hardness the metal will tend to fracture when worked and so it may be re-annealed periodically as the shape progresses. Annealing is stopped when the workpiece is near its final desired shape, and so the final product will have a desired stiffness and hardness. The technique of repoussé exploits these properties of copper, enabling the construction of durable jewelry articles and sculptures (including the Statue of Liberty).

For metal objects designed to flex, such as springs, specialized alloys are usually employed in order to avoid work hardening (a result of plastic deformation) and metal fatigue, with specific heat treatments required to obtain the necessary characteristics.

Devices made from aluminum and its alloys, such as aircraft, must be carefully designed to minimize or evenly distribute flexure, which can lead to work hardening and in turn stress cracking, possibly causing catastrophic failure. For this reason modern aluminum aircraft will have an imposed working lifetime (dependent upon the type of loads encountered), after which the aircraft must be retired.

Theory

Before work hardening, the lattice of the material exhibits a regular, nearly defect-free pattern (almost no dislocations). The defect-free lattice can be created or restored at any time by annealing. As the material is work hardened it becomes increasingly saturated with new dislocations, and more dislocations are prevented from nucleating (a resistance to dislocation-formation develops). This resistance to dislocation-formation manifests itself as a resistance to plastic deformation; hence, the observed strengthening.

In metallic crystals, irreversible deformation is usually carried out on a microscopic scale by defects called dislocations, which are created by fluctuations in local stress fields within the material culminating in a lattice rearrangement as the dislocations propagate through the lattice. At normal temperatures the dislocations are not annihilated by annealing. Instead, the dislocations accumulate, interact with one another, and serve as pinning points or obstacles that significantly impede their motion. This leads to an increase in the yield strength of the material and a subsequent decrease in ductility.

Such deformation increases the concentration of dislocations which may subsequently form low-angle grain boundaries surrounding sub-grains. Cold working generally results in a higher yield strength as a result of the increased number of dislocations and the Hall-Petch effect of the sub-grains, and a decrease in ductility. The effects of cold working may be reversed by annealing the material at high temperatures where recovery and recrystallization reduce the dislocation density.

A material's work hardenability can be predicted by analyzing a stress-strain curve, or studied in context by performing hardness tests before and after a process.

Elastic and Plastic Deformation

Work hardening is a consequence of plastic deformation, a permanent change in shape. This is distinct from elastic deformation, which is reversible. Most materials do not exhibit only one or the other, but rather a combination of the two. The following discussion mostly applies to metals, especially steels, which are well studied. Work hardening occurs most notably for ductile materials such as metals. Ductility is the ability of a material to undergo plastic deformations before fracture (for example, bending a steel rod until it finally breaks).

The tensile test is widely used to study deformation mechanisms. This is because under compression, most materials will experience trivial (lattice mismatch) and non-trivial (buckling) events before plastic deformation or fracture occur. Hence the intermediate processes that occur to the material under uniaxial compression before the incidence of plastic deformation make the compressive test fraught with difficulties.

A material generally deforms elastically under the influence of small forces; the material returns quickly to its original shape when the deforming force is removed. This phenomenon is called *elastic deformation*. This behavior in materials is described by Hooke's Law. Materials behave elastically until the deforming force increases beyond the elastic limit, which is also known as the yield stress. At that point, the material is permanently deformed and fails to return to its original shape when the force is removed. This phenomenon is called *plastic deformation*. For example, if one stretches a coil spring up to a certain point, it will return to its original shape, but once it is stretched beyond the elastic limit, it will remain deformed and won't return to its original state.

Elastic deformation stretches the bonds between atoms away from their equilibrium radius of separation, without applying enough energy to break the inter-atomic bonds. Plastic deformation, on the other hand, breaks inter-atomic bonds, and therefore involves the rearrangement of atoms in a solid material.

Dislocations and Lattice Strain Fields

In materials science parlance, dislocations are defined as line defects in a material's crystal structure. The bonds surrounding the dislocation are already elastically strained by the defect compared to the bonds between the constituents of the regular crystal lattice. Therefore, these bonds break at relatively lower stresses, leading to plastic deformation.

The strained bonds around a dislocation are characterized by lattice strain fields. For example, there are compressively strained bonds directly next to an edge dislocation and tensilely strained bonds beyond the end of an edge dislocation. These form compressive strain fields and tensile strain fields, respectively. Strain fields are analogous to electric fields in certain ways. Specifically, the strain fields of dislocations obey similar laws of attraction and repulsion; in order to reduce overall strain, compressive strains are attracted to tensile strains, and vice versa.

The visible (macroscopic) results of plastic deformation are the result of microscopic dislocation motion. For example, the stretching of a steel rod in a tensile tester is accommodated through dislocation motion on the atomic scale.

Increase of Dislocations and Work Hardening

Increase in the number of dislocations is a quantification of work hardening. Plastic deformation occurs as a consequence of work being done on a material; energy is added to the material. In addition, the energy is almost always applied fast enough and in large enough magnitude to not only move existing dislocations, but also to *produce* a great number of new dislocations by jarring or working the material sufficiently enough. New dislocations are generated in proximity to a Frank-Read source.

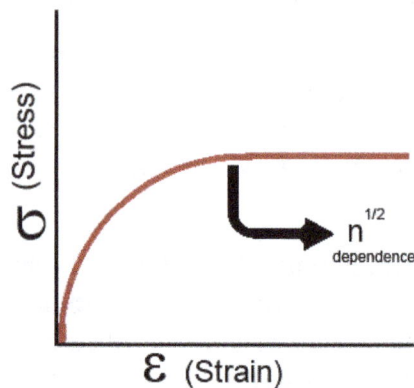

Figure 1: The yield stress of an ordered material has a half-root dependency on the number of dislocations present.

Yield strength is increased in a cold-worked material. Using lattice strain fields, it can be shown that an environment filled with dislocations will hinder the movement of any one dislocation. Because dislocation motion is hindered, plastic deformation cannot occur at normal stresses. Upon application of stresses just beyond the yield strength of the non-cold-worked material, a cold-worked material will continue to deform using the only mechanism available: elastic deformation, the regular scheme of stretching or compressing of electrical bonds (without dislocation motion) continues to occur, and the modulus of elasticity is unchanged. Eventually the stress is great enough to overcome the strain-field interactions and plastic deformation resumes.

However, ductility of a work-hardened material is decreased. Ductility is the extent to which a material can undergo plastic deformation, that is, it is how far a material can be plastically deformed before fracture. A cold-worked material is, in effect, a normal (brittle) material that has already been extended through part of its allowed plastic deformation. If dislocation motion and plastic deformation have been hindered enough by dislocation accumulation, and stretching of electronic bonds and elastic deformation have reached their limit, a third mode of deformation occurs: fracture.

Quantification of Work Hardening

The stress, τ of dislocation is dependent on the shear modulus, G, the magnitude of the Burgers vector, b, and the dislocation density, $\rho_\perp$:

$$\tau = \tau_0 + G\alpha b \rho_\perp^{1/2}$$

where τ_0 is the intrinsic strength of the material with low dislocation density and α is a correction factor specific to the material.

As shown in Figure and the equation above, work hardening has a half root dependency on the number of dislocations. The material exhibits high strength if there are either high levels of dislocations (greater than 10^{14} dislocations per m²) or no dislocations. A moderate number of dislocations (between 10^7 and 10^9 dislocations per m²) typically results in low strength.

Example

For an extreme example, in a tensile test a bar of steel is strained to just before the distance at which it usually fractures. The load is released smoothly and the material relieves some of its strain by decreasing in length. The decrease in length is called the elastic recovery, and the end result is a work-hardened steel bar. The fraction of length recovered (length recovered/original length) is equal to the yield-stress divided by the modulus of elasticity. (Here we discuss true stress in order to account for the drastic decrease in diameter in this tensile test.) The length recovered after removing a load from a material just before it breaks is equal to the length recovered after removing a load just before it enters plastic deformation.

The work-hardened steel bar has a large enough number of dislocations that the strain field interaction prevents all plastic deformation. Subsequent deformation requires a stress that varies linearly with the strain observed, the slope of the graph of stress vs. strain is the modulus of elasticity, as usual.

The work-hardened steel bar fractures when the applied stress exceeds the usual fracture stress and the strain exceeds usual fracture strain. This may be considered to be the elastic limit and the yield stress is now equal to the fracture toughness, which is of course, much higher than a non-work-hardened steel yield stress.

The amount of plastic deformation possible is zero, which is obviously less than the amount of plastic deformation possible for a non-work-hardened material. Thus, the ductility of the cold-worked bar is reduced.

Substantial and prolonged cavitation can also produce strain hardening.

Additionally, jewelers will construct structurally sound rings and other wearable objects (especially those worn on the hands) that require much more durability (than earrings for example) by utilizing a material's ability to be work hardened. While casting rings is done for a number of economical reasons (saving a great deal of time and cost of labor), a master jeweler may utilize the ability of a material to be work hardened and apply some combination of cold forming techniques during the production of a piece.

Processes

The following is a list of cold forming processes:

- Rolling
  - Swaging
  - Extrusion
  - Forging
  - Sizing
  - Riveting
  - Staking
  - Coining
  - Peening
  - Burnishing
  - Heading
  - Hubbing
  - Thread rolling
- Bending
  - Angle bending
  - Roll bending
  - Draw and compression
  - Roll forming
  - Seaming
  - Flanging
  - Straightening
- Shearing

- o   Slitting

- o   Blanking

- o   Piercing

- o   Lancing

- o   Perforating

- o   Notching

- o   Nibbling

- o   Shaving

- o   Trimming

- o   Cutoff

- o   Dinking

- Drawing

  - o   Tube drawing

  - o   Wire drawing

  - o   Spinning

  - o   Embossing

  - o   Stretch forming

  - o   Sheet metal drawing

  - o   Ironing

  - o   Superplastic forming

Techniques have been designed to maintain the general shape of the workpiece during work hardening, including shot peening and equal channel angular extrusion. The range of possible shapes is rather broad, including heads, threads, steps, knurls, chamfers, grooves, undercuts, and tapers.

Advantages and Disadvantages

Advantages:

- No heating required

- Better surface finish

- Superior dimensional control

- Better reproducibility and interchangeability

- Directional properties can be imparted into the metal

- Contamination problems are minimized

The increase in strength due to strain hardening is comparable to that of heat treating. Therefore, it is sometimes more economical to cold work a less costly and weaker metal than to hot work a more expensive metal that can be heat treated, especially if precision or a fine surface finish is required as well. The cold working process also reduces waste as compared to machining, or even eliminates with near net shape methods. The material savings becomes even more significant at larger volumes, and even more so when using expensive materials, such as copper, nickel, gold, tantalum, and palladium. The saving on raw material as a result of cold forming can be very significant, as is saving machining time. Production cycle times when cold working are very short. On multi-station machinery, production cycle times are even less. This can be very advantageous for large production runs.

During cold working the part undergoes work hardening and the microstructure deforms to follow the contours of the part surface. Unlike hot working, the inclusions and grains distort to follow the contour of the surface, resulting in anisotropic engineering properties.

Disadvantages:

- Greater forces are required

- Heavier and more powerful equipment and stronger tooling are required

- Metal is less ductile

- Metal surfaces must be clean and scale-free

- Intermediate anneals may be required to compensate for loss of ductility that accompanies strain hardening

- The imparted directional properties may be detrimental

- Undesirable residual stress may be produced

Due to the large capital costs required to set up a cold working process the process is usually only suitable for large volume productions.

Intermediate annealings may be required to reach the required ductility to continue cold working a workpiece, otherwise it may fracture if the ultimate tensile strength is exceeded. An anneal may also be used to obtain the proper engineering properties required in the final workpiece. Also, the distorted grain structure that gives the workpiece its superior strength can lead to residual stresses.

Cold worked items suffer from a phenomenon known as *springback*, or *elastic springback*. After the deforming force is removed from the workpiece, the workpiece springs back slightly. The amount a material springs back is equal to the yield strain (the strain at the yield point) for the material.

References

- Parr, Robert G.; Yang, Weitao (1989). Density-Functional Theory of Atoms and Molecules. New York: Oxford University Press. ISBN 0-19-509276-7.

- Music, D.; Geyer, R.W.; Schneider, J.M. (2016). "Recent progress and new directions in density functional theory based design of hard coatings". Surface & Coatings Technology. 286: 178. doi:10.1016/j.surfcoat.2015.12.021.

- Seidl, M.; Vuckovic, S.; Gori-Giorgi, P. (2016). "Challenging the Lieb–Oxford bound in a systematic way. Molecular Physics". Molecular Physics. 114 (7-8): 1076–1085. doi:10.1080/00268976.2015.1136440.

- Viraht, Xiao-Yin (2012). "Hohenberg-Kohn theorem including electron spin". Physical Review A. 86 (4): 042502. Bibcode:2012PhRvA..86d2502P. doi:10.1103/physreva.86.042502.

- A.J. Cohen, P. Mori-Sánchez and W. Yang (2012). "Challenges for Density Functional Theory". Chemical Reviews. 112 (1): 289–320. doi:10.1021/cr200107z. PMID 22191548.

Allied Fields of Materials Science

The allied fields of materials science are nanotechnology, polymer chemistry and radiation material science. Nanotechnology is the study of matter on atomic, molecular and supramolecular scale. This chapter helps the reader in developing an in depth understanding of all the fields that are allied to materials science.

Nanotechnology

Nanotechnology ("nanotech") is manipulation of matter on an atomic, molecular, and supramolecular scale. The earliest, widespread description of nanotechnology referred to the particular technological goal of precisely manipulating atoms and molecules for fabrication of macroscale products, also now referred to as molecular nanotechnology. A more generalized description of nanotechnology was subsequently established by the National Nanotechnology Initiative, which defines nanotechnology as the manipulation of matter with at least one dimension sized from 1 to 100 nanometers. This definition reflects the fact that quantum mechanical effects are important at this quantum-realm scale, and so the definition shifted from a particular technological goal to a research category inclusive of all types of research and technologies that deal with the special properties of matter which occur below the given size threshold. It is therefore common to see the plural form "nanotechnologies" as well as "nanoscale technologies" to refer to the broad range of research and applications whose common trait is size. Because of the variety of potential applications (including industrial and military), governments have invested billions of dollars in nanotechnology research. Until 2012, through its National Nanotechnology Initiative, the USA has invested 3.7 billion dollars, the European Union has invested 1.2 billion and Japan 750 million dollars.

Nanotechnology as defined by size is naturally very broad, including fields of science as diverse as surface science, organic chemistry, molecular biology, semiconductor physics, microfabrication, molecular engineering, etc. The associated research and applications are equally diverse, ranging from extensions of conventional device physics to completely new approaches based upon molecular self-assembly, from developing new materials with dimensions on the nanoscale to direct control of matter on the atomic scale.

Scientists currently debate the future implications of nanotechnology. Nanotechnology may be able to create many new materials and devices with a vast range of applications, such as in nanomedicine, nanoelectronics, biomaterials energy production, and consumer products. On the other hand, nanotechnology raises many of the same issues as any new technology, including concerns about the toxicity and environmental impact of nanomaterials, and their potential effects on global economics, as well as speculation about various doomsday scenarios. These concerns have led to a debate among advocacy groups and governments on whether special regulation of nanotechnology is warranted.

Origins

The concepts that seeded nanotechnology were first discussed in 1959 by renowned physicist Richard Feynman in his talk *There's Plenty of Room at the Bottom*, in which he described the possibility of synthesis via direct manipulation of atoms. The term "nano-technology" was first used by Norio Taniguchi in 1974, though it was not widely known.

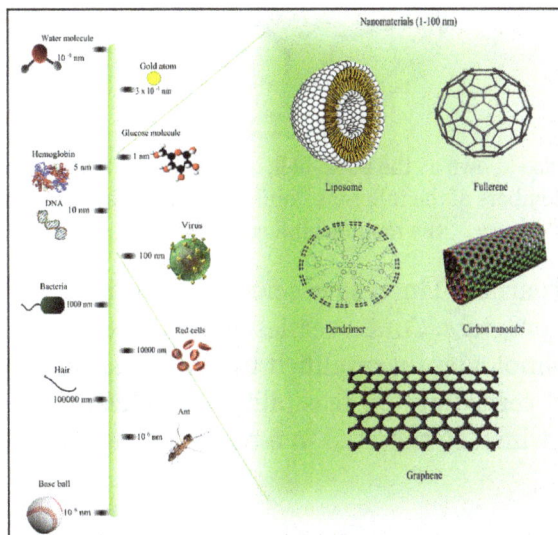

Comparison of Nanomaterials Sizes

Inspired by Feynman's concepts, K. Eric Drexler used the term "nanotechnology" in his 1986 book *Engines of Creation: The Coming Era of Nanotechnology*, which proposed the idea of a nanoscale "assembler" which would be able to build a copy of itself and of other items of arbitrary complexity with atomic control. Also in 1986, Drexler co-founded The Foresight Institute (with which he is no longer affiliated) to help increase public awareness and understanding of nanotechnology concepts and implications.

Thus, emergence of nanotechnology as a field in the 1980s occurred through convergence of Drexler's theoretical and public work, which developed and popularized a conceptual framework for nanotechnology, and high-visibility experimental advances that drew additional wide-scale attention to the prospects of atomic control of matter. In the 1980s, two major breakthroughs sparked the growth of nanotechnology in modern era.

First, the invention of the scanning tunneling microscope in 1981 which provided unprecedented visualization of individual atoms and bonds, and was successfully used to manipulate individual atoms in 1989. The microscope's developers Gerd Binnig and Heinrich Rohrer at IBM Zurich Research Laboratory received a Nobel Prize in Physics in 1986. Binnig, Quate and Gerber also invented the analogous atomic force microscope that year.

Second, Fullerenes were discovered in 1985 by Harry Kroto, Richard Smalley, and Robert Curl, who together won the 1996 Nobel Prize in Chemistry. C_{60} was not initially described as nanotechnology; the term was used regarding subsequent work with related graphene tubes (called carbon nanotubes and sometimes called Bucky tubes) which suggested potential applications for nanoscale electronics and devices.

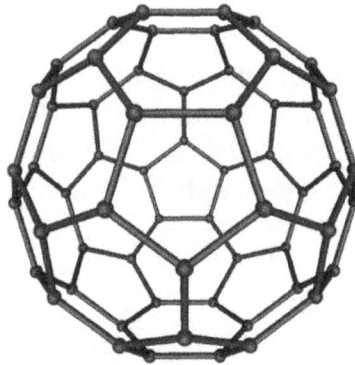

Buckminsterfullerene C60, also known as the buckyball, is a representative member of the carbon structures known as fullerenes. Members of the fullerene family are a major subject of research falling under the nanotechnology umbrella.

In the early 2000s, the field garnered increased scientific, political, and commercial attention that led to both controversy and progress. Controversies emerged regarding the definitions and potential implications of nanotechnologies, exemplified by the Royal Society's report on nanotechnology. Challenges were raised regarding the feasibility of applications envisioned by advocates of molecular nanotechnology, which culminated in a public debate between Drexler and Smalley in 2001 and 2003.

Meanwhile, commercialization of products based on advancements in nanoscale technologies began emerging. These products are limited to bulk applications of nanomaterials and do not involve atomic control of matter. Some examples include the Silver Nano platform for using silver nanoparticles as an antibacterial agent, nanoparticle-based transparent sunscreens, carbon fiber strengthening using silica nanoparticles, and carbon nanotubes for stain-resistant textiles.

Governments moved to promote and fund research into nanotechnology, such as in the U.S. with the National Nanotechnology Initiative, which formalized a size-based definition of nanotechnology and established funding for research on the nanoscale, and in Europe via the European Framework Programmes for Research and Technological Development.

By the mid-2000s new and serious scientific attention began to flourish. Projects emerged to produce nanotechnology roadmaps which center on atomically precise manipulation of matter and discuss existing and projected capabilities, goals, and applications.

Fundamental Concepts

Nanotechnology is the engineering of functional systems at the molecular scale. This covers both current work and concepts that are more advanced. In its original sense, nanotechnology refers to the projected ability to construct items from the bottom up, using techniques and tools being developed today to make complete, high performance products.

One nanometer (nm) is one billionth, or 10^{-9}, of a meter. By comparison, typical carbon-carbon bond lengths, or the spacing between these atoms in a molecule, are in the range 0.12–0.15 nm, and a DNA double-helix has a diameter around 2 nm. On the other hand, the smallest cellular lifeforms, the bacteria of the genus Mycoplasma, are around 200 nm in length. By convention, nanotechnology is taken as the scale range 1 to 100 nm following the definition used by the National

Nanotechnology Initiative in the US. The lower limit is set by the size of atoms (hydrogen has the smallest atoms, which are approximately a quarter of a nm diameter) since nanotechnology must build its devices from atoms and molecules. The upper limit is more or less arbitrary but is around the size below which phenomena not observed in larger structures start to become apparent and can be made use of in the nano device. These new phenomena make nanotechnology distinct from devices which are merely miniaturised versions of an equivalent macroscopic device; such devices are on a larger scale and come under the description of microtechnology.

To put that scale in another context, the comparative size of a nanometer to a meter is the same as that of a marble to the size of the earth. Or another way of putting it: a nanometer is the amount an average man's beard grows in the time it takes him to raise the razor to his face.

Two main approaches are used in nanotechnology. In the "bottom-up" approach, materials and devices are built from molecular components which assemble themselves chemically by principles of molecular recognition. In the "top-down" approach, nano-objects are constructed from larger entities without atomic-level control.

Areas of physics such as nanoelectronics, nanomechanics, nanophotonics and nanoionics have evolved during the last few decades to provide a basic scientific foundation of nanotechnology.

Larger to Smaller: a Materials Perspective

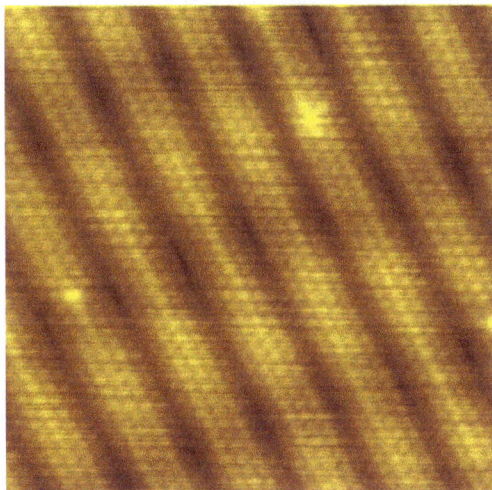

Image of reconstruction on a clean Gold(100) surface, as visualized using scanning tunneling microscopy. The positions of the individual atoms composing the surface are visible.

Several phenomena become pronounced as the size of the system decreases. These include statistical mechanical effects, as well as quantum mechanical effects, for example the "quantum size effect" where the electronic properties of solids are altered with great reductions in particle size. This effect does not come into play by going from macro to micro dimensions. However, quantum effects can become significant when the nanometer size range is reached, typically at distances of 100 nanometers or less, the so-called quantum realm. Additionally, a number of physical (mechanical, electrical, optical, etc.) properties change when compared to macroscopic systems. One example is the increase in surface area to volume ratio altering mechanical, thermal and catalytic properties of materials. Diffusion and reactions at nanoscale, nanostructures materials and nanodevices with fast ion transport are generally referred to nanoionics. *Mechanical* properties of nanosystems are of interest in the nanomechanics research. The catalytic activity of nanomaterials also opens potential risks in their interaction with biomaterials.

Materials reduced to the nanoscale can show different properties compared to what they exhibit on a macroscale, enabling unique applications. For instance, opaque substances can become transparent (copper); stable materials can turn combustible (aluminium); insoluble materials may become soluble (gold). A material such as gold, which is chemically inert at normal scales, can serve as a potent chemical catalyst at nanoscales. Much of the fascination with nanotechnology stems from these quantum and surface phenomena that matter exhibits at the nanoscale.

Simple to Complex: a Molecular Perspective

Modern synthetic chemistry has reached the point where it is possible to prepare small molecules to almost any structure. These methods are used today to manufacture a wide variety of useful chemicals such as pharmaceuticals or commercial polymers. This ability raises the question of extending this kind of control to the next-larger level, seeking methods to assemble these single molecules into supramolecular assemblies consisting of many molecules arranged in a well defined manner.

These approaches utilize the concepts of molecular self-assembly and/or supramolecular chemistry to automatically arrange themselves into some useful conformation through a bottom-up approach. The concept of molecular recognition is especially important: molecules can be designed so that a specific configuration or arrangement is favored due to non-covalent intermolecular forces. The Watson–Crick basepairing rules are a direct result of this, as is the specificity of an enzyme being targeted to a single substrate, or the specific folding of the protein itself. Thus, two or more components can be designed to be complementary and mutually attractive so that they make a more complex and useful whole.

Such bottom-up approaches should be capable of producing devices in parallel and be much cheaper than top-down methods, but could potentially be overwhelmed as the size and complexity of the desired assembly increases. Most useful structures require complex and thermodynamically unlikely arrangements of atoms. Nevertheless, there are many examples of self-assembly based on molecular recognition in biology, most notably Watson–Crick basepairing and enzyme-substrate interactions. The challenge for nanotechnology is whether these principles can be used to engineer new constructs in addition to natural ones.

Molecular Nanotechnology: a Long-term View

Molecular nanotechnology, sometimes called molecular manufacturing, describes engineered nanosystems (nanoscale machines) operating on the molecular scale. Molecular nanotechnology is especially associated with the molecular assembler, a machine that can produce a desired structure or device atom-by-atom using the principles of mechanosynthesis. Manufacturing in the context of productive nanosystems is not related to, and should be clearly distinguished from, the conventional technologies used to manufacture nanomaterials such as carbon nanotubes and nanoparticles.

When the term "nanotechnology" was independently coined and popularized by Eric Drexler (who at the time was unaware of an earlier usage by Norio Taniguchi) it referred to a future manufacturing technology based on molecular machine systems. The premise was that molecular scale biological analogies of traditional machine components demonstrated molecular machines were possible: by the countless examples found in biology, it is known that sophisticated, stochastically optimised biological machines can be produced.

It is hoped that developments in nanotechnology will make possible their construction by some other means, perhaps using biomimetic principles. However, Drexler and other researchers have proposed that advanced nanotechnology, although perhaps initially implemented by biomimetic means, ultimately could be based on mechanical engineering principles, namely, a manufacturing technology based on the mechanical functionality of these components (such as gears, bearings, motors, and structural members) that would enable programmable, positional assembly to atomic specification. The physics and engineering performance of exemplar designs were analyzed in Drexler's book *Nanosystems*.

In general it is very difficult to assemble devices on the atomic scale, as one has to position atoms on other atoms of comparable size and stickiness. Another view, put forth by Carlo Montemagno, is that future nanosystems will be hybrids of silicon technology and biological molecular machines. Richard Smalley argued that mechanosynthesis are impossible due to the difficulties in mechanically manipulating individual molecules.

This led to an exchange of letters in the ACS publication Chemical & Engineering News in 2003. Though biology clearly demonstrates that molecular machine systems are possible, non-biological molecular machines are today only in their infancy. Leaders in research on non-biological molecular machines are Dr. Alex Zettl and his colleagues at Lawrence Berkeley Laboratories and UC Berkeley. They have constructed at least three distinct molecular devices whose motion is controlled from the desktop with changing voltage: a nanotube nanomotor, a molecular actuator, and a nanoelectromechanical relaxation oscillator.

An experiment indicating that positional molecular assembly is possible was performed by Ho and Lee at Cornell University in 1999. They used a scanning tunneling microscope to move an individual carbon monoxide molecule (CO) to an individual iron atom (Fe) sitting on a flat silver crystal, and chemically bound the CO to the Fe by applying a voltage.

Current Research

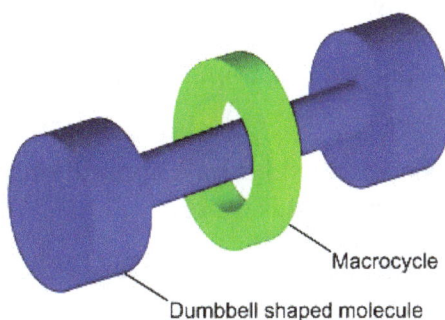

Graphical representation of a rotaxane, useful as a molecular switch.

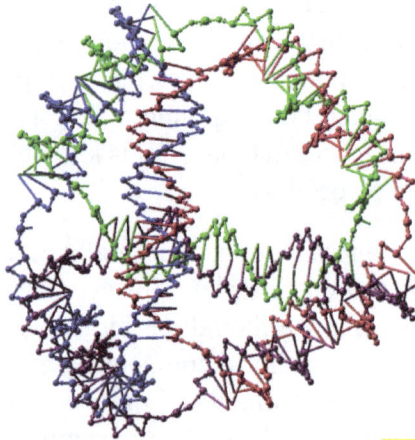

This DNA tetrahedron is an artificially designed nanostructure of the type made in the field of DNA nanotechnology. Each edge of the tetrahedron is a 20 base pair DNA double helix, and each vertex is a three-arm junction.

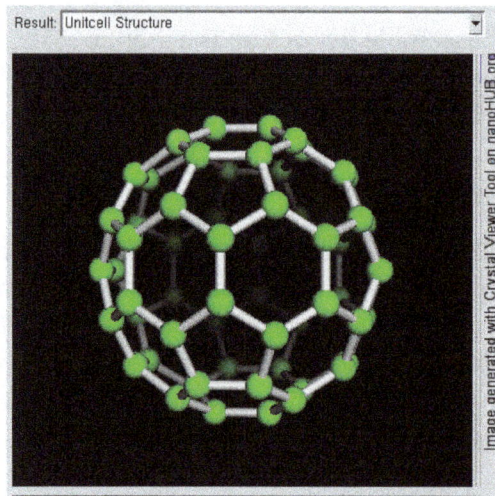

Rotating view of C_{60}, one kind of fullerene.

Nanomaterials

The nanomaterials field includes subfields which develop or study materials having unique properties arising from their nanoscale dimensions.

- Interface and colloid science has given rise to many materials which may be useful in nanotechnology, such as carbon nanotubes and other fullerenes, and various nanoparticles and nanorods. Nanomaterials with fast ion transport are related also to nanoionics and nanoelectronics.

- Nanoscale materials can also be used for bulk applications; most present commercial applications of nanotechnology are of this flavor.

- Progress has been made in using these materials for medical applications.

- Nanoscale materials such as nanopillars are sometimes used in solar cells which combats the cost of traditional Silicon solar cells.

- Development of applications incorporating semiconductor nanoparticles to be used in the next generation of products, such as display technology, lighting, solar cells and biological imaging.

- Recent application of nanomaterials include a range of biomedical applications, such as tissue engineering, drug delivery, and biosensors.

Bottom-up Approaches

These seek to arrange smaller components into more complex assemblies.

- DNA nanotechnology utilizes the specificity of Watson–Crick basepairing to construct well-defined structures out of DNA and other nucleic acids.

- Approaches from the field of "classical" chemical synthesis (Inorganic and organic synthesis) also aim at designing molecules with well-defined shape (e.g. bis-peptides).

- More generally, molecular self-assembly seeks to use concepts of supramolecular chemistry, and molecular recognition in particular, to cause single-molecule components to automatically arrange themselves into some useful conformation.

- Atomic force microscope tips can be used as a nanoscale "write head" to deposit a chemical upon a surface in a desired pattern in a process called dip pen nanolithography. This technique fits into the larger subfield of nanolithography.

Top-down Approaches

These seek to create smaller devices by using larger ones to direct their assembly.

- Many technologies that descended from conventional solid-state silicon methods for fabricating microprocessors are now capable of creating features smaller than 100 nm, falling under the definition of nanotechnology. Giant magnetoresistance-based hard drives already on the market fit this description, as do atomic layer deposition (ALD) techniques. Peter Grünberg and Albert Fert received the Nobel Prize in Physics in 2007 for their discovery of Giant magnetoresistance and contributions to the field of spintronics.

- Solid-state techniques can also be used to create devices known as nanoelectromechanical systems or NEMS, which are related to microelectromechanical systems or MEMS.

- Focused ion beams can directly remove material, or even deposit material when suitable precursor gasses are applied at the same time. For example, this technique is used routinely to create sub-100 nm sections of material for analysis in Transmission electron microscopy.

- Atomic force microscope tips can be used as a nanoscale "write head" to deposit a resist, which is then followed by an etching process to remove material in a top-down method.

Functional Approaches

These seek to develop components of a desired functionality without regard to how they might be assembled.

- Magnetic assembly for the synthesis of anisotropic superparamagnetic materials such as recently presented magnetic nanochains.

- Molecular scale electronics seeks to develop molecules with useful electronic properties. These could then be used as single-molecule components in a nanoelectronic device. For an example see rotaxane.

- Synthetic chemical methods can also be used to create synthetic molecular motors, such as in a so-called nanocar.

Biomimetic Approaches

- Bionics or biomimicry seeks to apply biological methods and systems found in nature, to the study and design of engineering systems and modern technology. Biomineralization is one example of the systems studied.

- Bionanotechnology is the use of biomolecules for applications in nanotechnology, including use of viruses and lipid assemblies. Nanocellulose is a potential bulk-scale application.

Speculative

These subfields seek to anticipate what inventions nanotechnology might yield, or attempt to propose an agenda along which inquiry might progress. These often take a big-picture view of nanotechnology, with more emphasis on its societal implications than the details of how such inventions could actually be created.

- Molecular nanotechnology is a proposed approach which involves manipulating single molecules in finely controlled, deterministic ways. This is more theoretical than the other subfields, and many of its proposed techniques are beyond current capabilities.

- Nanorobotics centers on self-sufficient machines of some functionality operating at the nanoscale. There are hopes for applying nanorobots in medicine, but it may not be easy to do such a thing because of several drawbacks of such devices. Nevertheless, progress on innovative materials and methodologies has been demonstrated with some patents granted about new nanomanufacturing devices for future commercial applications, which also progressively helps in the development towards nanorobots with the use of embedded nanobioelectronics concepts.

- Productive nanosystems are "systems of nanosystems" which will be complex nanosystems that produce atomically precise parts for other nanosystems, not necessarily using novel nanoscale-emergent properties, but well-understood fundamentals of manufacturing. Because of the discrete (i.e. atomic) nature of matter and the possibility of exponential growth, this stage is seen as the basis of another industrial revolution. Mihail Roco, one of the architects of the USA's National Nanotechnology Initiative, has proposed four states of nanotechnology that seem to parallel the technical progress of the Industrial Revolution, progressing from passive nanostructures to active nanodevices to complex nanomachines and ultimately to productive nanosystems.

- Programmable matter seeks to design materials whose properties can be easily, reversibly and externally controlled though a fusion of information science and materials science.

- Due to the popularity and media exposure of the term nanotechnology, the words picotechnology and femtotechnology have been coined in analogy to it, although these are only used rarely and informally.

Dimensionality in Nanomaterials

Nanomaterials can be classified in 0D, 1D, 2D and 3D nanomaterials. The dimensionality play a major role in determining the characteristic of nanomaterials including physical, chemical and biological characteristics. With the decrease in dimensionality, an increase in surface-to-volume ratio is observed. This indicate that smaller dimensional nanomaterials have higher surface area compared to 3D nanomaterials. Recently, two dimensional (2D) nanomaterials are extensively investigated for electronic, biomedical, drug delivery and biosensor applications.

Tools and Techniques

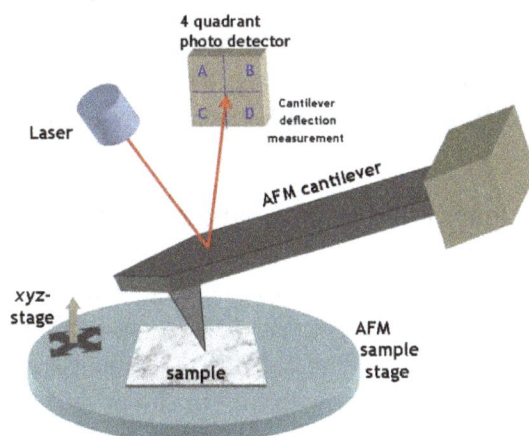

Typical AFM setup. A microfabricated cantilever with a sharp tip is deflected by features on a sample surface, much like in a phonograph but on a much smaller scale. A laser beam reflects off the backside of the cantilever into a set of photodetectors, allowing the deflection to be measured and assembled into an image of the surface.

There are several important modern developments. The atomic force microscope (AFM) and the Scanning Tunneling Microscope (STM) are two early versions of scanning probes that launched nanotechnology. There are other types of scanning probe microscopy. Although conceptually similar to the scanning confocal microscope developed by Marvin Minsky in 1961 and the scanning acoustic microscope (SAM) developed by Calvin Quate and coworkers in the 1970s, newer scanning probe microscopes have much higher resolution, since they are not limited by the wavelength of sound or light.

The tip of a scanning probe can also be used to manipulate nanostructures (a process called positional assembly). Feature-oriented scanning methodology may be a promising way to implement these nanomanipulations in automatic mode. However, this is still a slow process because of low scanning velocity of the microscope.

Various techniques of nanolithography such as optical lithography, X-ray lithography dip pen nanolithography, electron beam lithography or nanoimprint lithography were also developed. Lithography is a top-down fabrication technique where a bulk material is reduced in size to nanoscale pattern.

Another group of nanotechnological techniques include those used for fabrication of nanotubes and nanowires, those used in semiconductor fabrication such as deep ultraviolet lithography, electron beam lithography, focused ion beam machining, nanoimprint lithography, atomic layer deposition, and molecular vapor deposition, and further including molecular self-assembly techniques such as those employing di-block copolymers. The precursors of these techniques preceded the nanotech era, and are extensions in the development of scientific advancements rather than techniques which were devised with the sole purpose of creating nanotechnology and which were results of nanotechnology research.

The top-down approach anticipates nanodevices that must be built piece by piece in stages, much as manufactured items are made. Scanning probe microscopy is an important technique both for characterization and synthesis of nanomaterials. Atomic force microscopes and scanning tunneling microscopes can be used to look at surfaces and to move atoms around. By designing different tips for these microscopes, they can be used for carving out structures on surfaces and to help guide self-assembling structures. By using, for example, feature-oriented scanning approach, atoms or molecules can be moved around on a surface with scanning probe microscopy techniques. At present, it is expensive and time-consuming for mass production but very suitable for laboratory experimentation.

In contrast, bottom-up techniques build or grow larger structures atom by atom or molecule by molecule. These techniques include chemical synthesis, self-assembly and positional assembly. Dual polarisation interferometry is one tool suitable for characterisation of self assembled thin films. Another variation of the bottom-up approach is molecular beam epitaxy or MBE. Researchers at Bell Telephone Laboratories like John R. Arthur. Alfred Y. Cho, and Art C. Gossard developed and implemented MBE as a research tool in the late 1960s and 1970s. Samples made by MBE were key to the discovery of the fractional quantum Hall effect for which the 1998 Nobel Prize in Physics was awarded. MBE allows scientists to lay down atomically precise layers of atoms and, in the process, build up complex structures. Important for research on semiconductors, MBE is also widely used to make samples and devices for the newly emerging field of spintronics.

However, new therapeutic products, based on responsive nanomaterials, such as the ultradeformable, stress-sensitive Transfersome vesicles, are under development and already approved for human use in some countries.

Applications

Nanostructures provide this surface with superhydrophobicity, which lets water droplets roll down the inclined plane.

As of August 21, 2008, the Project on Emerging Nanotechnologies estimates that over 800 manufacturer-identified nanotech products are publicly available, with new ones hitting the market at a pace of 3–4 per week. The project lists all of the products in a publicly accessible online database.

Most applications are limited to the use of "first generation" passive nanomaterials which includes titanium dioxide in sunscreen, cosmetics, surface coatings, and some food products; Carbon allotropes used to produce gecko tape; silver in food packaging, clothing, disinfectants and household appliances; zinc oxide in sunscreens and cosmetics, surface coatings, paints and outdoor furniture varnishes; and cerium oxide as a fuel catalyst.

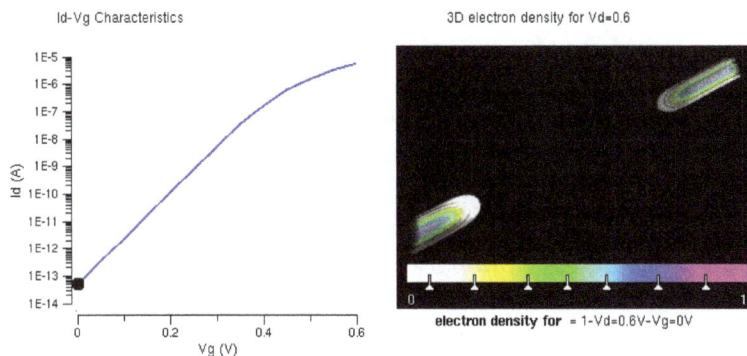

One of the major applications of nanotechnology is in the area of nanoelectronics with MOSFET's being made of small nanowires ~10 nm in length. Here is a simulation of such a nanowire.

Further applications allow tennis balls to last longer, golf balls to fly straighter, and even bowling balls to become more durable and have a harder surface. Trousers and socks have been infused with nanotechnology so that they will last longer and keep people cool in the summer. Bandages are being infused with silver nanoparticles to heal cuts faster. Video game consoles and personal computers may become cheaper, faster, and contain more memory thanks to nanotechnology. Nanotechnology may have the ability to make existing medical applications cheaper and easier to use in places like the general practitioner's office and at home. Cars are being manufactured with nanomaterials so they may need fewer metals and less fuel to operate in the future.

Scientists are now turning to nanotechnology in an attempt to develop diesel engines with cleaner exhaust fumes. Platinum is currently used as the diesel engine catalyst in these engines. The catalyst is what cleans the exhaust fume particles. First a reduction catalyst is employed to take nitrogen atoms from NOx molecules in order to free oxygen. Next the oxidation catalyst oxidizes the hydrocarbons and carbon monoxide to form carbon dioxide and water. Platinum is used in both the reduction and the oxidation catalysts. Using platinum though, is inefficient in that it is expensive and unsustainable. Danish company InnovationsFonden invested DKK 15 million in a search for new catalyst substitutes using nanotechnology. The goal of the project, launched in the autumn of 2014, is to maximize surface area and minimize the amount of material required. Objects tend to minimize their surface energy; two drops of water, for example, will join to form one drop and decrease surface area. If the catalyst's surface area that is exposed to the exhaust fumes is maximized, efficiency of the catalyst is maximized. The team working on this project aims to create nanoparticles that will not merge. Every time the surface is optimized, material is saved. Thus, creating these nanoparticles will increase the effectiveness of the resulting diesel engine catalyst—in turn leading to cleaner exhaust fumes—and will decrease cost. If successful, the team hopes to reduce platinum use by 25%.

Nanotechnology also has a prominent role in the fast developing field of Tissue Engineering. When designing scaffolds, researchers attempt to the mimic the nanoscale features of a Cell's microenvironment to direct its differentiation down a suitable lineage. For example, when creating scaffolds to support the growth of bone, researchers may mimic osteoclast resorption pits.

Researchers have successfully used DNA origami-based nanobots capable of carrying out logic functions to achieve targeted drug delivery in cockroaches. It is said that the computational power of these nanobots can be scaled up to that of a Commodore 64.

Implications

An area of concern is the effect that industrial-scale manufacturing and use of nanomaterials would have on human health and the environment, as suggested by nanotoxicology research. For these reasons, some groups advocate that nanotechnology be regulated by governments. Others counter that overregulation would stifle scientific research and the development of beneficial innovations. Public health research agencies, such as the National Institute for Occupational Safety and Health are actively conducting research on potential health effects stemming from exposures to nanoparticles.

Some nanoparticle products may have unintended consequences. Researchers have discovered that bacteriostatic silver nanoparticles used in socks to reduce foot odor are being released in the wash. These particles are then flushed into the waste water stream and may destroy bacteria which are critical components of natural ecosystems, farms, and waste treatment processes.

Public deliberations on risk perception in the US and UK carried out by the Center for Nanotechnology in Society found that participants were more positive about nanotechnologies for energy applications than for health applications, with health applications raising moral and ethical dilemmas such as cost and availability.

Experts, including director of the Woodrow Wilson Center's Project on Emerging Nanotechnologies David Rejeski, have testified that successful commercialization depends on adequate oversight, risk research strategy, and public engagement. Berkeley, California is currently the only city in the United States to regulate nanotechnology; Cambridge, Massachusetts in 2008 considered enacting a similar law, but ultimately rejected it. Relevant for both research on and application of nanotechnologies, the insurability of nanotechnology is contested. Without state regulation of nanotechnology, the availability of private insurance for potential damages is seen as necessary to ensure that burdens are not socialised implicitly.

Health and Environmental Concerns

A video on the health and safety implications of nanotechnology

Nanofibers are used in several areas and in different products, in everything from aircraft wings to tennis rackets. Inhaling airborne nanoparticles and nanofibers may lead to a number of pulmonary diseases, e.g. fibrosis. Researchers have found that when rats breathed in nanoparticles, the particles settled in the brain and lungs, which led to significant increases in biomarkers for inflammation and stress response and that nanoparticles induce skin aging through oxidative stress in hairless mice.

A two-year study at UCLA's School of Public Health found lab mice consuming nano-titanium dioxide showed DNA and chromosome damage to a degree "linked to all the big killers of man, namely cancer, heart disease, neurological disease and aging".

A major study published more recently in Nature Nanotechnology suggests some forms of carbon nanotubes – a poster child for the "nanotechnology revolution" – could be as harmful as asbestos if inhaled in sufficient quantities. Anthony Seaton of the Institute of Occupational Medicine in Edinburgh, Scotland, who contributed to the article on carbon nanotubes said "We know that some of them probably have the potential to cause mesothelioma. So those sorts of materials need to be handled very carefully." In the absence of specific regulation forthcoming from governments, Paull and Lyons (2008) have called for an exclusion of engineered nanoparticles in food. A newspaper article reports that workers in a paint factory developed serious lung disease and nanoparticles were found in their lungs.

Regulation

Calls for tighter regulation of nanotechnology have occurred alongside a growing debate related to the human health and safety risks of nanotechnology. There is significant debate about who is responsible for the regulation of nanotechnology. Some regulatory agencies currently cover some nanotechnology products and processes (to varying degrees) – by "bolting on" nanotechnology to existing regulations – there are clear gaps in these regimes. Davies (2008) has proposed a regulatory road map describing steps to deal with these shortcomings.

Stakeholders concerned by the lack of a regulatory framework to assess and control risks associated with the release of nanoparticles and nanotubes have drawn parallels with bovine spongiform encephalopathy ("mad cow" disease), thalidomide, genetically modified food, nuclear energy, reproductive technologies, biotechnology, and asbestosis. Dr. Andrew Maynard, chief science advisor to the Woodrow Wilson Center's Project on Emerging Nanotechnologies, concludes that there is insufficient funding for human health and safety research, and as a result there is currently limited understanding of the human health and safety risks associated with nanotechnology. As a result, some academics have called for stricter application of the precautionary principle, with delayed marketing approval, enhanced labelling and additional safety data development requirements in relation to certain forms of nanotechnology.

The Royal Society report identified a risk of nanoparticles or nanotubes being released during disposal, destruction and recycling, and recommended that "manufacturers of products that fall under extended producer responsibility regimes such as end-of-life regulations publish procedures outlining how these materials will be managed to minimize possible human and environmental exposure" (p. xiii).

The Center for Nanotechnology in Society has found that people respond to nanotechnologies differently, depending on application – with participants in public deliberations more positive about nanotechnologies for energy than health applications – suggesting that any public calls for nano regulations may differ by technology sector.

Polymer Chemistry

Polymer chemistry is a chemistry subdiscipline that deals with the structures, chemical synthesis and properties of polymers, primarily synthetic polymers such as plastics and elastomers. Polymer chemistry is related to the broader field of polymer science, which also encompasses polymer physics and polymer engineering.

The chemist Hermann Staudinger first proposed that polymers consisted of long chains of atoms held together by covalent bonds, which he called macromolecules. His work expanded the chemical understanding of polymers and was followed by an expansion of the field of polymer chemistry during which such polymeric materials as neoprene, nylon and polyester were invented.

According to the International Union of Pure and Applied Chemistry, macromolecules refer to the individual molecular chains and are the domain of chemistry. Polymers describe the bulk properties of polymer materials and belong to the field of polymer physics as a subfield of physics.

Polymers and Their Properties

Polymers are high molecular mass compounds formed by polymerization of monomers. The simple reactive molecule from which the repeating structural units of a polymer are derived are called monomer. A polymer is chemically described by its degree of polymerisation, molar mass distribution, tacticity, copolymer distribution, the degree of branching, by its end-groups, crosslinks, crystallinity and thermal properties such as its glass transition temperature and melting temperature. Polymers in solution have special characteristics with respect to solubility, viscosity and gelation.

Schematically polymers are subdivided into biopolymers and synthetic polymers according to their origin. Each one of these classes of compounds can be subdivided into more specific categories in relationship to their use, properties and physicochemical characteristics.

The biochemistry and industrial chemistry are disciplines that are interested in the study of the chemistry of polymers.

- Biopolymers: Polymers produced by living organisms:

    o Structural proteins: collagen, keratin, elastin.

    o Functional proteins: enzymes, hormones.

    o Structural polysaccharides: cellulose, chitin.

    o Reserve polysaccharides: starch, glycogen.

    o Nucleic acids: DNA, RNA.

- Synthetic polymers: these polymers are used for plastics, synthetic fibers, paints, building materials, furniture, mechanical parts and adhesives. These are divided into:

    o Thermoplastic polymers such as polyethylene, teflon, polystyrene, polypropylene, polyester, polyurethane, Poly(methyl methacrylate), vinyl chloride, nylon, rayon, cellulose, silicon, glass fiber, among others.

    o Thermoset plastics:vulcanized rubber, bakelite, Kevlar, polyepoxide.

History

The work of Henri Braconnot in 1777 and the work of Christian Schönbein in 1846 led to the discovery of nitrocellulose, which, when treated with camphor produced celluloid. Dissolved in ether or acetone, it is collodion, used as a wound dressing since the U.S. Civil War. Cellulose acetate was first prepared in 1865. In 1834, Friedrich Ludersdorf and Nathaniel Hayward independently discovered that adding sulfur to raw natural rubber (polyisoprene) helped prevent the material from becoming sticky. In 1844 Charles Goodyear received a U.S. patent for vulcanizing rubber with sulfur and heat. Thomas Hancock had received a patent for the same process in the UK the year before.

In 1884 Hilaire de Chardonnet started the first artificial fiber plant based on regenerated cellulose, or viscose rayon, as a substitute for silk, but it was very flammable. In 1907 Leo Baekeland invented the first synthetic polymer, a thermosetting phenol-formaldehyde resin called Bakelite. Around the same time, Hermann Leuchs reported the synthesis of N-carboxyanhydrides and their high molecular weight products upon reaction with nucleophiles, but stopped short of referring to these as polymers, possibly due to the strong views espoused by Emil Fischer, his direct supervisor, denying the possibility of any covalent molecule exceeding 6,000 daltons. Cellophane was invented in 1908 by Jocques Brandenberger who squirted sheets of viscose rayon into an acid bath.

In 1922 Hermann Staudinger (of Worms, Germany 1881-1965) was the first to propose that polymers consisted of long chains of atoms held together by covalent bonds. He also proposed to name these compounds macromolecules. Before that, scientists believed that polymers were clusters of small molecules (called colloids), without definite molecular weights, held together by an unknown force. Staudinger received the Nobel Prize in Chemistry in 1953. Wallace Carothers invented the first synthetic rubber called neoprene in 1931, the first polyester, and went on to invent nylon, a true silk replacement, in 1935. Paul Flory was awarded the Nobel Prize in Chemistry in 1974 for his work on polymer random coil configurations in solution in the 1950s. Stephanie Kwolek developed an aramid, or aromatic nylon named Kevlar, patented in 1966.

There are now a large number of commercial polymers, including composite materials such as carbon fiber-epoxy, polystyrene-polybutadiene (HIPS), acrylonitrile-butadiene-styrene (ABS), and other such materials that combine the best properties of their various components, including polymers designed to work at high temperatures in automobile engines.

In spite of the great importance of the polymer industry, it took a long time before universities introduced teaching and research programs in polymer chemistry. An "Institut fur Makromolekulare Chemie was founded in 1940 in Freiburg, Germany under the direction of Hermann Staudinger. In America a "Polymer Research Institute" (PRI) was established in 1941 by Herman Mark at the Polytechnic Institute of Brooklyn (now Polytechnic Institute of NYU). Several hundred graduates of PRI played an important role in the US polymer industry and academia. Other PRI's were founded in 1961 by Richard S. Stein at the University of Massachusetts, Amherst, in 1967 by Eric Baer at Case Western Reserve University, in 1982 at The University of Southern Mississippi, and in 1988 at the University of Akron.

Theories

Association theory is a discredited theory which tried to explain molecular structures of macromolecules.Some other theories related to polymers include

- Scheutjens–Fleer theory

- Flory–Huggins solution theory

- Cossee-Arlman mechanism

- Polymer field theory

- Hoffman Nucleation Theory

Examples of Polymers

Biopolymers

They are produced by living organisms:

- structural proteins: collagen, keratin, elastin...

- chemically functional proteins: enzymes, hormones, transport proteins...

- structural polysaccharides: cellulose, chitin...

- storage polysaccharides: starch, glycogen...

- nucleic acids: DNA, RNA

Synthetic Polymers

Examples of Synthetic polymers are plastics, fibers, paints, building materials, furniture, mechanical parts, adhesives:

- thermoplastics: polyethylene, Teflon, polystyrene, polypropylene, polyester, polyurethane, polymethyl methacrylate, polyvinyl chloride, nylon, rayon, celluloid, silicone...

- thermosetting plastics: vulcanized rubber, Bakelite, Kevlar, epoxy...

Radiation Material Science

Radiation materials science describes the interaction of radiation with matter: a broad subject covering many forms of irradiation and of matter.

Main Aim of Radiation Material Science

Some of the most profound effects of irradiation on materials occur in the core of nuclear power reactors where atoms comprising the structural components are displaced numerous times over the course of their engineering lifetimes. The consequences of radiation to core components includes changes in shape and volume by tens of percent, increases in hardness by factors of five or more, severe reduction in ductility and increased embrittlement, and susceptibility to environmentally induced cracking. For these structures to fulfill their purpose, a firm understanding of the effect of radiation on materials is required in order to account for irradiation effects in design, to mitigate its effect by changing operating conditions, or to serve as a guide for creating new, more radiation-tolerant materials that can better serve their purpose.

Radiation

The types of radiation that can alter structural materials consist of neutrons, ions, electrons and gamma rays. All of these forms of radiation have the capability to displace atoms from their lattice sites, which is the fundamental process that drives the changes in structural metals. The inclusion of ions among the irradiating particles provides a tie-in to other fields and disciplines such as the use of accelerators for the transmutation of nuclear waste, or in the creation of new materials by ion implantation, ion beam mixing, plasma assisted ion implantation and ion beam assisted deposition.

The effect of irradiation on materials is rooted in the initial event in which an energetic projectile strikes a target. While the event is made up of several steps or processes, the primary result is the displacement of an atom from its lattice site. Irradiation displaces an atom from its site, leaving a vacant site behind (a vacancy) and the displaced atom eventually comes to rest in a location that is between lattice sites, becoming an interstitial atom. The vacancy-interstitial pair is central to radiation effects in crystalline solids and is known as a Frenkel pair (FP). The presence of the Frenkel pair and other consequences of irradiation damage determine the physical effects, and with the application of stress, the mechanical effects of irradiation by the occurring of interstitial, phenomena, such as swelling, growth, phase transition, segregation, etc., will be effected. In addition to the atomic displacement, an energetic charged particle moving in a lattice also gives energy to electrons in the system, via the electronic stopping power. This energy transfer can also for high-energy particles produce damage in non-metallic materials, as so called ion tracks.

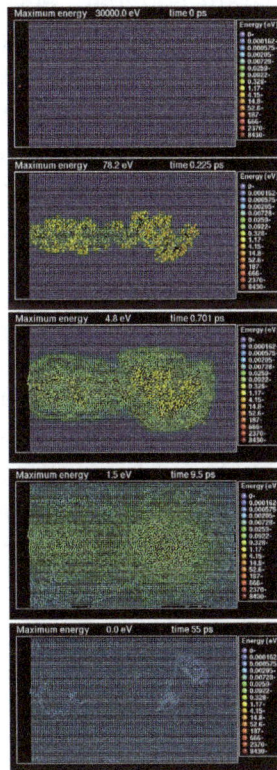

Image sequence of the time development of a collision cascade in the heat spike regime produced by a 30 keV Xe ion impacting on Au under channeling conditions. The image is produced by a classical molecular dynamics simulation of a collision cascade. The image shows a cross section of two atomic layers in the middle of a threedimensional simulation cell. Each sphere illustrates the position of an atom, and the colors show the kinetic energy of each atom as indicated by the scale on the right. At the end, both point defects and dislocation loops remain.

Radiation Resistant Materials

To generate materials that fit the increasing demands of nuclear reactors to operate with higher efficiency or for longer lifetimes, materials must be designed with radiation resistance in mind. In particular, Generation IV nuclear reactors operate at higher temperatures and pressures compared to modern pressurized water reactors, which account for a vast amount of western reactors. This leads to increased vulnerability to normal mechanical failure in terms of creep resistance as well as radiation damaging events such as neutron-induced swelling and radiation induced segregation of phases. By accounting for radiation damage, reactor materials would be able to withstand longer operating lifetimes. This allows reactors to be decommissioned after longer periods of time, improving return on investment of reactors without compromising safety. This is of particular interest in developing commercial viability of advanced and theoretical nuclear reactors, and this goal can be accomplished through engineering resistance to these displacement events.

Grain Boundary Engineering

Face centered cubic (FCC) metals such as austenitic steels and Ni-base alloys can benefit greatly from grain boundary engineering. Grain boundary engineering attempts to generate higher amounts of special grain boundaries, characterized by favorable orientations between grains. By increasing populations of low energy boundaries without increasing grain size, fracture mechanics of these FCC metals can be changed to improve mechanical properties given a similar displacements per atom (DPA) value versus non grain boundary engineered alloys. This method of treatment in particular yields better resistance to stress corrosion cracking and oxidation.

Materials Selection

By using advanced methods of material selection, materials can be judged on criteria such as neutron-absorption cross sectional area. Selecting materials with minimum neutron-absorption can heavily minimize the number of DPA that occur over a reactor material's lifetime. This slows the radiation embrittlement process by preventing mobility of atoms in the first place, proactively selecting materials that do not interact with the nuclear radiation as frequently. This can have a huge impact on total damage especially when comparing the materials of modern advanced reactors of zirconium to stainless steel reactor cores, which can differ in absorption cross section by an order of magnitude from more optimal materials.

Example values for thermal neutron cross section are shown in the table below.

Element	Thermal Neutron Cross Section (barns)
Magnesium	0.059
Lead	0.17
Zirconium	0.18
Aluminum	0.23

Iron	2.56
Austenitic Stainless Steel	3.1
Nickel	4.5
Titanium	6.1
Cadmium	2520

Short Range Order (SRO) Self-organization

For nickel-chromium and iron-chromium alloys, short range order can be designed on the nano-scale (<5 nm) that absorbs the interstitial and vacancy's generated by PKA events. This allows materials that mitigate the swelling that normally occurs in the presence of high DPAs and keep the overall volume percent change under the ten percent range. This occurs through generating a metastable phase that is in constant, dynamic equilibrium with surrounding material. This metastable phase is characterized by having an enthalpy of mixing that is effectively zero with respect to the main lattice. This allows phase transformation to absorb and disperse the point defects that typically accumulate in more rigid lattices. This extends the life of the alloy through making vacancy and interstitial creation less successful as constant neutron excitement in the form of displacement cascades transform the SRO phase, while the SRO reforms in the bulk solid solution.

Resource

- Fundamentals of Radiation Material Science, Gary S. Was, Springer-Verlag Berlin Heidelberg 2007

- R. S. Averback and T. Diaz de la Rubia (1998). "Displacement damage in irradiated metals and semiconductors". In H. Ehrenfest and F. Spaepen. Solid State Physics 51. Academic Press. pp. 281–402.

- R. Smith, ed. (1997). Atomic & ion collisions in solids and at surfaces: theory, simulation and applications. Cambridge University Press. ISBN 0-521-44022-X.

References

- Drexler, K. Eric (1992). Nanosystems: Molecular Machinery, Manufacturing, and Computation. New York: John Wiley & Sons. ISBN 0-471-57547-X.

- Allhoff, Fritz; Lin, Patrick; Moore, Daniel (2010). What is nanotechnology and why does it matter?: from science to ethics. John Wiley and Sons. pp. 3–5. ISBN 1-4051-7545-1.

- Lapshin, R. V. (2011). "Feature-oriented scanning probe microscopy". In H. S. Nalwa. Encyclopedia of Nanoscience and Nanotechnology (PDF). 14. USA: American Scientific Publishers. pp. 105–115. ISBN 1-58883-163-9.

- Murray R.G.E. (1993) Advances in Bacterial Paracrystalline Surface Layers. T. J. Beveridge, S. F. Koval (Eds.). Plenum Press. ISBN 978-0-306-44582-8. pp. 3–9.

- Kafshgari MH, Voelcker NH, Harding FJ (2015) Applications of zero-valent silicon nanostructures in biomedicine. Nanomedicine (Lond) 10 (16):2553-71. doi:10.2217/nnm.15.91 PMID 26295171

- Cassidy (2014). "Nanotechnology in the Regeneration of Complex Tissues". Bone and Tissue Regeneration Insights: 25. doi:10.4137/BTRI.S12331.

Permissions

We would like to thank the editorial team for lending their expertise to make the book truly unique. They have played a crucial role in the development of this book. Without their invaluable contributions this book wouldn't have been possible. They have made vital efforts to compile up to date information on the varied aspects of this subject to make this book a valuable addition to the collection of many professionals and students.

This book was conceptualized with the vision of imparting up-to-date and integrated information in this field. To ensure the same, a matchless editorial board was set up. Every individual on the board went through rigorous rounds of assessment to prove their worth. After which they invested a large part of their time researching and compiling the most relevant data for our readers.

The editorial board has been involved in producing this book since its inception. They have spent rigorous hours researching and exploring the diverse topics which have resulted in the successful publishing of this book. They have passed on their knowledge of decades through this book. To expedite this challenging task, the publisher supported the team at every step. A small team of assistant editors was also appointed to further simplify the editing procedure and attain best results for the readers.

Apart from the editorial board, the designing team has also invested a significant amount of their time in understanding the subject and creating the most relevant covers. They scrutinized every image to scout for the most suitable representation of the subject and create an appropriate cover for the book.

The publishing team has been an ardent support to the editorial, designing and production team. Their endless efforts to recruit the best for this project, has resulted in the accomplishment of this book. They are a veteran in the field of academics and their pool of knowledge is as vast as their experience in printing. Their expertise and guidance has proved useful at every step. Their uncompromising quality standards have made this book an exceptional effort. Their encouragement from time to time has been an inspiration for everyone.

The publisher and the editorial board hope that this book will prove to be a valuable piece of knowledge for students, practitioners and scholars across the globe.

Index

www.ingramcontent.com/pod-product-compliance
Lightning Source LLC
Chambersburg PA
CBHW061319190326
41458CB00011B/3842